Build Your Own Test Equipment

Homer L. Davidson

TAB Books
Division of McGraw-Hill
New York San Francisco Washington, D.C. Auckland Bogotá
Caracas Lisbon London Madrid Mexico City Milan
Montreal New Delhi San Juan Singapore
Sydney Tokyo Toronto

TAB Books is a division of McGraw-Hill.

pbk 8 9 10 11 12 13 14 FGR/FGR 9 9 8
hd 1 2 3 4 5 6 7 8 9 FGR/FGR 9 9 8 7 6 5 4 3 2 1

Library of Congress Cataloging-in-Publication Data

Davidson, Homer L.
Build your own test equipment / by Homer L. Davidson
p. cm.
Includes index.
ISBN 0-8306-8475-1 (h) ISBN 0-8306-3475-4 (p)
1. Electronic instruments—Design and construction—Amateurs' manuals. I. Title.
TK9965.D38 1991 90-27915
621.381'542—dc20 CIP

Acquisitions Editor: Roland S. Phelps
Technical Editor: Andrew Yoder
Production: Katherine G. Brown
Book Design: Jaclyn J. Boone

Contents

Introduction

Why should anyone try to build their own test instruments when so many commercial units are available? The answer is simple: to build test equipment that is not currently on the market, to learn how test instruments work, to locate defective circuits, and to save some money. Besides, it's fun. With a low-priced digital multimeter (DMM) and the test instruments in this book, you can service many different units in the entertainment field.

The 32 test instruments found in this book have actually been constructed, tested and placed in service. Some of the test instruments are very simple and can be built in one evening, and others might take a day or two to construct. The book shows how to build various test units used on the service bench by the electronic technician, the electronic activist, the beginner, and the homeowner.

Throughout the book are complete descriptions of how to connect wire, and solder each test instrument. It is very discouraging to work several hours on a project, only to find that it does not operate. Information about testing and servicing possible troubles and pitfalls is at the end of the text for each test instrument. The complete parts lists include exact part numbers, and where the critical parts can be found and ordered. Of course, some of these components might be found in your own electronic junk box.

The electronic technician uses many different test instruments to service amplifiers, cassette decks, CD players, camcorders, radio receivers, microwave ovens, and television sets. This book is divided to cover five different types of test equipment: that used by the novice, beginner, homeowner, electronic student, and electronic technician.

How to select the correct cabinet and chassis, and how to make perf and etched boards is covered in chapter 1. The project, the experimenter's powered breadboard, is a universal solderless breadboard upon which you can design and experiment with the various electronic circuits.

Seven simple test instruments that anyone can build are found in chapter 2. Although these test instruments are easy to construct, several might wind up on the electronic technician's test bench. The diode, light bulb, and fuse testers can be used in every home. The low-voltage motor tester can be used to locate the defective low-voltage toy motors or expensive CD and camcorder motors (FIG. I-1).

Chapter 3 provides instructions and data on how to build fine test instruments for the novice and beginner, starting with a transistor tester that connects to a regular DMM (digital multimeter). A diode/SCR/triac and continuity checker and noise generator should be found upon any test bench; both instruments can be built from information in this chapter.

Seven test instruments for the advanced or electronic enthusiast are included in chapter 4. The low-voltage power supply, audio amp checker, sine/square-wave generator, hand-held signal injector, and tracer are ideal test instruments to learn how to build and handy to have around when building projects (FIG. I-2). The radio amateur or electronic technician can use the crystal checker to locate defective crystals in ham gear or saw filters and crystals in the TV chassis.

The laser/infrared CD diode checker, infrared remote-control tester, and deluxe duoregulated power supply can be used every week by the

I-1 This deluxe motor tester has a 0 to 5-volt supply to test camcorder, CD, and battery motors. A fixed 5-, 10-, 12-, and 15-volt regulated power supply provides up to 2-amp operation for larger cassette, toy, and small dc motors.

I-2 Two sine and square waveform generator testers: one operates on batteries and the other is powered with an ac transformer.

I-3 This deluxe, dual-powered voltage supply has two separate voltage sources: a variable 1.2-to-27 volt, 1.5-amp regulated power supply, and four fixed voltage sources—10, 15, 20, and 24 volts dc.

I-4 This dummy load provides 4- and 8-ohm load with up to 100 watts of power for use when servicing stereo amps, large power amplifiers, or PA systems, for example.

electronic technician (FIG. I-3). Since tuner subbers are hard to find, here is a project that can determine if the tuner or chassis is defective in the TV receiver. Although most of the test instruments found in chapter 5 are for the electronic technician's service bench, several can be used by the telephone lineman, electrician, audio/radio technician, audio enthusiast, and homeowner (FIG. I-4).

Make test instruments shockproof and easy to operate. Be careful when soldering or working near power-line equipment. Exercise extreme care when servicing microwave ovens and television receivers. Always use an isolation power transformer, to plug the TV set or unit into, before attaching the test instrument. Remember, safety comes first in test-instrument construction. Do not use old ac cords or exposed power-line wires in your favorite test instruments.

Of course, not everyone will want to build every project in the book, but several of these test instruments are for everyone. Just select the test instrument you would like to build and get started.

Transistor/IC terminal connections

TRANSISTORS

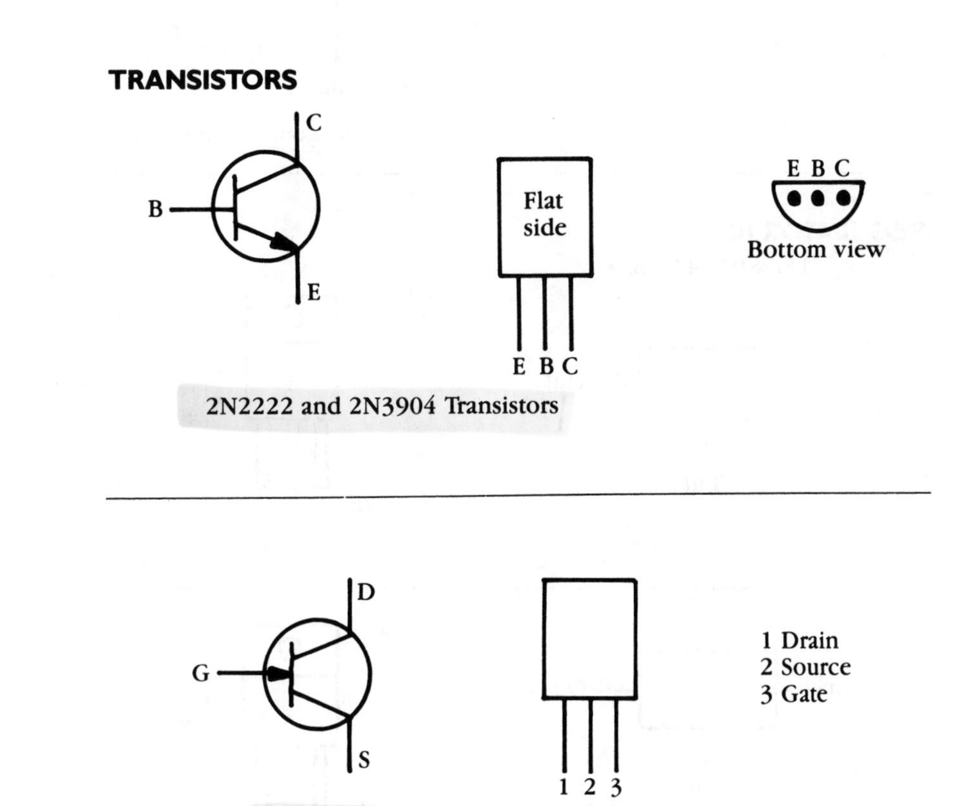

2N2222 and 2N3904 Transistors

MPF102

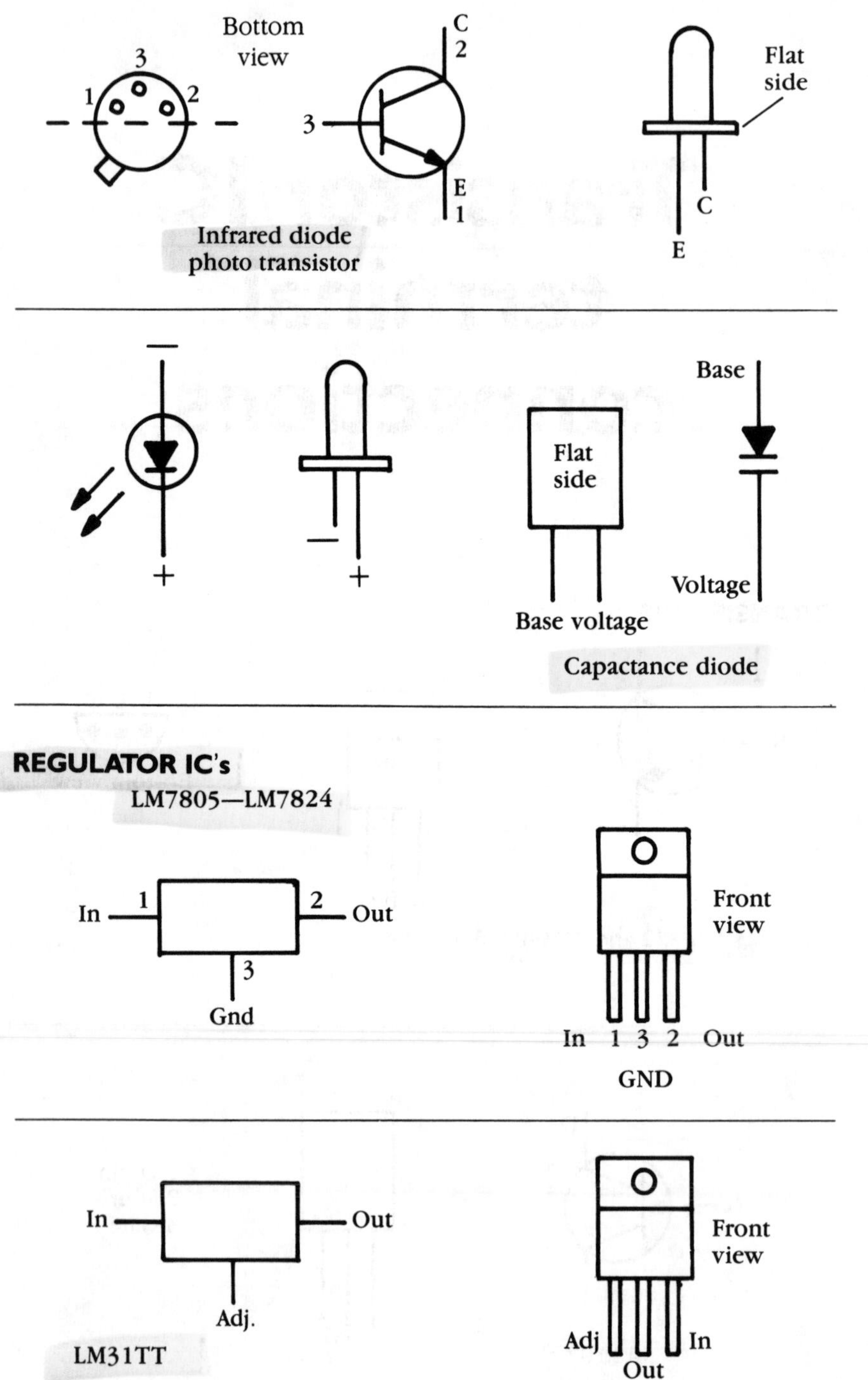
Bottom
view
3
1
2
C
2
3
E
1
Infrared diode
photo transistor
Flat
side
C
E
—
+
—
+
Flat
side
Base voltage
Base
Voltage
Capactance diode
REGULATOR IC's
LM7805—LM7824
In
1
2
Out
3
Gnd
Front
view
In 1 3 2 Out
GND
In
Out
Adj.
LM31TT
Front
view
Adj
In
Out

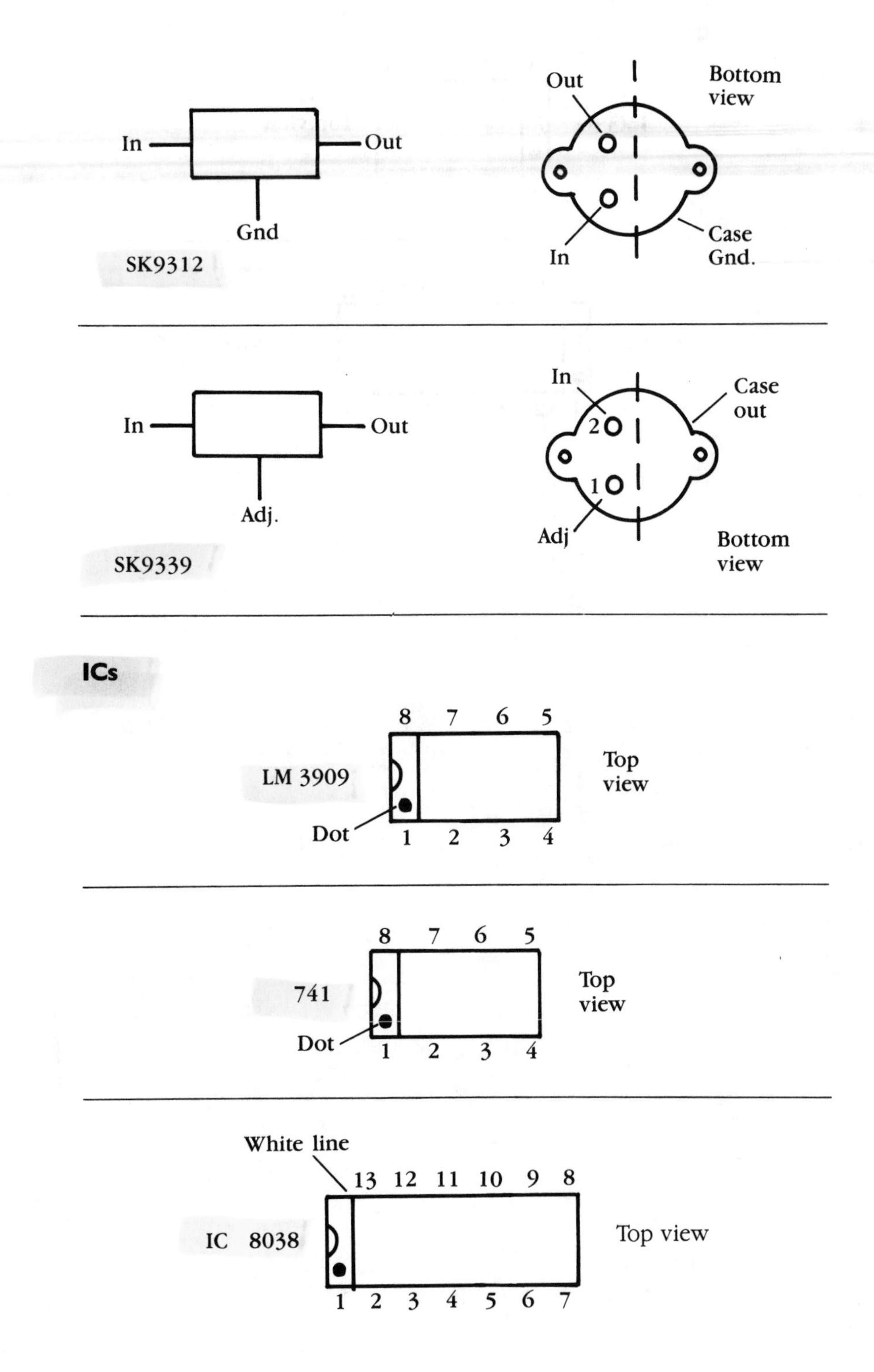
In
Out
Gnd
SK9312
Out
Bottom view
In
Case Gnd.
In
Out
Adj.
SK9339
In
Case out
2
1
Adj
Bottom view
ICs
8 7 6 5
LM 3909
Top view
Dot
1 2 3 4
8 7 6 5
741
Top view
Dot
1 2 3 4
White line
13 12 11 10 9 8
IC 8038
Top view
1 2 3 4 5 6 7

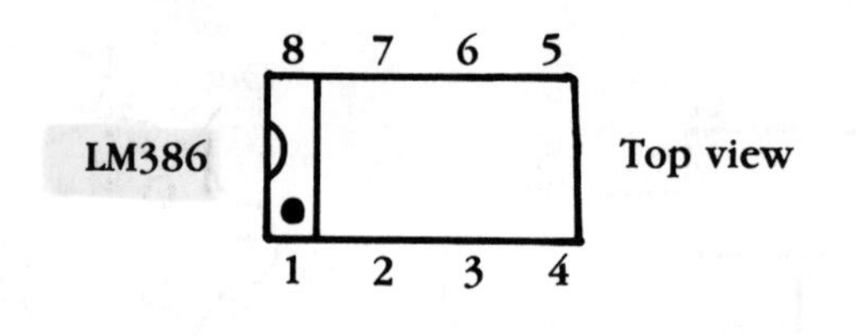
8 7 6 5
LM386
Top view
1 2 3 4

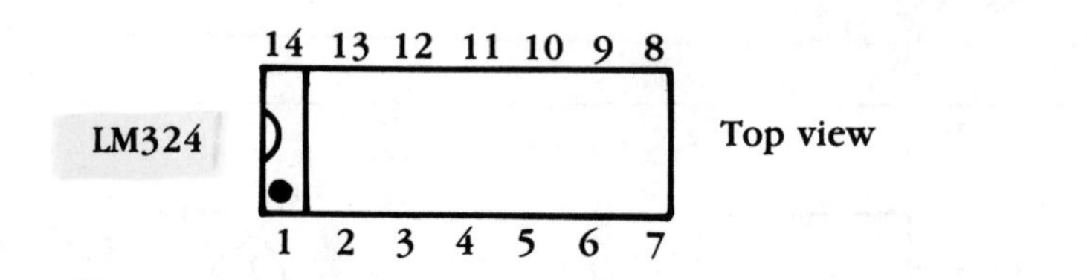
14 13 12 11 10 9 8
LM324
Top view
1 2 3 4 5 6 7

Chapter 1

Chassis, cabinets, and boards

When building electronic projects, choose the correct cabinet, chassis, and mounting boards. The nicer the appearance, the greater the acceptance. Place your electronic test instruments in a case or cabinet just like those used for commercial test instruments (FIG. 1-1). Build like a professional; make your projects snappy-looking and rugged.

SELECTING THE CABINET

Choose a cabinet large enough to support the chassis and required components. It's best to choose a large cabinet than one that is too small. Do not try to squeeze the parts close together; this procedure increases the chance of shorting or overheating. Notice how professional manufacturers of test equipment package their finished products. Often, the cabinet is twice the size or even much larger than the chassis.

Many types of cabinets and cases are available for building test equipment. The cabinet can be plastic, steel, aluminum, or a combination of all three (FIG. 1-2). Naturally, the plastic and aluminum cabinets and chassis are the easiest to drill, cut, and work with. A cracked-finish steel cabinet combined with a bottom aluminum chassis, however, can provide a professional-looking test instrument.

The larger the cabinet, the higher the cost. For large test instruments, choose deluxe boxes, cabinets, and containers. The professional project case or cabinet provides a durable, nice appearance and a great finish to your project. You can purchase professional instrument cases and cabinets for meters, preamps, amplifiers, power supplies, and signal genera-

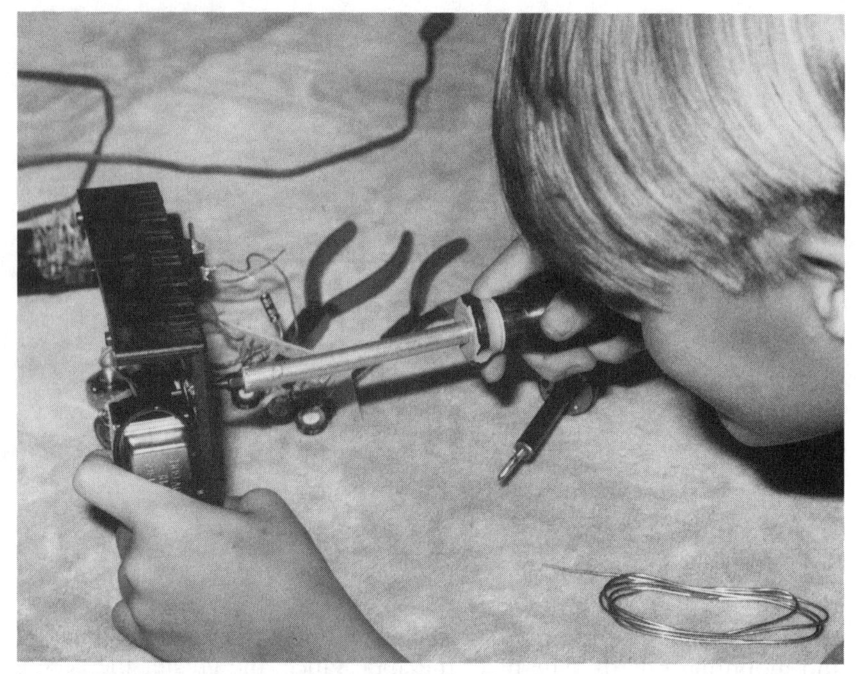

1-1 Building your own test equipment can be fun. Even youngsters can get into the act.

1-2 Cabinets and cases come in many sizes and shapes. Choose the cabinet that will make your test equipment look the most professional.

tors. Do not hesitate to spend as much money for the case as you did for all the required electronic parts (FIG. 1-3).

Sometimes these small electronic test instruments do not require large, expensive cabinets, so you can choose the plastic economy cases.

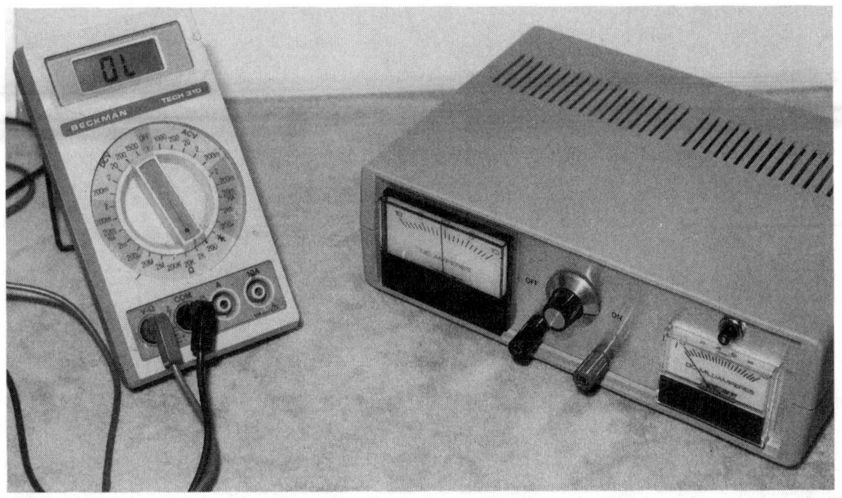

1-3 A good rule of thumb is to spend as much money for the cabinet as the total cost of all the parts.

Some of these hobby cases contain a board. Just add your own test circuit to the printed circuit board with predrilled holes for components and ICs. In other case/pc combos, a pc board with IC grid pattern, an aluminum cover, and vinyl feet are included.

You might be able to order or find surplus test equipment cabinets or cases at various electronic stores. Most of these cabinets are brand new with various holes and cut-outs. Usually, these are from end-of-year lots, purchased at a discount. Cases constructed for other electronic equipment can be salvaged for your electronic project. Most of these cabinets can be modified for your own test equipment. Check chapter 8 for source information on cabinets, cases, and electronic parts.

CHOOSING THE CORRECT CHASSIS

Yesterday, the chassis was either steel or aluminum. Today, the chassis can be an etched pc or perfboard (FIG. 1-4). Since tubes are no longer used in construction articles, the transistor and IC components are mounted on pc or perfboard circuits. The pc board is etched with etchant solution and the perfboard can be purchased predrilled with or without copper-ringed holes.

Perfboard construction

Perfboard is available in large or small sheets (FIG. 1-5). The perfboard might have IC or component holes (.042″) prepunched in a standard .100 × .100 grid. Predrilled grid boards accept dip IC sockets and headers. Columns and rows are indexed to identify component pin outs with solder-ringed holes.

1-4 Today's chassis are made from perfboard and pc boards. Cut a piece from a larger size board to capitalize on size and cost.

1-5 Perfboards do not have foil but do have holes in a grid pattern for pushing part leads through. They are easily cut with a hacksaw or saber saw.

Some of these universal pc boards accept ICs with up to 40 pins. Regular IC/LS1 boards are made in the same manner. You can cut a piece of perfboard to the correct size with a hack saw, saber saw, or scroll saw. The perfboard is easily drilled and cut. You can make a quick chassis for small test instruments.

The multipurpose board with copper-ringed holes can have two bus strips at the center for common or ground terminals. The dual IC board is made especially for IC components. Some of these IC boards can be snapped in half if only one section is needed. The predrilled pc perfboard is easy to work with and has copper foil and ringed holes for quick soldering. These boards can be isolated from a metal chassis with pc board stand-offs using screws or bolts to prevent components from grounding on metal chassis (FIG. 1-6).

Etched pc boards

The copper-clad pc board is available in large or small sheets with single- or double-sided copper. The double-sided copper board was designed so that circuits can be etched on both sides of the pc board. You can easily cut these boards with a hack saw or saber saw (FIG. 1-7).

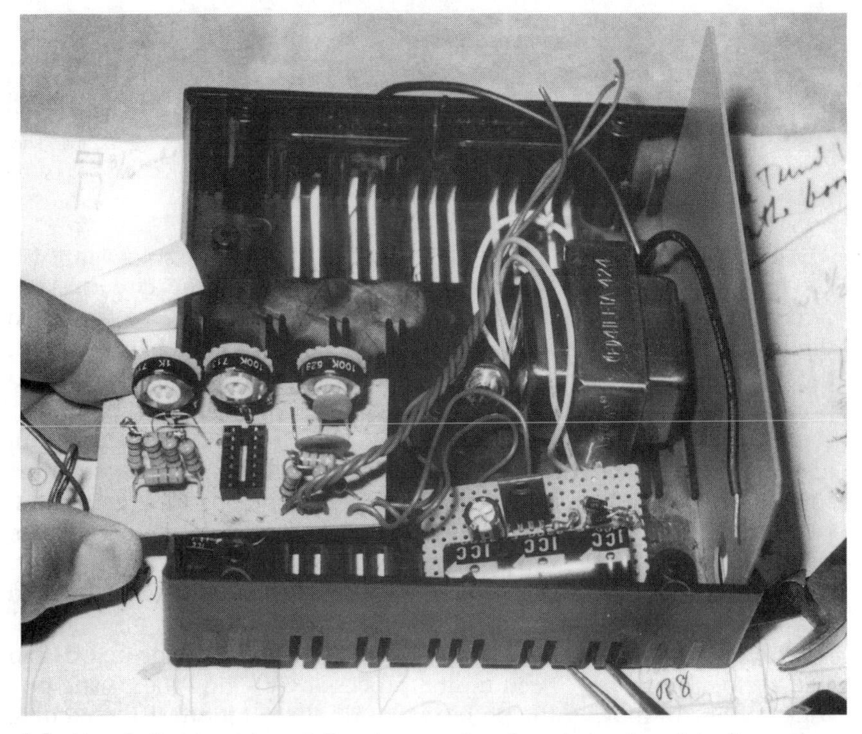

1-6 Use plastic or metal standoffs to keep small perfboard or pc boards in place and prevent them from shorting out.

1-7 Printed circuit boards can have copper foil on one or both sides of the board. The double-sided board is used when the circuit cannot be fitted on one side of the board.

The copper circuit is left on the pc board by applying solution-resistant ink or black layout strips, circles, and holes for the circuit. Also, direct-dry etch transfers, consisting of circles, strips and IC pads applied directly on the copper foil can be used. When placed in etchant solution, the copper is eaten away and only the wiring circuit remains. Purchase a printed-circuit (PC) board kit to etch your own circuits.

Often, the pc board kit consists of small printed boards, a solution-resistant ink pen, a plastic bottle of etchant solution, solution-resistant ink solvent, a scratch pad, a .16 drill bit, and one plastic box for etching (FIG. 1-8). Scrub the board in soap and water to remove oxidization and lacquer. Design the circuit upon the pc copper side with the pen, circles, strips, and direct-dry transfers. The etched chassis in this book can be traced with a piece of carbon paper placed upon a pc board (FIG. 1-9).

Before etching, make sure the ink is dry. Etching small pc boards can take 20 to 30 minutes. Speed up the process by rocking the plastic box back and forth, moving the pc board inside the solution. If the etching process takes over one hour, old etching or over-used etchant solution is probably the cause.

1-8 Design and build your own etched and drilled printed circuit boards with a pc board etching kit.

1-9 You can tin the remaining copper wiring with solder to protect it. Once you etch the circuit on the pc board, drill out the required holes for the parts.

TESTING

Before firing up the test instrument, recheck the wiring at least twice. Start at the front of the circuit and trace each wire and component. Then, start at the output of the test instrument circuit and work towards the front, crossing off each wire connection. Inspect all soldering connections with a small hand magnifying glass.

Voltage measurements

Take accurate voltage measurements from the IC or transistor to locate a defective part or connection with the digital multimeter (DMM). Voltages other than those found in the schematic indicate possible trouble. Low supply voltage at the VCC pin of the IC component might indicate a leaky IC or defective power source.

High voltage at the collector pin of a transistor might indicate an open transistor junction. Low voltage at the collector terminal might reveal a leaky transistor. Practically the same voltage on all three transistor terminals might indicate a leaky transistor. Improper voltages at wiring connections might indicate a poorly soldered or broken connection. The various power supply voltages can be checked at the output terminals (FIG. 1-10).

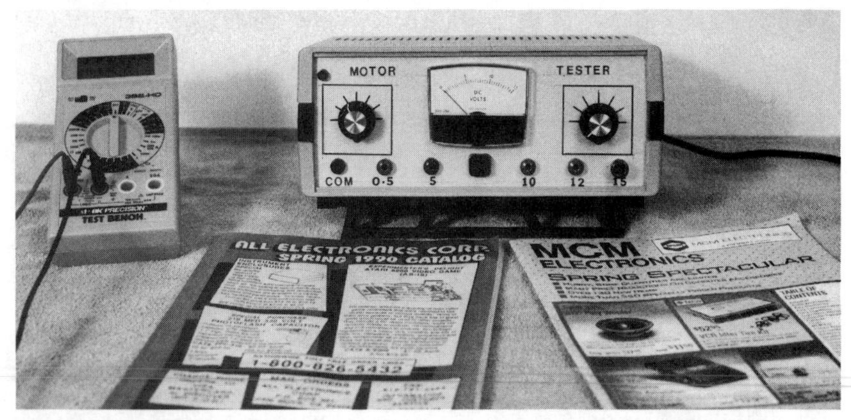

1-10 Check the voltage of transistors and IC components with a multimeter; pictured at left is a digital multimeter. A motor tester is shown at right.

Resistance measurements

Rotate the ohmmeter to the 20-ohm scale for shorted connections. Poor connections or soldering joints can be located with low ohmmeter readings. Trace the circuit for possible broken wires or connections from one component to the next with low-ohmmeter test probes (FIG. 1-11). A poor ground connection can be located with the ohmmeter. Besides low ohmmeter measurements, excessive current might reveal a leaky component or misplaced connection.

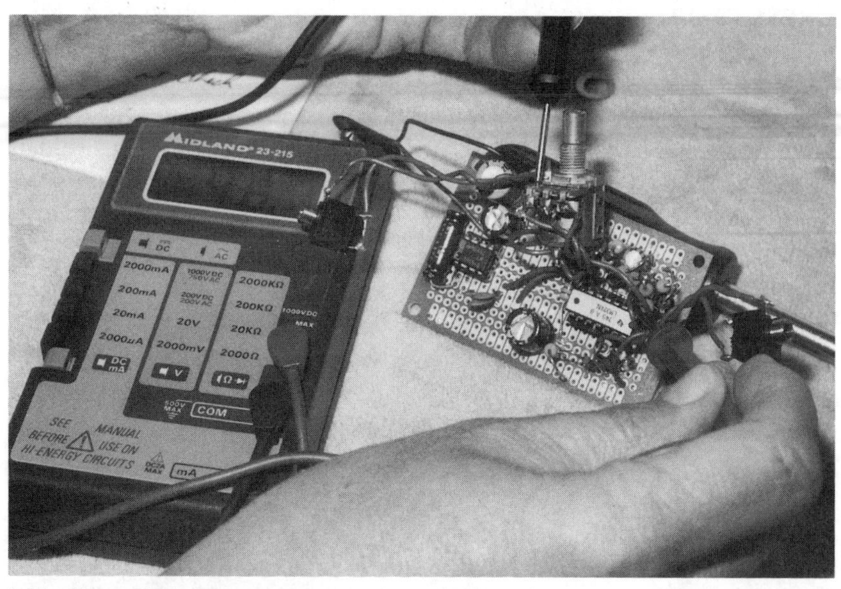

1-11 Take resistance measurements to locate broken wires or traces, incorrect wiring, or poorly soldered connections.

Current tests

Check the total current pull of the test instrument if the voltage is low and the project does not operate. Clip the current meter across the switch with the switch off (FIG. 1-12). A heavy current reading indicates a leaky component or improper wiring. No current measurement might indicate an open transistor, IC, regulator, or circuit. To measure correct current, place the meter probes in series with the suspected component.

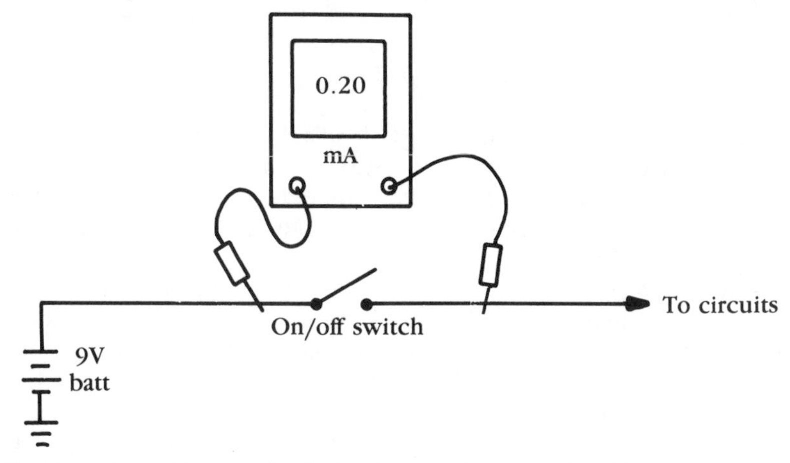

1-12 Check to see if your project is drawing excessive current by taking a current reading across the on/off switch. You should read the entire current draw with the switch off.

DRESS IT UP

Now dress up your cabinet with a few decals and letters. Project labels are available as rub-on letters, symbols, and calibration marks (FIG. 1-13) from electronic or stationery stores. Most of these rub-on labels will stick to plastic or metal surfaces. Place dark letters and decals on light cabinets and white symbols on dark cabinets. Protect decals and lettering with lacquer or clear spray.

1-13 Dress up your project with dry-transfer panel labels. Remember to first clean the surface with mild solvent before applying labels.

A tapewriter kit with labeling tapes will adhere to almost any smooth, clean, dry surface. These tapes are corrosion-proof, water and moisture resistant, and will withstand extreme temperatures up to 150°F. Several different colored rolls of tape might be interchangeable. Some tapewriters can print both vertically and horizontally.

Simply select the correct letter or number of the label and press halfway down on the trigger. Select each letter or number in order. Advance the tape until the last character of the message is in line with the cut-off indicator arrow. Now, cut tape by firmly pulling down cut-off lever. Clean the surface with a mild solvent, such as rubber-cement thinner. Do not use soap on bare aluminum. Remove the back side and press the label firmly against the instrument surface (FIG. 1-14). Make sure the label is lined up correctly.

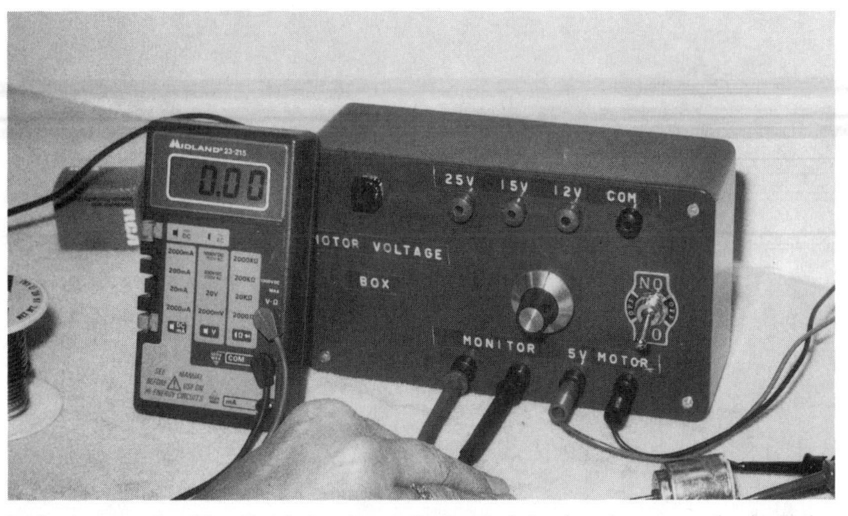

1-14 A tapewriter kit with labeling tape adheres labels to almost any smooth, clean, dry surface.

PARTS LIST

Most electronic projects include a complete parts list. For the following test instruments, a part list is included, complete with Radio Shack part numbers (FIG. 1-15). Of course, these part numbers are not critical and equivalent components can be substituted. You might have parts that can be used from the junk box or other electronic stores. A complete list of mail order firms are in chapter 8.

EXPERIMENTER'S POWERED BREADBOARD

How would you like to design and test your own electronic circuits in minutes? Just build this regulated breadboard project and you will be in business (FIG. 1-16). Underneath the universal breadboard is a voltage-regulated power supply with 5-, 8-, and 12-volt sources. Besides designing your own circuits, you can test the test equipment circuits in this book.

The two bottom rows of the breadboard have a common ground with all voltage sources and is connected to the black terminal post. The 5-volt source is connected to the first red post with the top row right holes. Connect the 8-volt source to the second row of holes on the right side. Wire both rows of holes on the left side of the breadboard to the 12-volt source. Now you can provide wire jumpers to any voltage source and circuit.

Power supply circuits

The power supply circuits consist of a 12.6-volt ac step-down transformer. D1 provides full-wave rectification (FIG. 1-17). The 12.6 volts ac is

1-15 A parts list is included with each test equipment project in this book. Radio Shack and other part numbers are given, although you can use equivalent replacement parts.

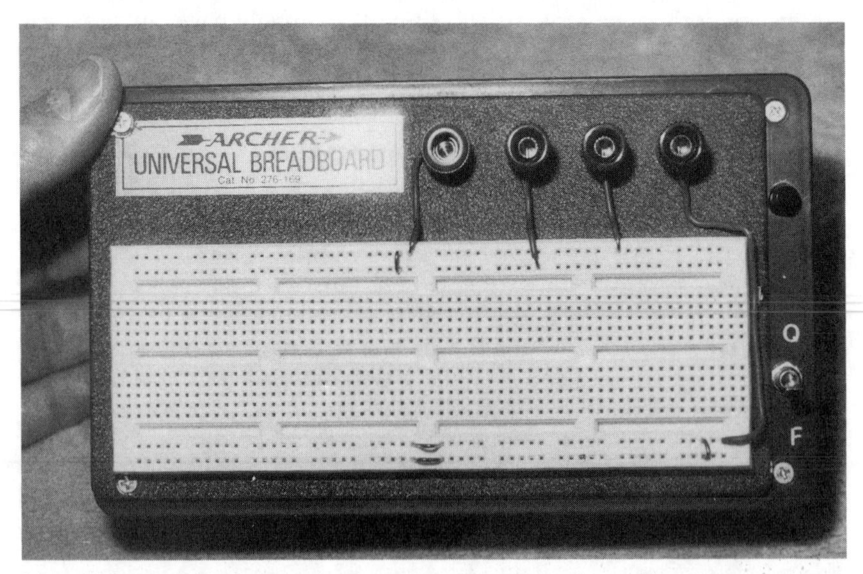

1-16 You can quickly design and put together your own projects with an ac-powered universal breadboard.

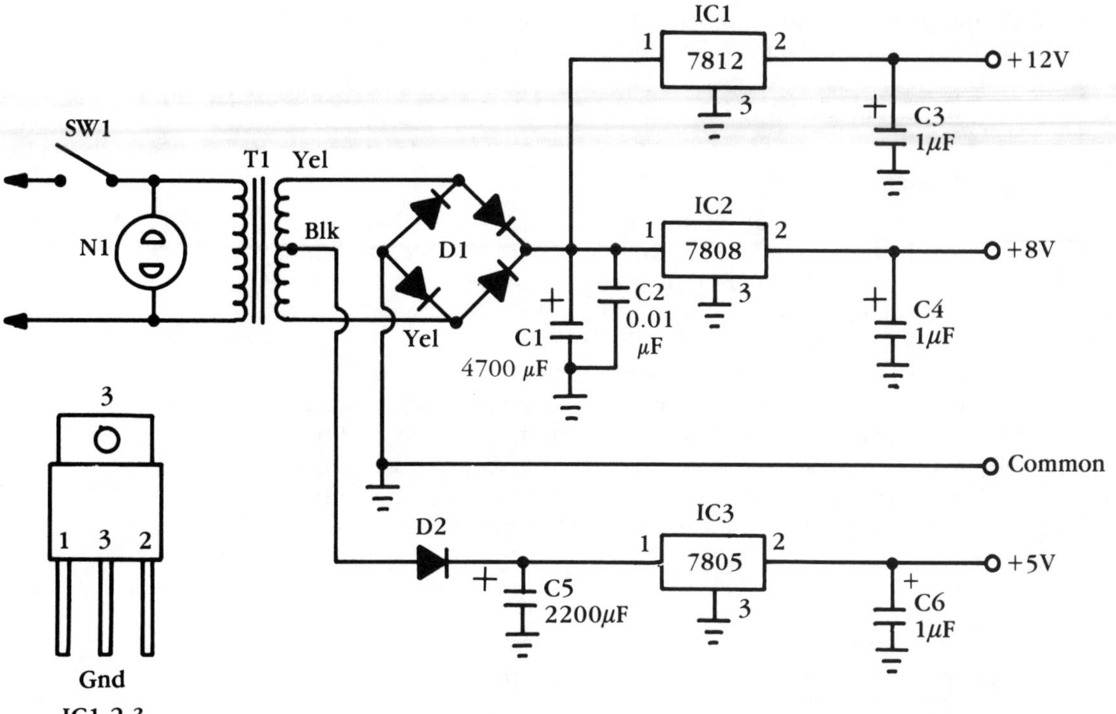

1-17 The power supply consists of a stepdown transformer, bridge rectifier, filters, and voltage regulators. IC1, IC2, and IC3 provide regulated 12, 8, and 5 volts.

applied across the two sine-wave terminals of the 4-amp rectifier. If they are handy, you can use four 3-amp diodes in place of D1. Wire the common-ground terminal post to the negative terminal of D1. The positive terminal (+) of D1 should measure 19 volts dc.

The 5-volt dc source is taken from the black center-top terminal of T1. D2 provides half-wave rectification with C5 operating as a filter capacitor. IC3 provides voltage regulation at +5 volts. All IC regulator terminals are the same. Remember, the center and tab terminals are at ground potential.

Only the 8- and 12-volt dc sources provide full-wave rectification to C1, IC1, and IC2. IC1 is a 12-volt IC regulator, 7812. The 8-volt source is regulated by IC2 (7808). C3, C4, and C6 (1 μF) electrolytic smoothing capacitors are connected to each output voltage source.

Preparing breadboard

Since only three posts are found on the original breadboard (Radio Shack 276-169), another must be added. Remove the four rubber feet from the bottom of commercial breadboard with a pocket knife. Drill an 11/32-inch hole for another red terminal post. This post must be insulated from the metal chassis. A plastic terminal post works nicely here. Double check the terminal at ground with the low-ohm scale of the DMM.

Preparing the top panel

Lay the breadboard assembly upon the top-panel lid of the plastic case. Mark and match the present holes that hold the top lid in place. With this size of plastic case, the breadboard's right edge must be in line with the case edge. This design allows room to the left for the on/off switch (SW1) and the neon indicator light.

Drill three 1/8-inch holes in the universal breadboard assembly. The right two holes will match with the top-panel lid screw holes. Countersink the holes in the metal chassis so the same lid-hole screws will hold the lid and breadboard chassis in position (FIG. 1-18). Cut a section in the top plastic lid so that the four terminal posts will fit inside. Now, drill a 15/64-inch hole for N1 and a 13/64-inch hole for SW1 in the plastic top lid.

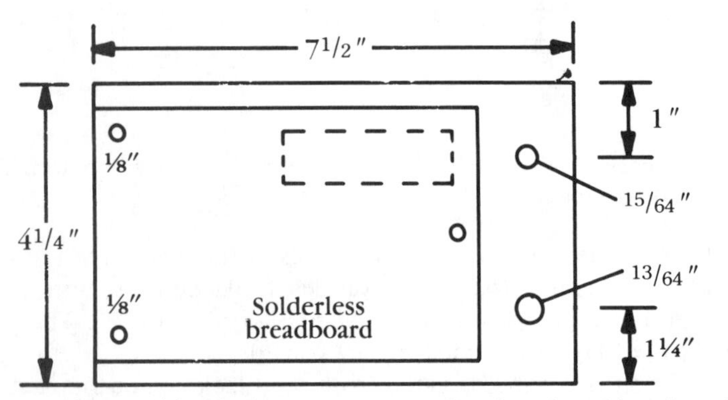

1-18 Lay out the top of the plastic case to accommodate the universal breadboard, neon indicator, and ac toggle switch.

Preparing the bottom case

The power transformer is mounted in the opposite end of the bottom case from the terminal posts. Drill two 1/8-inch holes to mount the transformer. Drill a 15/64-inch hole for rubber grommet and ac cord. Match the top breadboard chassis and mark the left mounting hole. Bolt the center hole of breadboard to the top panel. Now, the two right holes should match with the bottom case. Do not fasten the bottom case until the power supply is tested and ready.

Cut a 4 1/2-x-3 1/2-inch piece of perfboard. This chassis will hold all small components, except those on the front panel and T1. Drill two 1/8-inch holes in the center for board mounting. Try to keep all parts away

1-19 Close-up view of the perfboard chassis. Each regulator has its own heatsink.

from these holes so that the board can be mounted to the bottom case (FIG. 1-19).

Notice that each IC regulator has its own heatsink. Bolt the heatsink to each IC before mounting. Dab silicone heatsink grease behind each IC before bolting it to the metal sink.

Wiring

Mount all small components on the perfboard chassis as you solder them into the circuit. Don't worry where the various parts are mounted. Mount them close enough so that the terminal leads will connect with the corresponding circuits. Use a common bare ground wire to tie all grounds into the circuit.

Start by positioning D1 near the end of the perfboard. Bend ac connections toward the transformer. Likewise, bend the positive and negative leads toward the center of the perfboard. Mount C1 behind D1. Wrap the positive and negative terminal of D1 around the respective terminals of C1. Mark the positive and negative terminals of C1 on the bottom of chassis with a felt-tip pen. Recheck the polarity of C1.

Face each IC regulator outward on perfboard chassis. Tie the input terminals of IC1 and IC2 together to positive side of C1. Connect both center terminals to the common ground wire. Solder a piece of standard hookup wire to pin 2 of each IC regulator that will tie into the respective terminal posts. Solder C3 and C4, observing correct polarity.

Leave one end of D2 stick out towards the transformer and solder to terminal 1 of IC3 (7805). Connect capacitor C5 after D2. Recheck C5 polarity terminal connections. Ground the center terminal of IC3 to the common bare ground wire. Solder a four-inch flexible hookup wire to terminal 2 (output) and to the positive terminal of C6, to tie on the 5-volt terminal post.

Double-check

Before going any further, double-check the wiring of perfboard. Recheck the polarity of each capacitor and of D1. Make sure the input terminal (1) of each IC regulator ties to its respective rectifier and filter capacitor. Make sure the output terminal (2) of each IC regulator leads to the correct voltage terminal post.

Wiring connections

After the perfboard has been completely wired, connect the long, stranded hookup wire to each voltage source (FIG. 1-20). Make sure each voltage is connected to the correct post. Solder these wires to the bottom terminal posts underneath the breadboard. Run another ground lead to the bare common ground wire and the black terminal post. Recheck for poorly soldered connections, since the originals have large brass posts.

1-20 Connecting and soldering the wires from perfboard chassis to the top breadboard.

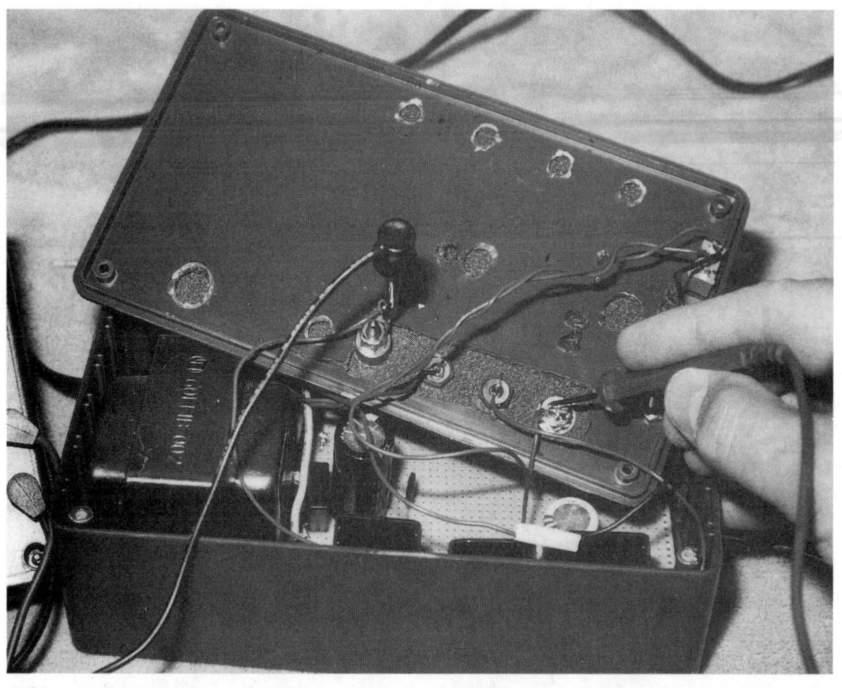

1-21 Checking current and voltage measurements on the breadboard project. Insert the current meter between the breadboard pins and the terminal post.

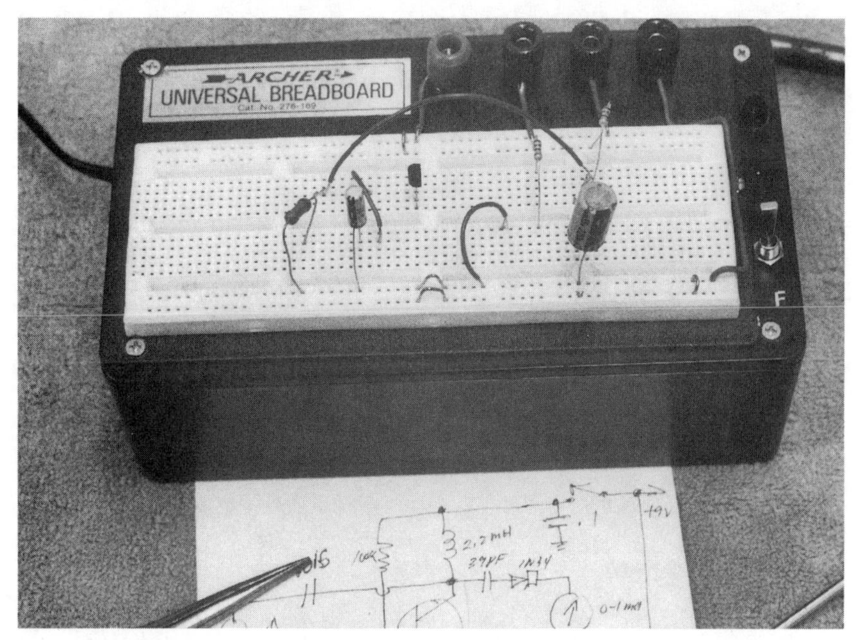

1-22 Check the finished breadboard project by assembling a simple test circuit.

Parts list

SW1	SPST push on/push off, 275-617 or equiv.
N1	Red neon 120-V light assembly, 272-712 or equiv.
T1	18-V C.F. power transformer 1.2-A, 273-1515 or equiv.
D1	4-A full-wave bridge Rec. 100 V, 276-1171 or equiv.
D2	1-A silicon 100 V diode.
C1	4700-μF 35-V electrolytic capacitor.
C2	.01-μF 100-V ceramic capacitor.
C3, C4, C6	1-μF 35-V electrolytic capacitor.
C5	2200-μF 35-V electrolytic capacitor.
IC1	12-V regulator, 7812.
IC2	8-V regulator, 7808.
IC3	5-Volt regulator, 7805.
Case	7½″-X-4¼″-X-2¼″ economy plastic case, 270-224.
Universal breadboard	4″ × 7″, 276-169 or equivalent.
Terminal post	Multipurpose post, 274-661 or equivalent.
Misc.	Ac cord, perfboard grommet, hoodup wire, solder, etc.

Connect one black primary lead (T1) to the on/off switch. Tie a knot in the ac cord and solder one lead to the other switch terminal. Solder the other transformer primary directly to the remaining wire of the ac cord. The two yellow leads of T1 can be soldered to the D1 input terminals.

Testing

Now, recheck all wiring connections. Plug the ac cord in and flip on SW1. N1 should light. Test each terminal post for correct voltage (FIG. 1-21). If some posts have the correct voltage and one does not, suspect improper connections or the IC regulator.

Measure the output voltage across C1. Improper voltage here indicates a poor or incorrect hookup of D1. Recheck the wiring connections. After a complete checkup of the power supply, start your next project on the breadboard (FIG. 1-22).

Simple test instruments

This chapter features seven simple test instruments that can be used around the house and on the electronic technician's service bench. The light-bulb tester, simple fuse tester, and power-line polarity tester can be used for testing around the house or garage. The simple diode/LED tester, microwave-oven tester, cheap diode/LED tester, low-voltage motor tester, and lamp ac triac tester can be used on the service bench.

SIMPLE DIODE/LED TESTER

Sometimes the most simple test gear can be more helpful than a large test instrument. This little diode/LED tester can check different diodes—power silicon rectifiers, switching barriers, zeners, small-signal diodes (1N34), and bridge rectifiers (FIG. 2-1). Also, the tester can be used to check suspected LEDs.

Only five major components are used to complete the diode/LED tester, so the total cost should be under $10.00 (much less with parts found in the junk box). The tester is built around a dual-LED with red and green colors. The dc voltage flows in one direction to trigger the red and in the opposite direction to light the green.

The circuit

A 12.6-volt ac step-down transformer supplies ac voltage through R1 to the dual LED. The circuit contains no ac switch because very little current is used until the diode is to be tested (FIG. 2-2). Simply pull the plug when not in use. R1 can be any resistance from 470 to 720 ohms. The lower the

2-1 Plug in the tester and place the suspect diode in the fuse holder. Use the test leads to check the defective diode in the circuit.

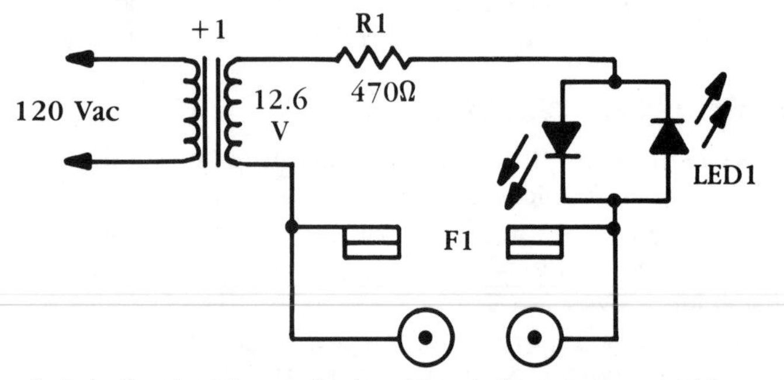

2-2 The diode testing circuit is very simple, using only five components. All components in the loop of the transformer secondary (12.6 V) are wired in series.

resistance, the brighter the LED. The green color is not as bright as the red, so a 470-ohm resistor was used for R1. LED 1 has only two leads and it makes no difference how the component is wired into the circuit (276-012).

Use a single fuse clip holder to quickly test the suspected diode. Plug the test leads into the jacks for in-circuit diode tests. Notice that the secondary of T1 and all other parts are wired in series—the circuit will be completed when a suspected diode is placed in the circuit.

Construction

All components are mounted inside the case, except the fuseholder. First, drill all holes in the front cover of a 4-×-2¼-×-1⁵⁄₁₆-inch plastic case. Center all holes. Drill a 7/32-inch hole for the LED and a 15/64-inch hole for the test jacks. The center hole for fuseholder should be 1/8-inch and the holes for the fuse clip terminals (FIG. 2-3) should be 5/64-inch.

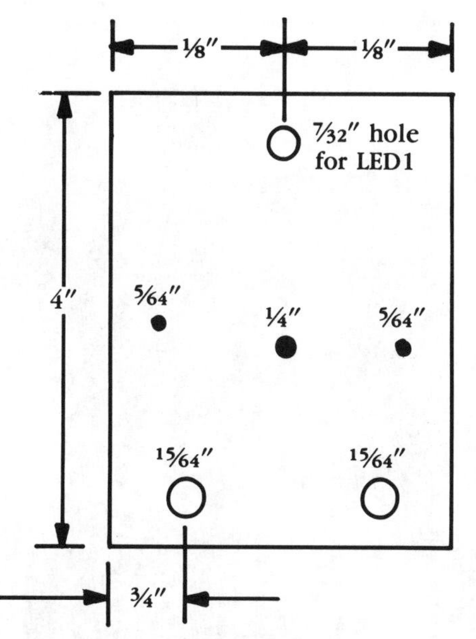

2-3 Drill the front panel holes with several small drill bits.

Since T1 fits snug inside the plastic case, mount it near the top end so it will not obstruct the fuseholder and test jacks. Drill two 8/32-inch holes to mount the power transformer in the bottom case. Center and drill a 7/32-inch hole at the top for the ac cord.

Mount the power transformer and place an insulated 3- or 4-terminal strip at the bottom bolt and nut to tie component leads together. Tie a knot in the ac cord and leave four inches inside the case. Mount the tip jacks and fuse block to the top plastic panel. Push the LED through the top hole. Place rubber silicone cement or epoxy over the end of the LED to hold it in position. Let the cement set for a few hours before wiring.

Solder

Connect one side of the ac cord and the black lead of T1 to the outside terminal lug. Solder the other black transformer wire and the ac lead to the second terminal lug. Now the ac power line is isolated from possible shock hazards. Place the other two yellow (12.6 V) leads of T1 in the remaining terminals.

Solder R1 to one of the terminal lugs and a short piece of hookup wire to LED1. Connect the other end of the LED directly to one side of the fuse clip and terminal post. The polarity of LED1's terminals is not important. Solder the red and black terminal posts to both sides of the fuse clip before connecting it to the inside components. All secondary parts are wired in series (FIG. 2-4).

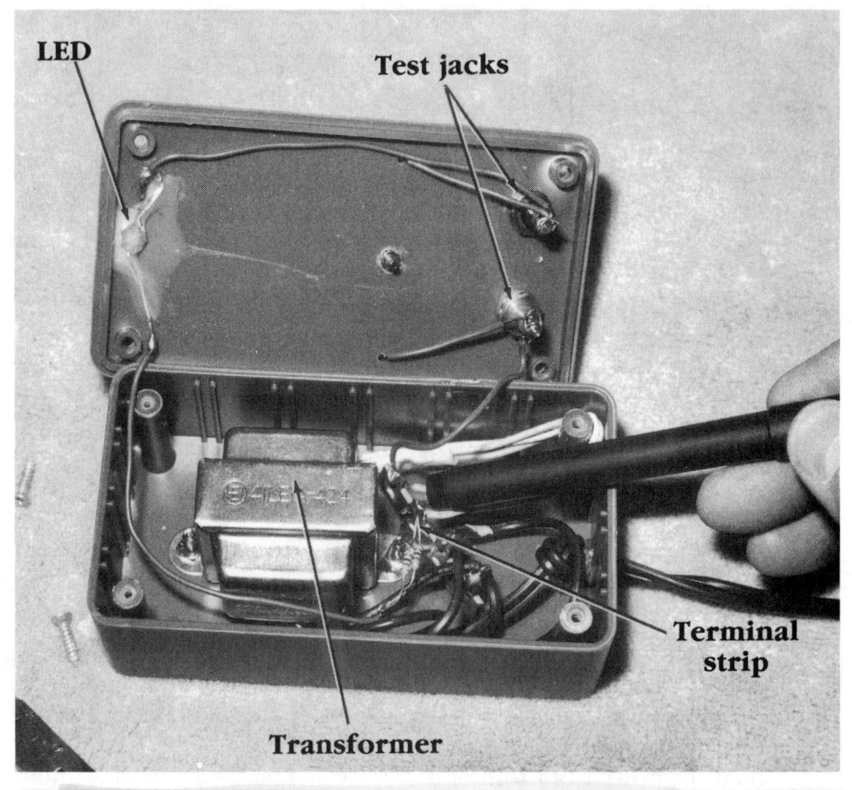

2-4 All components are soldered to a common, four-lug terminal strip. Mount the transformer in the bottom half of case.

Testing

Simply short the fuse clip with the blade of a screwdriver and the LED should light. If not, double-check all wiring and connections, making sure all are good and clean. Measure the ac voltage across the secondary winding (12.6 V) of T1 with a DMM or VOM. Still if no voltage registers, check the ac plug, cord, and soldered connections. Make sure the secondary (yellow) leads do not have a direct short when they are connected to the terminal strip. Insert the test leads and short the terminals. The LED should light.

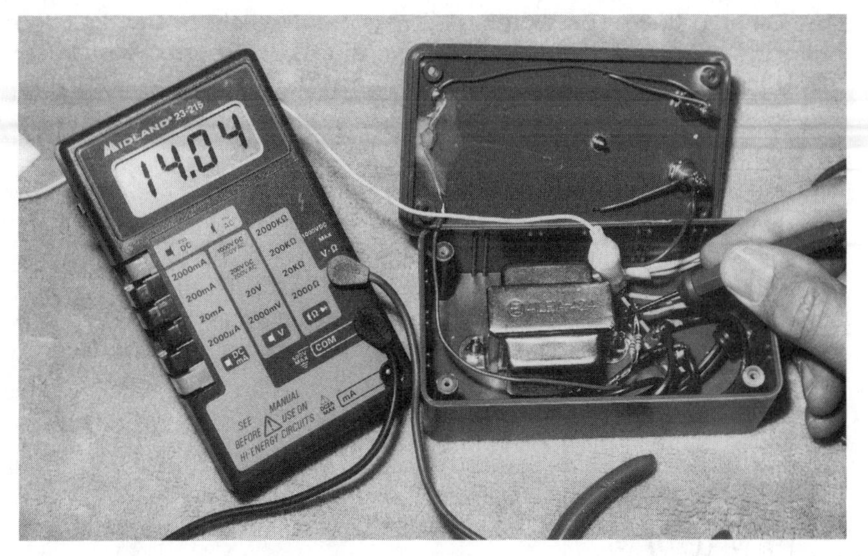

2-5 Check all wiring if the unit does not operate. Measure the voltage at the pin terminals and the secondary of power transformer (12.6 V).

How to use

Place the suspected diode across the fuse clip for an indication from the LED. Bend both clip pieces together to form a small "V" to lay the rectifiers in for testing. The test leads can be used to check diodes in the chassis.

A working diode lights either red or green. Do not worry about diode polarity. Reverse the diode and the other color lights for a normal diode. For instance, when the diode is placed across the fuse clip, the red color lights and when diode leads are reversed, the green color lights showing that the diode is good.

When both colors light with the diode test, the diode is shorted. Likewise, when the test leads are shorted together, both colors light with a yellowish cast. A very dim light indicates a high junction or resistance

Parts list

T1	12.6-V 450 mA power transformer, 273-1365 or equiv.
R1	470Ω 1-W resistor.
LED 1	Dual-color LED, 276-012.
TP1, TP2	Red and white banana terminal jacks, 274-725.
F1	Single-fuse clip holder, 270-739 or equiv.
TR1	4-lug terminal strip.
Case	Plastic economy case, 4″-X-2$\frac{1}{4}$″-X-1$\frac{5}{16}$″, 270-231 or equiv.
Misc.	Two $\frac{8}{32}$″ nuts and bolts, ac cord, hookup wire, solder, etc.

connection of the suspected diode. Replace it. No light in any direction indicates an open diode. Most defective diodes are shorted with some resistance. You can see both colors light up at the base of the LED.

LIGHT-BULB BLOCK TESTER

This little tester can test continuity in 120-volt power-line circuits. Just clip the tester across a suspected fuse in the fuse box. If the fuse is open, the bulb will light. Clip the tester terminals across the motor terminals of the table saw when the motor will not rotate to see if 120 volts is being applied. Place the tester leads across any ac switch that is open or intermittent and notice if bulb lights intermittently (FIG. 2-6).

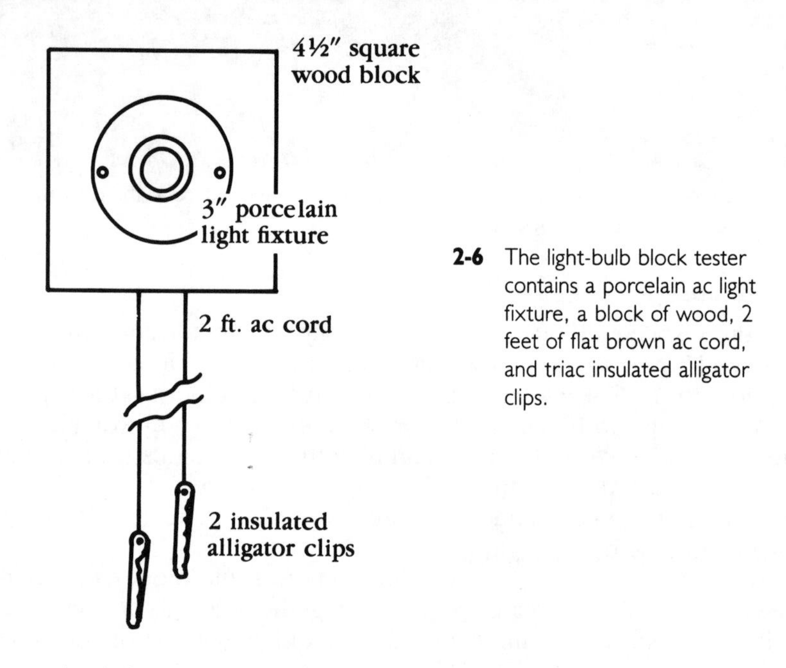

2-6 The light-bulb block tester contains a porcelain ac light fixture, a block of wood, 2 feet of flat brown ac cord, and triac insulated alligator clips.

Check fuses, triacs, and transformers in the microwave oven with this block tester. The tester can be used in series with an ac television chassis to indicate overloading. When the light goes out, you have located the defective component. The tester can be used in just about any power-line circuit.

Construction

The light-bulb tester was constructed on clean, dry, 2-×-6-inch pieces of wood. A three-inch porcelain socket is used to hold the light bulb. Cut a 4 1/2-inch square piece of wood. This piece can be a scrap of lumber or it can be cut from the end of a 2-×-6 stud. Round all corners with a sander or sanding block (FIG. 2-7).

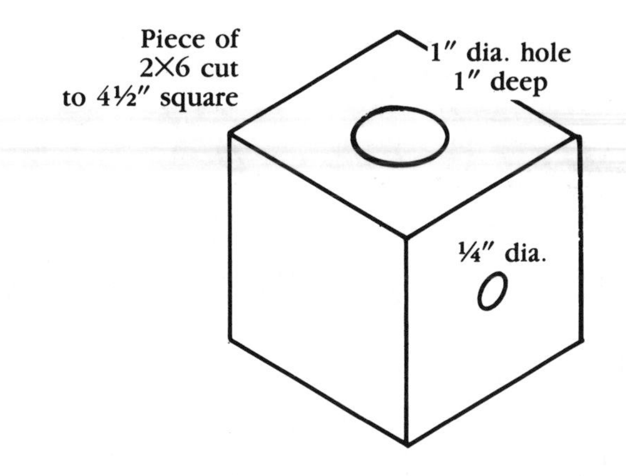

2-7 Cut the 2 x 6 piece of wood 4 1/2 inches square with a 1-inch hole in the middle. Drill a 1/4-inch hole in one end for the ac cord.

Drill a one-inch hole in the middle and top area of the wood. Do not drill through, but only down about one inch. Now, drill a 1/4-inch hole in one end, through to the middle hole. The ac cord will run through this hole. Make sure the block is clean. Spray on two coats of clear lacquer and let each coat dry for a few hours.

Wiring

Feed the end of a 24-inch piece of brown flat ac cord through the block and up to the light socket. Remove the light socket from the porcelain housing. Tie a knot in the cord, leaving 4 inches, so the wire cord will not pull out. Connect the wire ends to the two brass or bronze screws. It doesn't matter which wire goes where. It's best to solder each stranded wire end for a better connection before attaching it to the screws. Double-check each screw and wire for tightness.

Reassemble the porcelain light fixture and pull any extra wire out. Screw the light fixture to the board. Now solder on two insulated sharp-pointed alligator clips for easy application. Screw in a 25-watt bulb for most 120-volt continuity tests. A 100-watt bulb can be used to check for overloaded circuits in the TV chassis.

Testing

The light-bulb tester can be checked by clipping the leads across a power outlet receptacle. Next time that a house fuse blows and you don't know which one is open, clip the tester across each fuse until the bulb lights (FIG. 2-8). Be careful! Remember, you are working with voltages that can shock or kill you.

When a fuse continues to blow after you replace it, clip the tester across the house fuse terminals. Pull out each ac appliance, TV, and floor

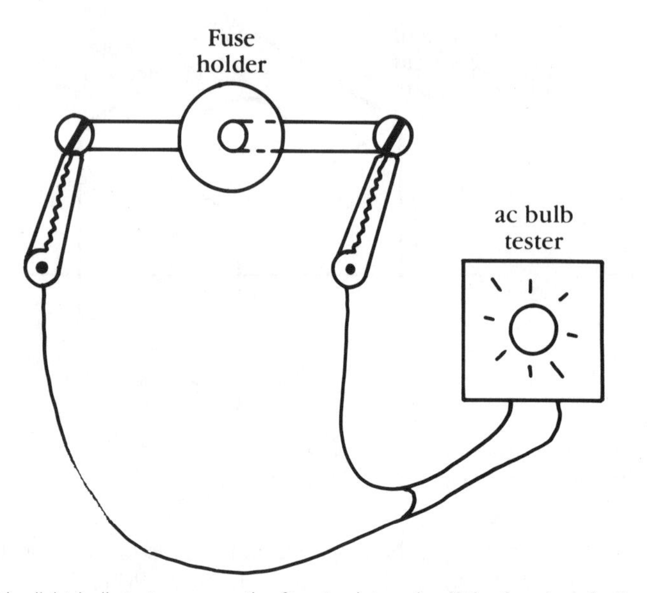

2-8 Clip the light bulb tester across the fuse to determine if the fuse is defective or open. You can also use it as monitor when the fuse keeps blowing.

lamp until the light goes out. When the light goes out, you have located the defective or shorted appliance or fixture.

If you have a power saw or planer that does not turn on, clip the tester across the switch terminals to see if switch is open. If the light stays on with the switch in any position, the switch needs to be replaced. Clip the tester across the motor terminals with the switch off (FIG. 2-9). Now, turn on the switch to see if voltage is found at the motor. The bulb should light. A grounded motor will indicate when one side of the tester is applied to the metal motor and to one motor terminal. The motor is defective if the 120-volt light bulb is lit.

Many TV technicians use a 100-watt bulb in series with the power cord to determine if the chassis is overloaded. In this situation, either the fuse keeps blowing or the circuit breaker will not hold. Clip the tester across the fuse or circuit-breaker terminals. The light will be bright if there is a leaky or shorted component in the TV chassis. With a normal chassis, the light bulb will be either half-lit or very dim. There might be

Parts list

One 3½″ diameter porcelain light fixture.
One 4½″ square piece of wood.
One 2′ piece of brown ac cord.
Two insulated/covered alligator clips.

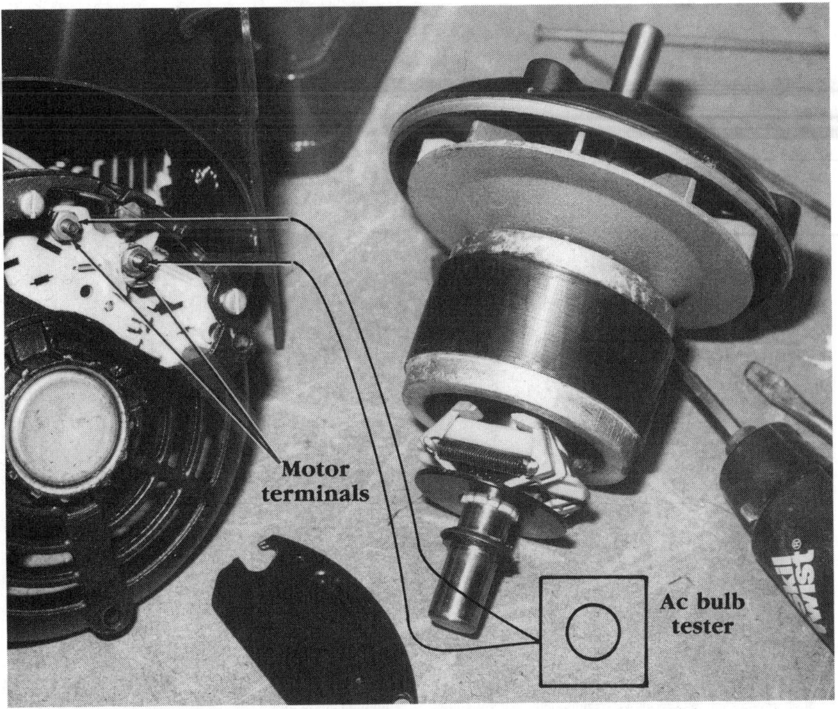

2-9 Clip the tester across motor terminals to see if they are receiving power.

several cases each year when you can use this tester, so why not build one to have around the house or on the workbench?

SIMPLE FUSE TESTER

You are fumbling around in the dark basement trying to locate a dead fuse in the fuse box after a heavy thunderstorm. The TV and living room lights went out during an NCAA basketball game. None of the fuses are marked and all you have is a small flashlight. Where do you go from here? To prevent this situation from happening, build this small fuse tester that can check most fuses found in the home (FIG. 2-10).

The circuit

All the fuses can be tested on the top side of a plastic box. You can place a fuse in the slot and test it with only one hand to avoid fumbling around in the dark. The jumbo LED indicator should be bright on each fuse test. Just take one fuse out at one time, test it, and replace the bad one.

The circuit is quite simple. All parts are wired in series and there are no switch or test cords to fool with. If you only have to use it once or twice a year, the batteries should last a few years. Choose alkaline C-batteries for longer life.

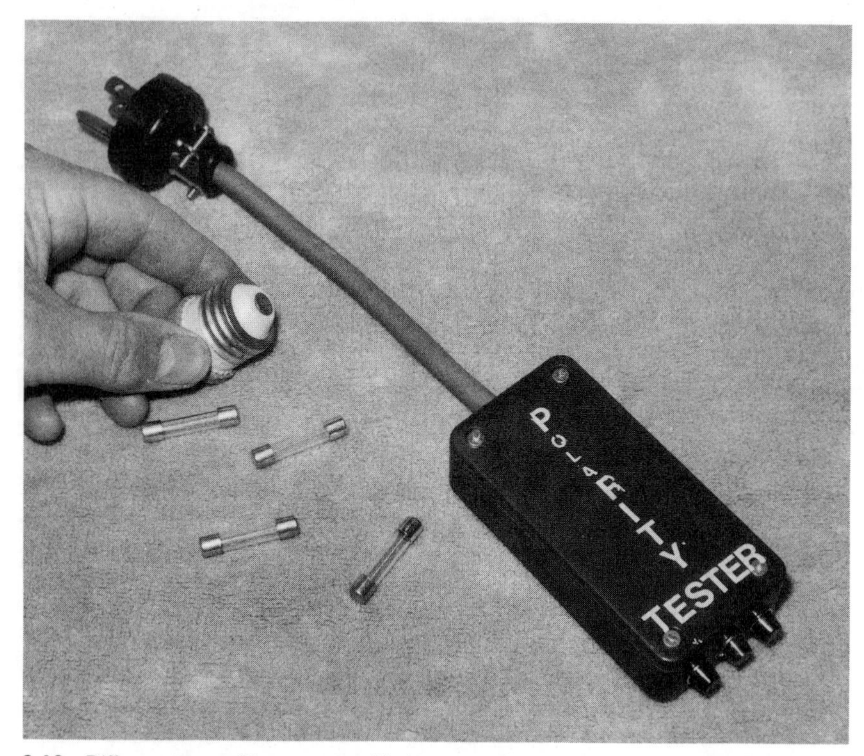

2-10 Different types of fuses are used in houses and consumer entertainment equipment.

A large red jumbo LED (276-065) with three leads is used here. The positive terminal of the battery should connect to the positive terminal of the LED (FIG. 2-11). Solder the two outside terminals together and use the center wire for the positive terminal. Place the two C-cells in a battery holder so they can easily be replaced.

Select a small fuse clip for the slender and cartridge fuses. This little tester will check fuses from your TV set, cassette deck, CD player, car radio, amplifier, and microwave oven. A homemade gizmo for checking round house fuses can be made from long and short bolts and nuts. Secure the long bolt, with a nut and star washer on top, and a nut below, to hold it in position (FIG. 2-12). Dab white enamel paint or rubber silicone on the bottom nut.

Construction

Choose a small plastic box that will hold the two C-cells in the bottom side. All the other parts are mounted on the top plastic cover. Mount the two bolts toward the top of cover. Place the bolts 1/2 inch apart. Drill two 1/8-inch holes for the bolts.

Next, mount the fuse clip assembly in the middle of the top cover.

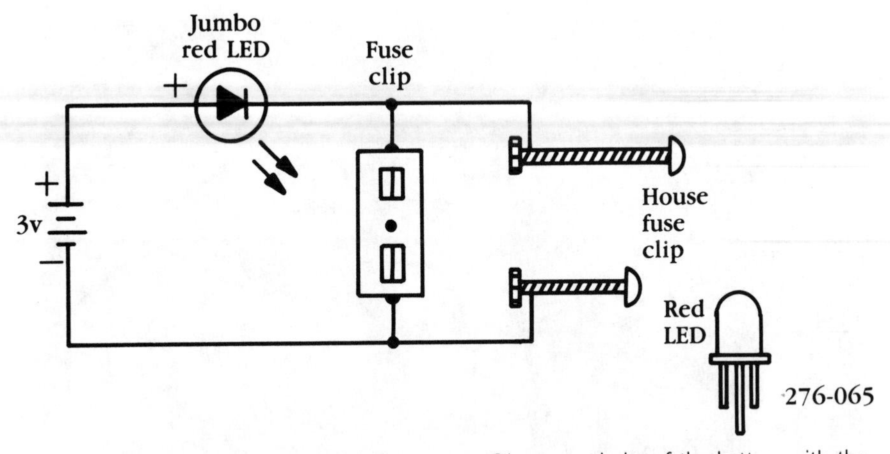

2-11 Circuit diagram of the simple fuse tester. Observe polarity of the battery with the LED.

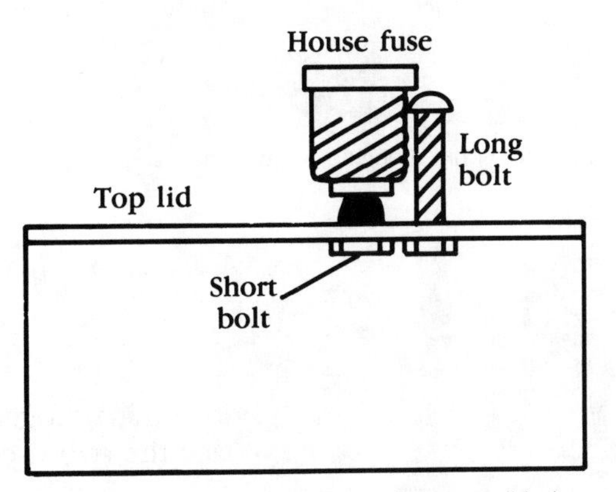

2-12 The house fuse clip is made out of two small bolts and nuts.

Drill one 1/8-inch hole in the middle and two 1/16-inch holes at each end for the hookup wires. Mount the jumbo LED at the bottom. Drill a hole just big enough for the tip of the LED to come through the top cover. The plastic rim will hold it inside. Cement the LED to the lid with rubber silicone. Mount all parts before soldering.

Wiring

First wire the two fuse clips in parallel. Solder the LED in series with the fuse clips and batteries. Make sure that the connections are correct on the batteries and the LED. Connect the negative side of the batteries to the two fuse clips. Double-check the wiring and LED polarity.

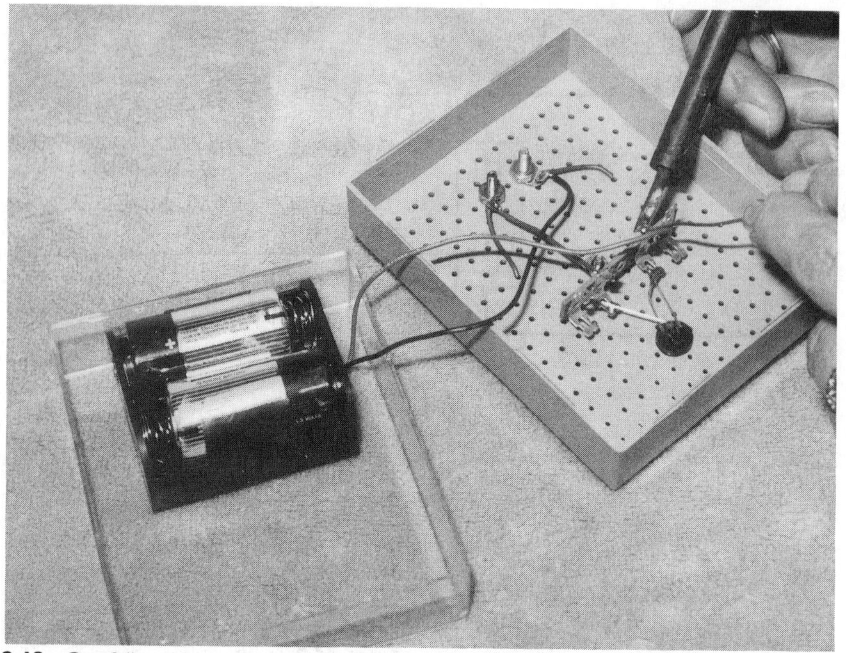

2-13 Carefully solder each connection and component in series. Wire the two fuse holders in parallel.

2-14 Testing a suspect house fuse. The fuse is open if the LED does not light.

Parts list

LED	Jumbo-type, 276-065 or equiv.
Battery holder	For 2 C-cells, 270-385 or equiv.
Fuse clip	Chassis mount, 270-739 or equiv.
Case	6″-X-3³⁄₁₆″-X-1⅞″ plastic economy box, 270-223.
Batt.	2 alkaline C-batteries.
Misc.	Bolts, nuts, solder, hookup wire, etc.

Usage

Before screwing the top lid to the bottom plastic box, short the fuse clip with a screwdriver. The LED should be very bright. Now, place a good fuse across the fuse terminals. If the fuse is normal, the LED will light. Check a house fuse on the bolt connections. If the LED does not light, the fuse is open. Now, place the tester in a spot where you will not forget it the next time a fuse blows.

POWER LINE POLARITY TESTER

All electrical outlets and appliances should be connected for safety to follow local and national electrical codes. Just plug this little tester into the service bench or into any three-prong outlet and it immediately shows if the receptacle is correctly wired. Protect yourself and others by grounding all power tools and electrical outlets.

On any electrical outlet, the positive side is on the right (FIG. 2-15). Usually, the bronze screw should match up with the smaller slot. The left side of the outlet is connected to the negative wire with the large slot. The negative screw can be either brass or white in color. The round hole at the bottom and metal-plate screw holder should be at ground potential.

The red light will indicate if the outlet is hooked up backwards or if power is present. The green light indicates a poor ground, and the amber bulb lights to indicate improper connections to the hot wire. No light from the amber bulb is normal.

The circuit

The test circuit is very simple and it uses 120-volt neon indicators (FIG. 2-16). Looking at the end view of the 3-wire plug, the red neon indicator is wired across the two top terminals. Connect the green bulb to the small spade (hot) and other lead to the round ground terminal. Wire one lead of the amber bulb to the ground plug (round) and to the neutral receptacle terminal.

2-15 The ac outlet should have the positive (black) wire connected to the right side or bronze screws of the receptacle.

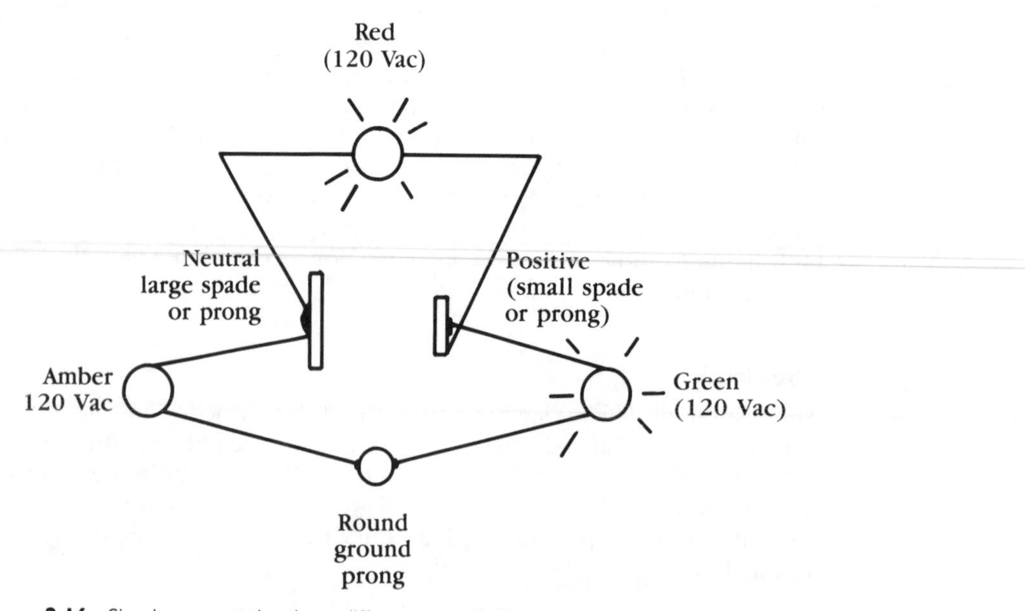

2-16 Simply connect the three different neon indicators across the ac plug.

Construction

Select a small plastic box to hold the three neon bulb assemblies. If the neon indicators are placed side by side in the end of the plastic box, a 4-×-2-×-1 3/16-inch box is ideal. Place the red bulb on the left, the amber in center, and the green to the right. Drill a 5/16-inch hole for the red and green indicators. Drill a 9/32-inch hole for the amber bulb assembly (FIG. 2-17). Now, drill a 17/64-inch hole in the red side of the box for the round three-wire cable.

2-17 Inside view of the power line polarity tester.

Wiring

Mount a 4-lug terminal strip at the bottom of the plastic box. Solder the black and white wires of the cable to the two outside terminal lugs. The black lead terminal is the hot lead. Connect the green ground wire to the center terminal. All ground or neutral wires should be connected to this terminal.

Start with the red light assembly and connect one wire to the hot and other to the neutral terminal. Solder the amber indicator across the neutral white wire and other lead to the green ground wire lug (FIG. 2-18). Connect the green bulb to the common ground and the other lead to the hot black lug wire. It does not matter which wires on the neon lamp are connected as long as both leads are black.

Slip the three-prong plug over the cable and tin each end. Connect the black wire to the small spade and the white wire to the large terminal. Wrap the green ground wire to the round ground terminal of the ac plug. Doublecheck each terminal and lamp to ensure correct hookup.

Testing

Now, give it a try by plugging into a three-prong receptacle. When the red bulb is lit, 120-volts ac is applied across the receptacle. If the red bulb is not lit, check for improper wiring or a blown fuse. Under normal conditions, the green bulb should be lit, indicating a good ground. If the green

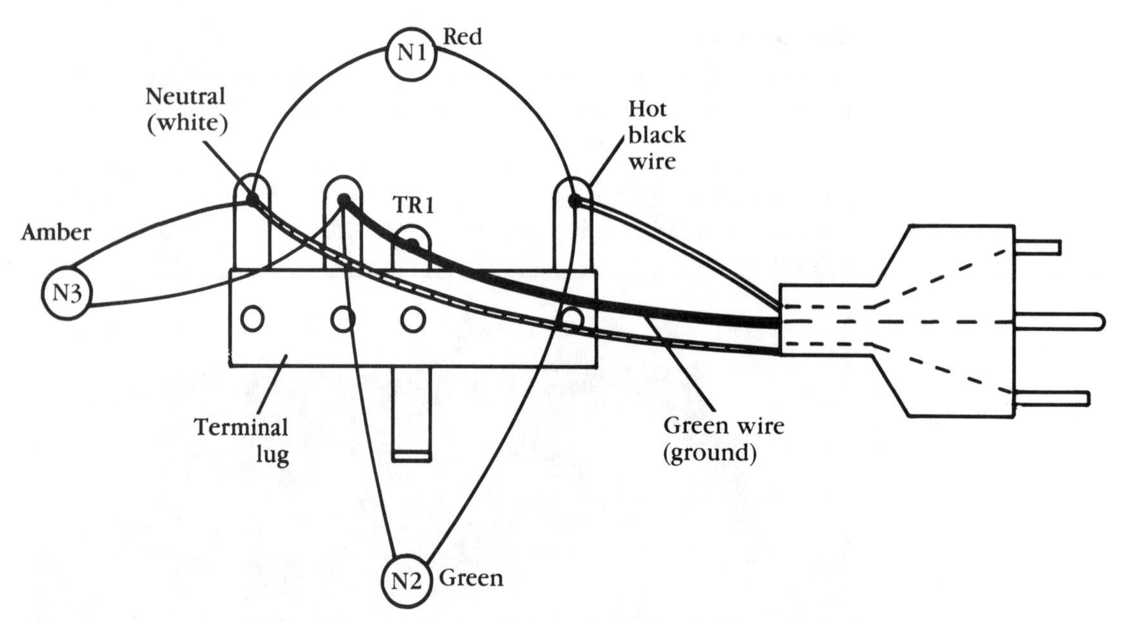

2-18 A pictorial drawing of how each part is wired into the circuit.

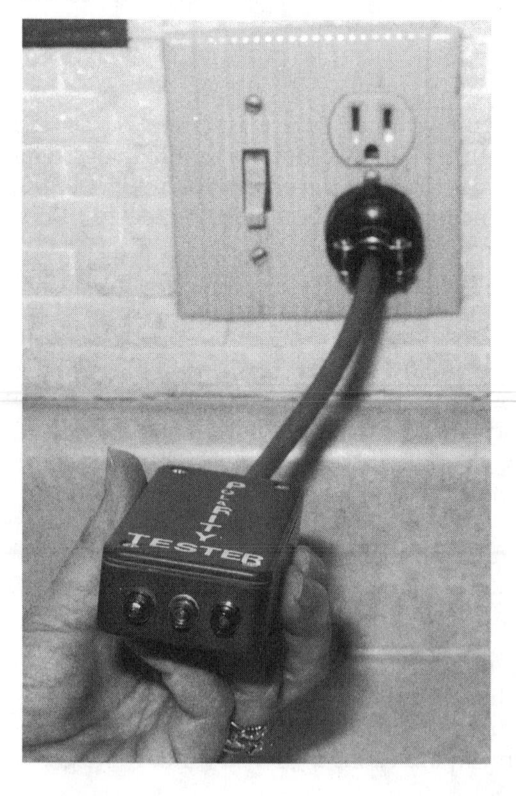

2-19 Using the polarity tester to check a kitchen outlet.

Parts list

N1	Red 120-volt neon bulb assembly, 272-712 or equiv.
N2	Green 120-volt neon bulb assembly, 272-708 or equiv.
N3	Amber 120-volt neon bulb assembly, 272-707 or equiv.
TR1	3-lug terminal strip.
Plug	3-prong polarity plug.
Case	4-×-2-×-.844-inch project case, 270-220 or equiv.
Misc.	Bolt, nut, one foot of three-wire cable, one rubber grommet, hookup wire, solder, etc.

light flickers or goes out, a poor ground exists at the receptacle or fuse box (FIG. 2-19).

The amber bulb should never light. If the amber bulb lights, the receptacle is not wired correctly. The hot wire is on the left side of the receptacle and the neutral wire is connected to the right side. Reverse the two wires in the receptacle after removing a fuse or turning power off at that fuse box. If the amber bulb is barely lit, it might indicate a poor ground between the neutral terminal of the outlet and the green ground wire. Check the ground wire of the outlet and the fuse box. If all three lights are on, power is present with a poor ground.

Conclusion

The small power line polarity tester should have no connections or bolts outside the plastic box for safety hazards. This small tester can quickly check power-line voltage and proper grounds. Solder all connections to be safe. Keep the tester handy for power voltage or thunderstorm breakdowns.

MICROWAVE OVEN TESTER

The microwave oven tester is a handy gadget to take along on a microwave oven service call or to use on the workbench. Place the small tester inside a plastic box to prevent electrical shock. Only four major components are needed to build this shockproof microwave oven tester.

The circuit

Actually, the transformer and 12-volt bulb take the place of an ordinary 120-volt light bulb. The only parts exposed to the power line circuits are the two alligator clips and the ac cord. You can lay this small tester along the side of the metal oven without arc-over or electric shock hazards. The tester can sit on top of the oven and not break down.

The insulated alligator clips are soldered to a two-foot piece of flat rubber ac power cord. Use strong enough clips so they will cling to the various sizes of terminals and wires (FIG. 2-20). Use an insulated strip for ac cord and transformer wire ends or wire them direct. Solder and carefully tape the wires. Tie a knot in the cord so it will not pull out.

2-20 The simple microwave tester circuit has only a few working components.

The power transformer steps down the power line voltage to 12.6 volts. Solder the secondary leads directly to the 12-volt lamp assembly. Choose the enclosed jumbo red lamp assembly with solder tabs (Radio Shack 272-336). If one of these assemblies is not available, choose an E-10 lamp base with solder lug connections. This screw-type base will hold an E-10 threaded bulb—12- or 14-volt type (FIG. 2-21).

Plastic box construction

Select a small plastic box to house the largest component (T1). The miniature power transformer should be rated at 12.6 volts with 300 mA's (Radio Shack 273-1385). The power transformer can be mounted at one end with the ac cord running out of the other. Place the bulb assembly in the middle/top area of the top cover.

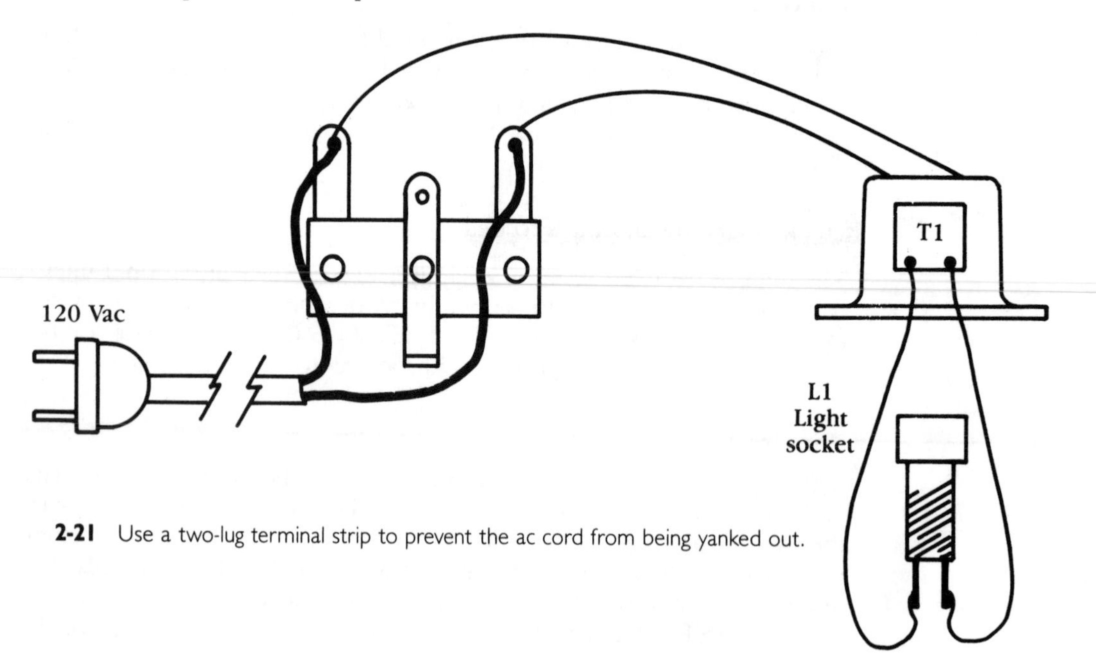

2-21 Use a two-lug terminal strip to prevent the ac cord from being yanked out.

Drill two 1/8-inch holes to mount the transformer. Drill a 11/32-inch hole in the other end of the plastic box and place a rubber grommet in the hole for the ac cord to come through.

How to use

Select a pair of alligator claw clips that will hold onto wires and component terminals (FIG. 2-22). Always clip the alligator clips to the oven before turning the oven on. Also, discharge the high-voltage capacitor after the microwave oven is turned off.

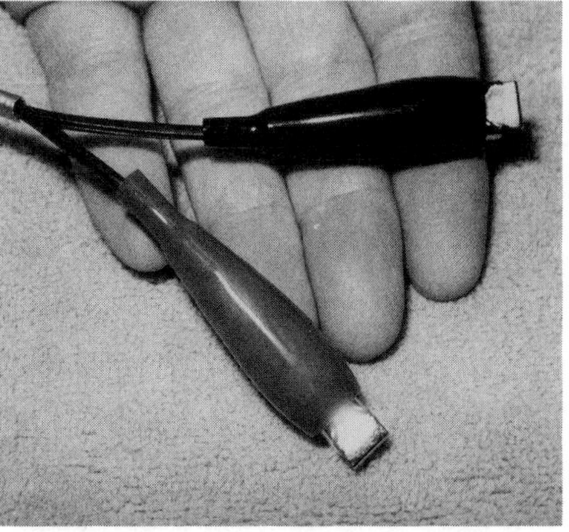

2-22 Place heavy-duty claw-type alligator clips on the end of the 2 foot cord for clamping securely on wires and connections.

The ac clips can be clipped across most any component in a microwave oven to determine if the part is open or defective. The small microwave tester can be clipped across interlock switches, cavity lights, thermal cutouts, monitor switches, various motors, switch points, oven relays, triacs and the primary winding of power transformers (FIG. 2-23). The tester can be used as a voltage monitor at the primary winding of high-voltage transformers to determine if voltage is applied to the high-voltage circuits.

Before clipping the test instrument to an oven component, make sure the high-voltage capacitor is discharged. If the high-voltage capacitor is not discharged, you could be shocked or injured while working around the microwave oven. A person with a weak heart could be killed. So take it slow and keep your mind on your work. Discharge the high-voltage (hv) capacitor by placing two long screwdriver blades across the capacitor terminals (FIG. 2-24). Always use screwdrivers with insulated handles. Sometimes the capacitor is mounted where one screwdriver metal blade can be used to discharge the capacitor. You should hear a loud snap or pop when the capacitor is discharged.

Fuse check If you have no way of testing a thermal-type oven fuse, clip the tester across the fuse terminals. Remember, the thermal fuse has a

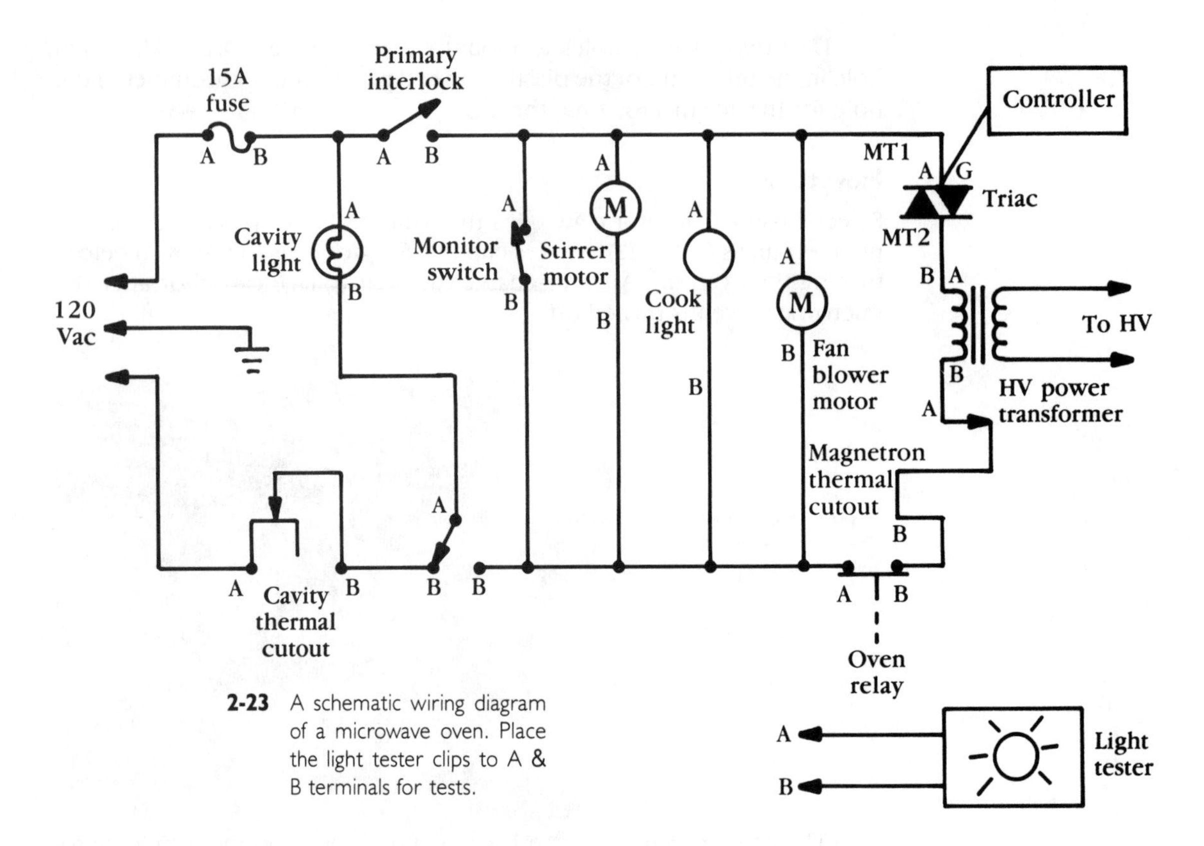

2-23 A schematic wiring diagram of a microwave oven. Place the light tester clips to A & B terminals for tests.

solid ceramic cover and you cannot see if the fuse is open or normal. The tester will light if the fuse is open.

Interlock switches Two or more interlock switches may be inside the microwave oven. These switches are to protect the operator and the oven from possible oven radiation. If the interlock switch does not open, the oven can operate when the door is open. Some of these interlock switches have more than one position and might be stacked.

Clip the two leads across the interlock switch. Now open and close the door. If the switch is in the normally closed position, the bulb will light when the door is closed. The light should go out when the door is opened. The monitor interlock switch is closed when the oven door is closed and the tester light is off. The tester should light when the door opens, indicating that the monitor switch is operating. If the test light flickers, suspect poor or dirty switch contacts.

Oven thermal cavity cutout All thermal cutouts are normally in the closed position. These cutouts will open if the oven is getting too hot. The cavity thermal cutout is usually located at the top of the oven. Clip the tester across the cavity thermal cutout (A & B). If the oven is cool and the test light is on, the unit is defective and should be replaced. Sometimes these thermal units overheat and cause poor circuit connections.

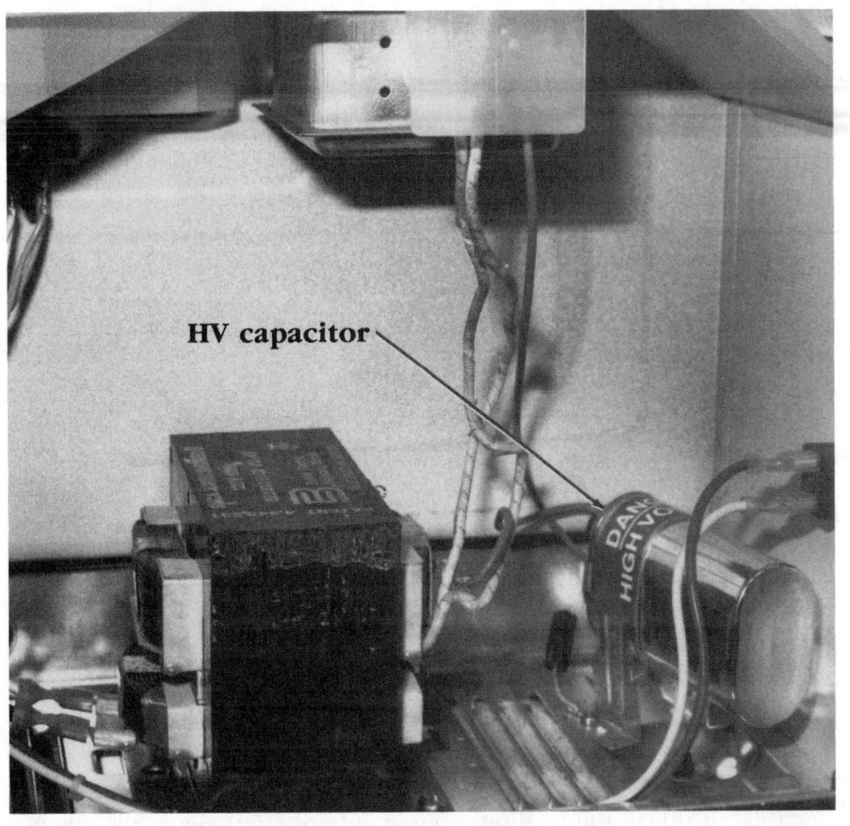

2-24 The HV capacitor in the microwave oven is located on the bottom panel of this oven.

Stirrer motor When the stirrer or turntable motor does not rotate and the oven appears to be cooking, suspect a defective motor or poor wire connections. Clip the light tester across the motor terminals. The light should be on when the oven is cooking. Check for an open fuse or a defective interlock switch with a dead test lamp.

Fan motor The fan motor keeps the magnetron cool and circulates air in the oven during the cooking process. Check for voltage across the fan motor with the light tester. If voltage is present, shut off the oven, discharge the hv capacitor, and rotate the fan by hand. The fan motor bearings might be sluggish or dry, preventing the fan from coming on. Lubricate them with light machine oil. Replace the fan motor if it does not rotate after oiling and if voltage is found at the motor terminals.

Oven relay The oven relay applies 120-volts ac to the primary winding of the hv transformer. The relay could operate from a separate oven switch or an electronic controller. Clip the light tester across the oven switch terminals. If the bulb lights with the oven turned on, suspect a defective relay. Clip the light tester across the solenoid winding terminals (if it is a 120-volt relay) to determine if power is getting to the relay.

Parts list

T1	Miniature 120-V power transformer with 12.6-V secondary at 300 MA, 273-1385 or equiv.
Lamp	12-V lamp assembly—jumbo red 105 mA, 272-336 or equiv.
Clips	2 heavy-duty insulated claw clips, 270-349 or equiv.
Box	Plastic economy case or box 4″ × 2⁷⁄₁₆″ × 1¹⁄₁₆″, 270-221 or equiv.
Misc.	Rubber grommet, 2 ft. flat rubber ac cord, bolts, nuts, two-lug terminal strip, solder, etc.

Triac assembly A triac assembly is used in place of the oven relay in some microwave ovens. If the oven does not start cooking with the fan blower operating, suspect that the triac is open or that the controller is defective. Clip the light tester across terminals MT1 and MT2. The test bulb will light indicating that the triac is defective. If the bulb does not light, the triac might be good, with a missing gate voltage.

Transformer monitor Clip the light tester across the primary winding of the hv transformer. The test light should be on with a normal oven. If the light is off, the circuit is open; possibly a defective triac, oven relay points, interlock, or fuse is to blame. The light clipped across the transformer can be used as a monitor for intermittent oven operation.

INEXPENSIVE CONTINUITY TESTER

The low-priced continuity tester consists of only four electronic components. This tester comes in handy to locate or check door bell wires, switches, to locate correct wires in the wall, motors, buzzers, transformers, and much more. The small LED will light when continuity is present. Just clip the alligator clips across the part to see if it is open or has continuity. The indicator blinks and the piezo buzzer sounds.

The circuit

The continuity circuit consists of a 3-volt battery, a blinking red LED, a piezo buzzer, and test leads. All components are wired in series except the LED and buzzer (FIG. 2-25). Place the two parts in parallel with each other and in series with the battery and test leads. When the component to be tested is connected to the alligator clips, all parts will be in series.

Construction

Choose a small plastic cabinet large enough to house the battery holder. Place the "C" batteries in one end of the box and mount the LED buzzer at the other end. Both the LED and the piezo buzzer are mounted on the top plastic cover.

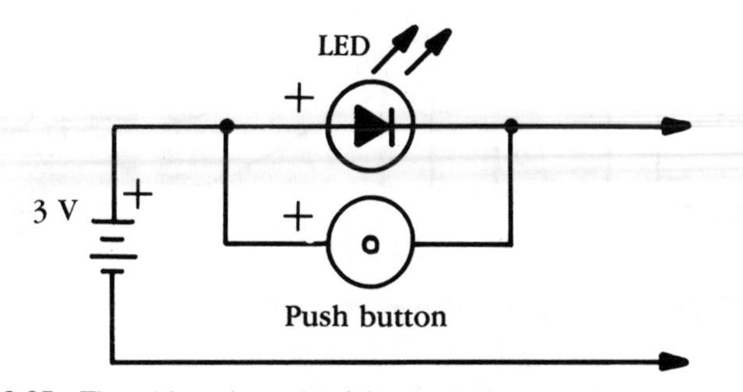

2-25 The wiring schematic of the continuity tester is very simple.

Drill two 1/8-inch holes in the bottom side of the box to secure the battery holder (FIG. 2-26). Lay the piezo buzzer on the top lid so it will not strike the batteries after it has been mounted. Mark two round holes where the pc mounts will go. Drill two 1/16-inch holes for the buzzer terminals and a 3/16-inch hole for the LED. Center two 1/8-inch holes on each side of the buzzer for the test leads. Cement the LED and buzzer to the top lid with rubber silicone.

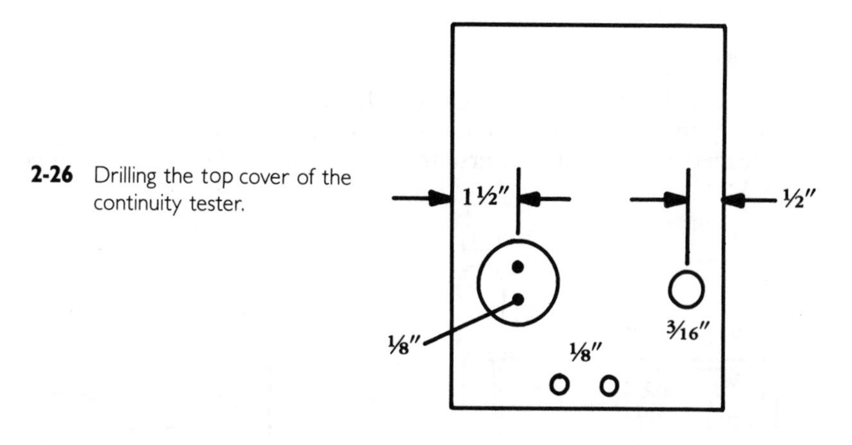

2-26 Drilling the top cover of the continuity tester.

Soldering

Since very few components need to be connected, perfboard or pc wiring is not required. Instead, bolt a three-lug terminal strip to the top panel to tie the test leads into the circuit (FIG. 2-27). Feed one end of the test leads through the front cover holes, remove the insulation, twist the flexible wires, and tin the ends with solder. Solder the two leads into the outside post of the terminal strip.

Connect the two terminals of the LED and pc buzzer together. Be careful when placing the hot iron on the buzzer leads, because too much heat can destroy the soldered connection inside. Apply enough heat to make a good connection. Form a round circle with the solid hookup wire

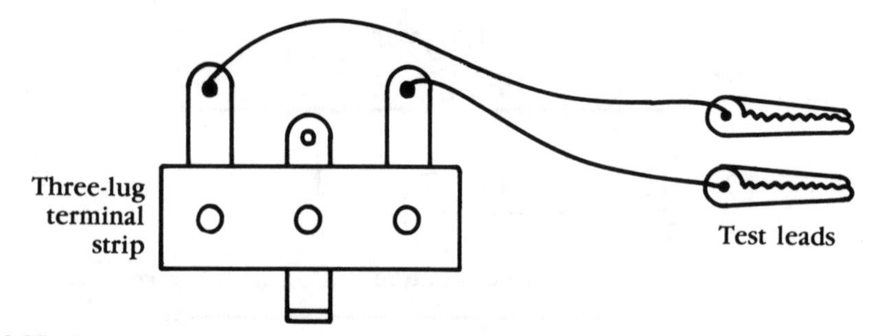

2-27 Bolt a three-terminal strip to the underside area of cabinet. Tin the ends of the test leads and solder them directly to the two outside terminals.

and loop it over the pc terminal before soldering. Connect all positive terminals together.

Solder the hookup wire to the positive battery terminal post and connect it to the positive terminal of the LED and to the piezo buzzer. Solder the negative terminal of the batteries directly to the other test lead (FIG. 2-28). Connect the remaining test lead to the negative side of the LED and the buzzer.

Testing

Simply clip the two alligator clips together and the LED should light intermittently and the buzzer should sound. If not, check the polarity of both the buzzer and the LED. Reverse the leads of the LED if the buzzer works

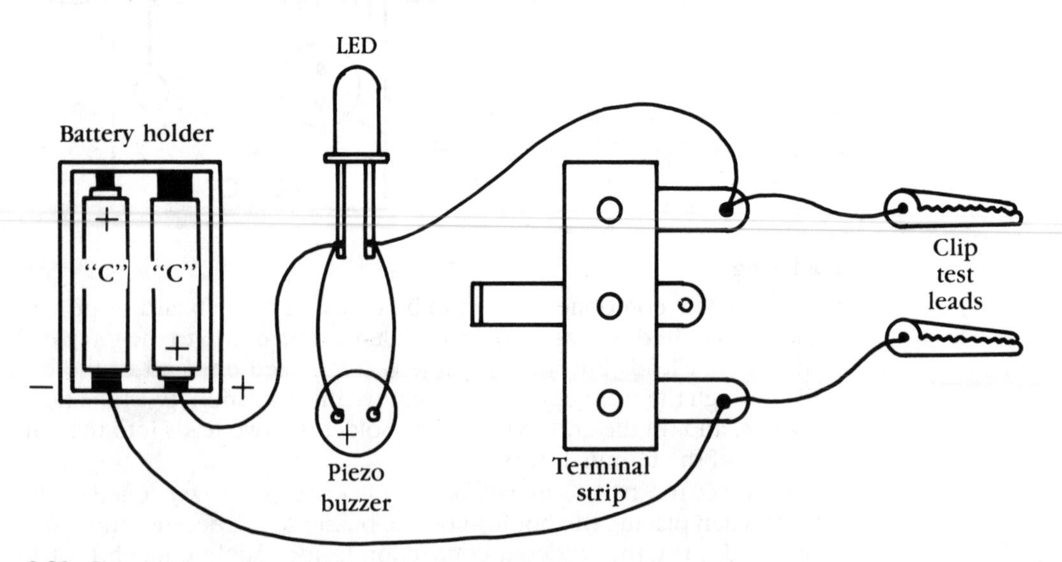

2-28 Pictorial diagram of soldering together the four components. Be careful not to leave the soldering iron on the buzzer leads too long.

Parts list

LED	Red blinking LED, 276-036 or equiv. (20 mA at 3V).
PB	Piezo buzzer (2800 Hz) 3 to 20 Vdc, 273-065 or equiv.
Batt.	2 C-cell alkaline batteries.
Holder	1 C-cell holder, 270-385 or equiv.
Box	Plastic economy case, 4¾″ × 2½″ × 1¹⁄₁₆″, 270-222 or equiv.
Misc.	One set of test leads, terminal strips, hookup wire, solder, etc.

and the LED does not. Check for a poor solder joint or a broken connection if neither action works.

Besides continuity tests, the cheap LED tester can be used to check other LEDs, diodes, power transformer windings, and coils (FIG. 2-29). Do not use the tester to check junction leads of transistors—the applied voltage might destroy low-signal transistors.

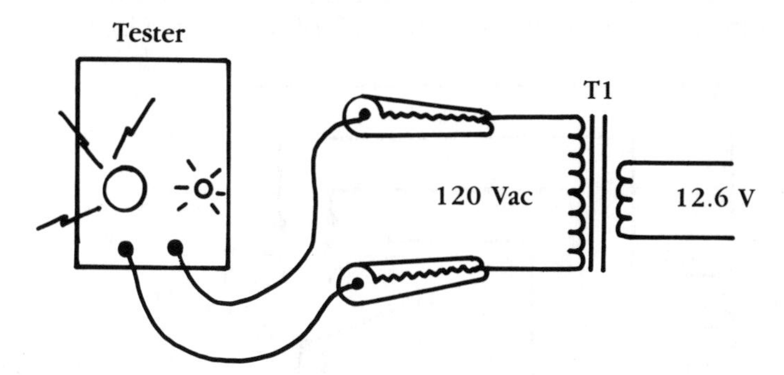

2-29 This little continuity tester can check transformers, coils, LEDs, diodes, and many other devices.

LOW-VOLTAGE DC MOTOR TESTER

The low-voltage motor tester can be used to check the condition of small dc motors in CD players and camcorders. These motors require from less than one volt to as much as 6 volts to operate. Most of these dc motors are controlled directly by a driver IC and a control microprocessor (FIG. 2-30). The dc voltage is clipped across the motor winding to see if the motor rotates. The various switched voltages change the speed of the motor.

The circuit

The low-voltage tester is a simple battery supply circuit that has its voltage varied by a rheostat. The small wattage control used in audio and linear circuits will not work in this application. A higher-wattage rheostat is needed because the dc motor pulls greater current. Different voltages are switched into the circuit, then varied with R1.

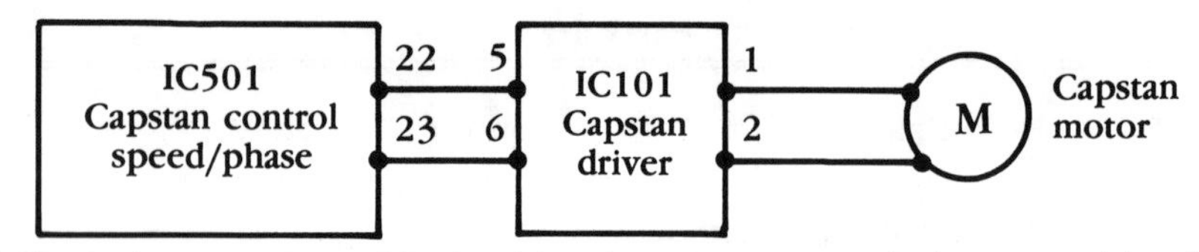

2-30 In the camcorder or CD player the microprocessor controls the capstan driver (IC 101), which in turn rotates the small capstan motor.

Wire six D-cells in series to produce six volts. Each battery is tapped and switched into the circuit. The voltage sources vary to the terminal jacks, which connect to the motor under test. Use a 10- or 25-watt control rheostat. You can find these controls in surplus or electronic parts mail-order catalogs.

The five-position rotary switch applies the various voltages to the motor. Position one shuts everything off—if not, the batteries will soon run down. All four batteries are wired in series (FIG. 2-31). Use D-cell alkaline batteries for longer life and greater current capacity. Each battery voltage is tapped and applied to succeeding positions of SW1. Check the voltage applied to the motor with the pair of monitoring jacks.

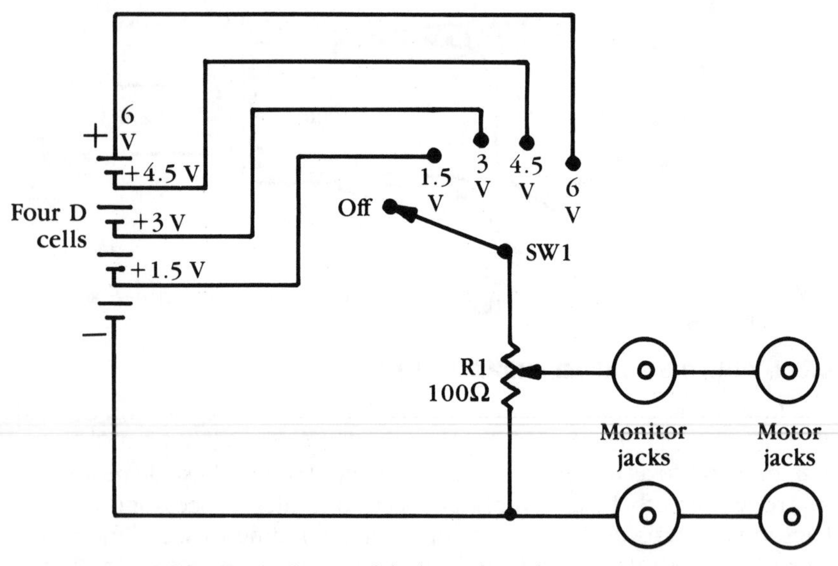

2-31 Circuit diagram of the low-voltage dc motor tester.

Construction

Select a plastic box large enough to contain the battery holder and the rheostat, the two largest components. In this project, a 7½-×-4¼-×-2¼-inch plastic case does the job. Mount the four banana jacks at the bottom edge of the front panel. Place the selecting switch and control to the right of the jacks.

Drill a $^3/_8$-inch hole for the 5-position switch (or use a 6- or 10-position switch, if you have it). Drill a $^{13}/_{32}$-inch hole for the 10-watt rheostat. Center and space the four $^1/_4$-inch holes at the left and bottom side of the top lid (FIG. 2-32). Before mounting SW1, spray the switch contacts with cleaning fluid, then bolt the battery holder to the bottom box area.

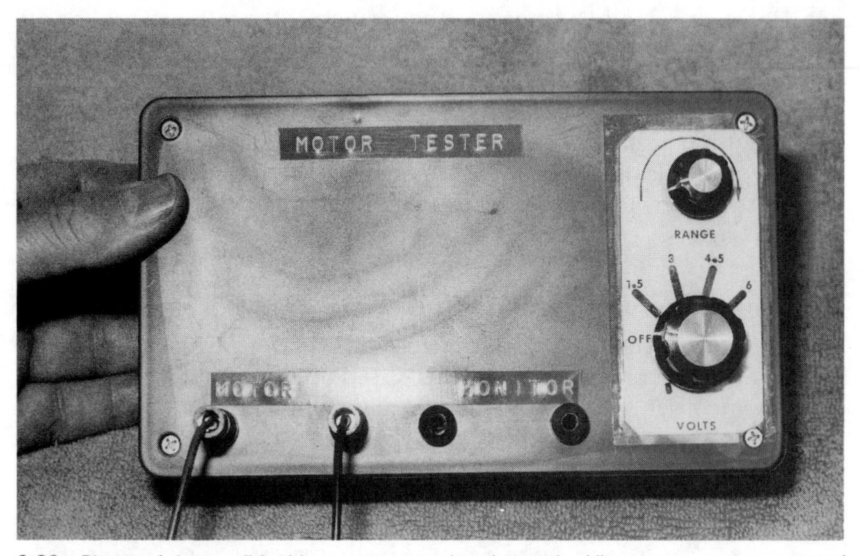

2-32 Photo of the top lid with parts mounted underneath. All components are mounted on the lid except the batteries and battery holder.

The four-cell battery holder must be modified to connect each voltage terminal to SW1. Form or cut three shiny strips of metal that will slip in between each battery cell. Cut the pieces of metal, $^1/_2$-inch diameter with a terminal lug formed at one end. Solder a piece of hookup wire to the end that will connect to SW1. After you place the batteries in the holder, slide the terminal strip between each battery with hookup wire and solder to SW1.

Wiring

After the large components are mounted, solder the parts that should be placed on the top lid. Start with SW1 and connect each battery terminal wire to the rotary switch. Solder the 1.5-volt wire to terminal 2 of SW1. Remember to leave position 1 open in order to turn the motor tester off. Check terminal 2 with the ohmmeter if in doubt. Likewise, solder each voltage wire, starting with the lowest voltage. Place the 6-volt wire of battery holder to the fifth position of SW1.

Connect SW1 switching terminal to the high side of R1. Run the negative battery terminal to the bottom side of R1 and to the black jack terminals. Connect a red wire from the center terminal of R1 to the two red jacks. Make sure that the red jacks are connected together. Use the first set of banana jacks for voltage monitoring with an outside DMM or voltmeter.

Testing

Plug a pair of voltmeter probes into the monitor jacks and check the voltage. No voltage should be measured at position 1; the tester should be off. Switch SW1 to the number 1 position; rotate R1 to highest level and you should measure 1.5 volts. Check each switch position for the correct voltage—terminal 3 should have 3 volts, etc.

Insert a pair of banana plugs into the voltage jacks for motor tests. Select a pair of flexible leads with banana plugs at one end and small hook clips at the other. These hook clips are easily connected to the motor terminals or wires.

Checking camcorder and CD motors

The defective motor in the camcorder or CD player might have become open, shorted, erratic, or frozen (FIG. 2-33). For instance, if a zoom motor in the camcorder operated to one end and stopped, the motor might be jammed in that position. Check the small gears that rotate to make the zoom lens assembly focus. Suspect a defective motor if the gear assembly appears normal.

Measure the voltage applied to the zoom motor. If voltage is present, the motor might be defective. If no voltage is found at the motor terminals, suspect a defective drive IC or control processor. To make sure the

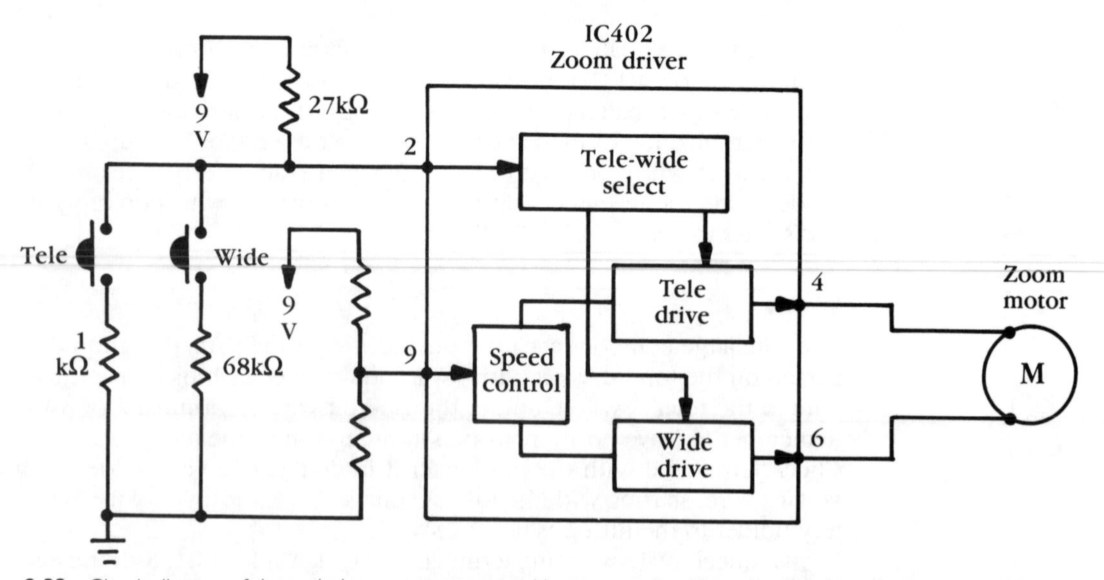

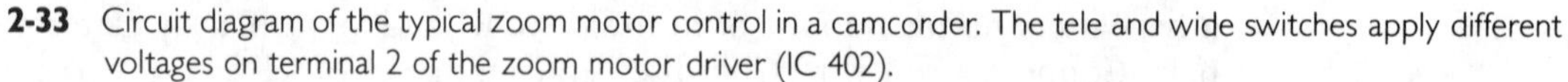
2-33 Circuit diagram of the typical zoom motor control in a camcorder. The tele and wide switches apply different voltages on terminal 2 of the zoom motor driver (IC 402).

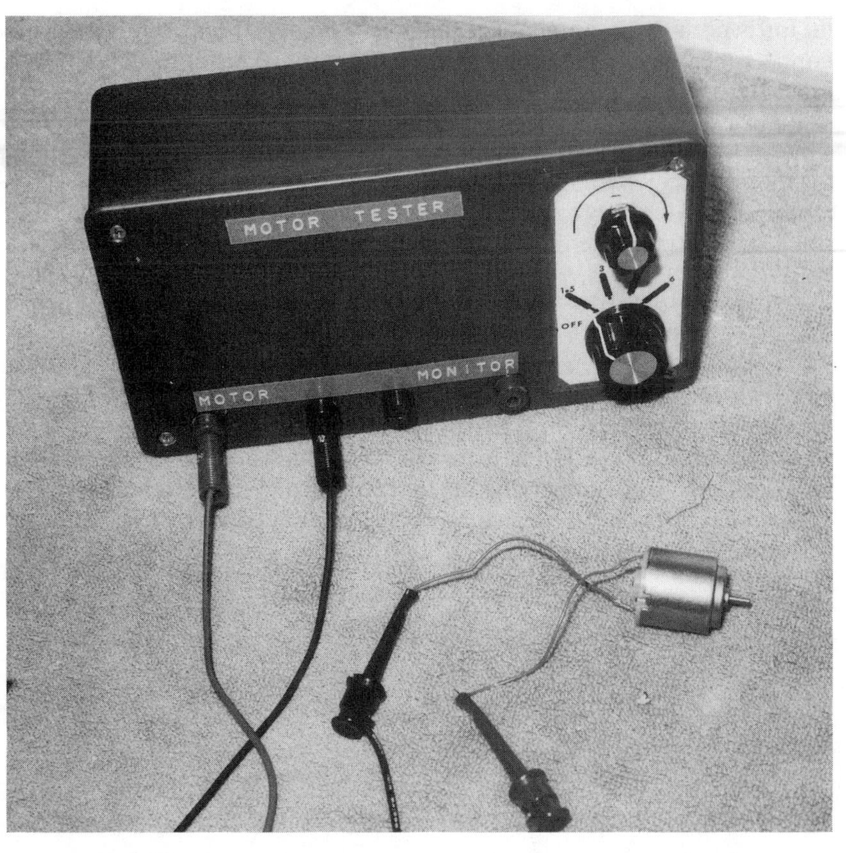

2-34 Test remote control model motors with the motor tester.

Parts list

SW1	5-position rotary switch, MRS-60 All Around Corp.
R1	100-Ω 10-W rheostat, RHE-100 All Around Corp.
Battery holder	4 D-cells, 270-396 or equiv., or 2 to 4 D-cells wired in series.
Case	Economy plastic box 7½″ × 4¼″ × 2¼″, 270-224 or equiv.
Jacks	2 sets of red and black banana jacks, 274-725.
Batt.	4 alkaline or heavy-duty D-cells.
Test leads	Mini-clip test leads with hook tips at one end, 278-1160 or equiv.
Misc.	Bolts, nuts, hookup wire, solder, etc.

motor is normal, disconnect the positive wire at the motor terminal and connect the positive wire of the low-voltage motor tester to the removed motor terminal. Clip the black wire to the other side of motor.

Always switch the motor tester dc voltage source to the lowest voltage (1.5 V). Make sure R1 is toward the ground or negative terminal. Monitor the dc voltage with a DMM or VOM. Slowly raise the voltage and the motor should start to rotate. As more voltage is applied, the motor will run faster. Notice if the motor is erratic or will not turn over at times (FIG. 2-34).

Suspect a defective motor when the motor will not operate on 1.5 volts. Switch SW1 to the 3-volt range. If it will not operate on this setting, the motor is open or defective. Use an exact replacement if the motor is erratic or broken. Always disconnect one side of the motor (high) from the circuit so that you do not damage the driver IC or wiring. Some of these motors unplug entirely from the pc board.

Besides testing or applying dc voltage to camcorders and CD players, you can check battery-operated motors in remote cars, trucks, and boats with this low-voltage dc tester.

Chapter 3

Novice test instruments

This chapter starts with a transistor switcher that provides easy transistor testing—in and out of the circuit—with the DMM. For the stereo listener is a balance indicator that will balance both channels or troubleshoot the various stereo circuits. With the diode/scr/triac/continuity checker, you can test important components before installing them in your favorite electronic project. The CB RF power tester shows how to check the signal output to see if the flea-powered walkie-talkie is even operating. Last, but not least, this chapter contains a noise generator project that can be used when signal tracing audio in the front-end circuits of radios and amplifiers.

Most of these electronic test instruments require only a few parts and can be constructed within a few hours. Schematic diagrams and important photos illustrate the layout of the parts and hookup for each project.

TRANSISTOR JUNCTION SWITCHER TESTER

You do not have to fumble, probe the wrong element, or accidentally short the probes with this transistor switcher. Simply clip the transistor to the collector (red), emitter (yellow), and base (green) alligator clips. Then, read the DMM for the results.

A DMM with a diode test can check the junctions of any transistor or diode. Rotate the DMM to the diode test position. Clip the red probe to the base and the black lead to the collector terminal to make a junction test between the base and the collector. Simply leave the red probe on the base and connect the black lead to the emitter terminals. The two measurements should be quite close for normal transistor tests (FIG. 3-1). If

3-1 The transistor switcher is connected to a power transistor to make junction tests.

you have a DMM with diode test, you can check transistors with the switch.

The transistor can be checked for shorted or leaky junctions. Most transistors leak between the collector and emitter. Large-, medium-, or small-signal transistors can be tested on this junction switcher tester with help from the DMM.

The circuit

The junction switcher eliminates changing probes on the three transistor terminals. Clip the red, green, and yellow alligator clips to the respective transistor terminals. SW3 selects the npn or pnp transistor (FIG. 3-2). SW1 stays in the first (B) position during normal npn transistor tests. When reverse junction tests are made, SW1 will switch to the second position (C) and (E) with SW2 in the last position or base (B) terminal.

Short or leakage tests are made with both SW1 and SW2 from one element to another. Remember, when both switches are on the same letter, the DMM will show a short which is normal. Remember, leakage tests are made from one element to another.

Preparing the top cover

Draw a line through the length of the lid. Drill a 1/4-inch hole at the top for a toggle switch and two 1/8-inch meter holes for probe leads (FIG. 3-3). Both rotary switches require a 3/8-inch hole in the top cover. At the bottom, drill three 1/8-inch holes for transistor test leads. Now, mount the three switches in their respective holes.

Wiring

To start, run the red positive lead to the red banana plug. Connect a 5-inch piece of red test lead to the center terminal of npn/pnp toggle switch.

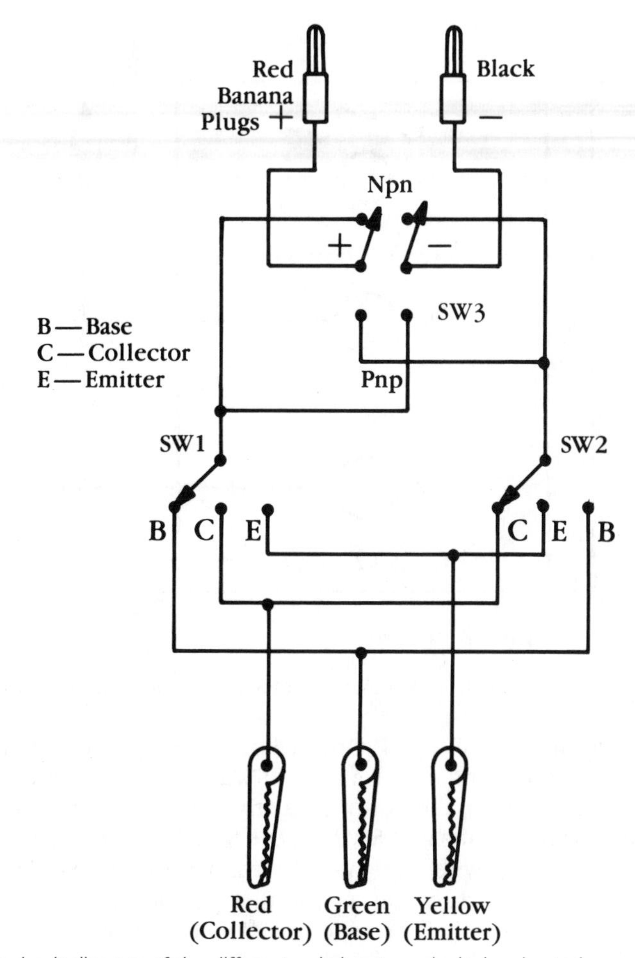

3-2 The circuit diagram of the different switches to make leaky, shorted, open, or normal transistor tests.

Likewise, solder a black flexible test lead to the center terminal of SW3, opposite the red terminal. Check the npn switch; the toggle should point toward the npn position. Now, use the ohmmeter to find which terminal is switched to the red lead.

The two bottom terminals are switched when the toggle of SW3 is flipped to npn position. Solder a piece of red hookup wire from the red terminal of SW3 to the common terminal of SW1. Rotate SW1 to start with position 1, the B-(base) switch terminal. Check with the ohmmeter to see which terminal is switched by SW1 on the first position. If you want to test the transistors in circuit, allow a two-foot length of test clip leads.

Solder the green wire from the transistor test lead to this terminal of SW1. Tie a knot in the test lead so it doesn't pull out. Now, solder a green

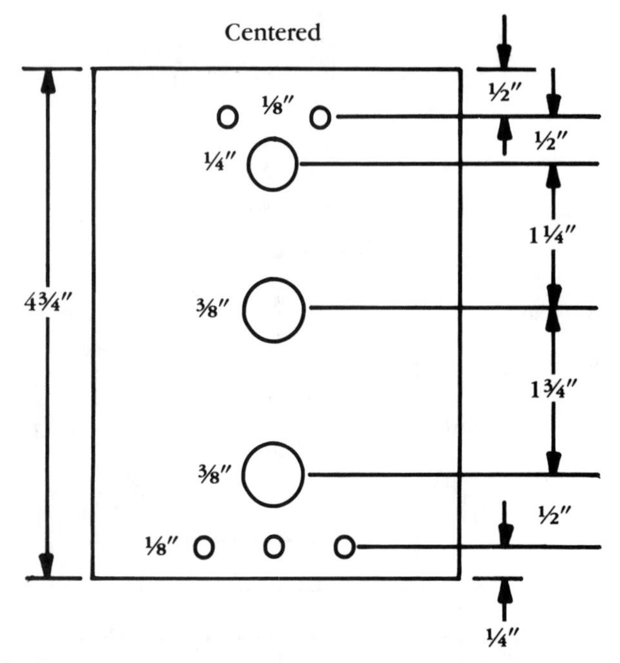

3-3 The required mounting-holes layout for the top of the project case.

wire from B terminal of SW1 to B terminal of SW2. Again, check the short in terminal with the common terminal of SW2 in third position.

Connect a wire from the bottom negative (black) terminal of SW3 to the common switch terminal of SW2. Rotate SW2 to the first position and check for the correct switching contact terminal of SW2. Solder the red transistor (collector) alligator clip to the first position (C) of SW2. Tie a knot in the test clip lead before soldering. Check each switched connection with the low ohmmeter scale.

Rotate SW1 to the second position (C). See which switch terminal ties in with the common terminal and solder the lead from SW1 to the red first terminal of SW2 (C).

Slip the yellow clip lead through the top cover hole and tie a knot in it. Rotate SW1 to the third position (E). Rotate SW2 to the second position (E). Now, solder the yellow (emitter) to these switched terminals. Note that each switch terminal is in order when starting with the first position. Now doublecheck the wiring for poor connections or misplaced wires.

Preparing banana plugs

Select a red and black banana plug and connect each colored flexible wire. These two plugs will be inserted into the DMM jacks. Most DMMs have banana-type jacks. If yours does not, pick up a couple of correct test plugs for the meter.

Cut the two flexible leads to the same length and tin both wire ends. Slip the plastic piece of plug over each wire end and solder the metal ends of the plugs. Use a hot soldering iron to heat the metal probe end and melt solder into end hole. Hold the metal probe with a pair of long-nose pliers. Keep excess solder away from the screw threads.

Testing

Check the wiring by flipping SW3 to npn, SW1 to the first position (B), and switch SW2 to the first position (C) (FIG. 3-4). Plug the banana plugs into the DMM (red and black) to the Ohm and Comm terminals. Switch the DMM to the low 200-ohm scale. Short the red and green (collector and base) clips together; the ohmmeter should indicate a short. Switch SW1 to the second position (C) and leave SW2 in the first position (C). Again, the DMM will show a short. Likewise, when SW1 and SW2 are both switched to the E positions, a short is noted. Now, switch SW1 to E and SW2 to B and short the yellow and green test clips. After the short-circuit test, the transistor junction tester is ready to go.

3-4 Close up view of the finished project making transistor tests.

How to test transistors

Plug the red and black plugs into the DMM. Turn the selector switch to the diode test position—remember, the digital multimeter must have a diode test in order to use this transistor switcher. Clip the colored leads to the correct transistor terminals: red to the collector, green to the base, and emitter to the yellow.

Today, most transistors are npn types, so switch SW3 (toggle) to the upward position (FIG. 3-5). Set SW1 and SW2 in the first position. You are now checking the base and collector junctions. Low-signal transistors have a higher reading (near 0.755), and power and output transistors have a lower measurement. Rotate SW2 to the E position and leave SW1 set at the B position. Notice that this reading is a little higher, (near 0.775). Use this test to check the base and emitter junctions.

Normal test	
SW1	B
SW2	C & E
Reverse tests	
SW1	C
SW2	B
SW1	E
SW2	B
Leakage tests	
SW1 C	SW2 E
SW1 B	SW2 C
SW1 E	SW2 B
Reverse leakage	
SW1 E	SW2 C
SW1 C	SW2 B
SW1 B	SW2 E

3-5 The switching chart used to test the suspected transistor for open, leaky, and normal conditions. Tape the chart to one side of tester for easy reference.

Test power audio and horizontal output transistors in the same manner, except that normal transistors produce lower measurements (0.573). When switched to the base and emitter positions, the reading increases (0.585). While the transistor is connected, check for leaky conditions. Each transistor will have a different resistance measurement, but they should be quite close to above readings.

Reverse the procedure to see if leakage is noted in the other direction. Switch SW1 to the collector (C) position and SW2 to the base (B) position. No measurement reading indicates no leakage. Likewise, rotate SW1 to the emitter (E) position and SW2 to the base (B) position. No measurement reading indicates that the transistor is normal. The pnp transistor can be tested in the same manner, just flip SW3 to the pnp position (downward).

High junction resistance

If one measurement is 0.755 in one direction and 0.1075 in the switched direction, suspect a high resistance joint in the reverse direction. For instance, if the base-to-collecter measurement is 0.775 and the base-to-emitter is 0.1075, the base-to-emitter junction is poor—discard the transistor. Remember, these junction test readings will be close. Very low measurements in either direction indicate a leaky transistor.

Leaky transistors

The leaky transistor will show a low resistance measurement in both directions. Connect the suspected transistor to the color-coded clips. If the transistor is leaky, the measurement might be below 0.2, and when shorted, .05 or less. The readings in both directions will be the same. The transistor can be shorted from base-to-collector, base-to-emitter, or emitter-to-collector. Most power transistors short between the emitter and the collector.

Check the transistor for leakage between the collector and emitter with SW1 in the second position (C) and SW2 in the second position (E). No measurement readings indicate a normal transistor. A low measurement might indicate a leaky transistor between the collector and emitter terminals. If for instance, the measurement was low (0.1) and was the same in the other direction (SW1 in the E and SW2 in the C position), the transistor is shorted (0.1). A shorted transistor will measure the same in both directions. Remember, if both switches are in the same lettered position, the DMM will show a shorted condition.

Testing unknown transistors

Check the transistor closely for some type of identification. The transistor, if numbered, can be referenced in a semiconductor manual for type and lead connections. If it has no numbers or identification, check the bottom lead line. Determine the base terminal with a reading from the other two terminals.

When you find a measurement that is quite close in both directions, you have identified all three terminals. For example, you might have a measurement between two transistor terminals with SW1 and B and SW2 at C. Leave the same clip lead and SW1 in the B setting and switch SW2 to E. If two close measurements are found, you have located the correct terminals. Remember, the base terminal must be found and clipped to the green lead with SW1 set in the B position. Interchange clips on the transistor until you find a measurement in both directions.

If you have a transistor and you do not know if it is npn or pnp, con-

Parts list

PL2	Red and black male banana plugs
SW1, SW2	2-pole 3-position switch, 275-1386.
SW3	DPDT flat-lever toggle switch, 275-636 or equiv.
Test clips	One of each red: yellow, and green small alligator clips.
Case	4⅝″-X-2⁹⁄₁₆″-X-1⁹⁄₁₆″ delux project case, 270-222 or equiv.
Misc.	Small flexible test-lead wire, nuts, bolts, solder, hook up wire, etc.

nect the transistor to the clip leads. Check for "CBE" on the flat side of the transistor. Follow the bottom transistor layout to determine which clip goes where. When set in the npn position with no measurement, flip SW3 to the pnp position. The transistor is a pnp if a measurement is noted. Check the pnp transistor in the same manner as the npn.

Open transistors

Any junction can be open from base-to-collector or base-to-emitter. Clip the colored leads to the respective transistor terminals. If there is no measurement with SW1 in the B position and with SW2 in the C position, suspect that the transistor is open from base-to-collector. Now, switch SW3 to the other type of transistor. If a measurement is noted, the transistor is not open, but the transistor is open if no measurement is found from B-to-C and yet B-to-E appears normal. In this case the open junction is from base to collector—throw away the transistor.

Diode tests

Diode tests can be made with switches connected to the DMM. Connect the suspect diode to the green and red alligator clips. Reverse the diode terminals if there is no measurement. The working diode will produce a normal measurement if the green clip is connected to the negative terminal and red clip to the positive terminal of the diode. If a low measurement is found in both B-to-C and C-to-B, the diode is leaky. No measurement in both directions indicates the diode is open.

Testing transistors and diodes with the transistor junction switcher is easy by following the small chart. Tape it to the side of the switcher.

STEREO BALANCE TESTER

When one of your stereo's channels becomes weak, you can use the stereo balance tester to troubleshoot it. The tester can be used to accurately balance the channels (FIG. 3-6). The speaker portion indicates if

3-6 The front view of the stereo-balance meter tester.

either channel is operating. Audio surplus meters can be used in this output circuit.

The signal circuit

With this small tester, you can plug the small audio generator directly into the audio inputs or clip them into the circuit. The audio oscillator is built around an inexpensive chip; the LM 3909 is a LED IC powered by only one flashlight cell with minimum current. C2, R1, and R2 create a frequency near 1 kHz. Changing the values of R1 and R2 will change the frequency of the circuit (FIG. 3-7).

Only $1^1/_2$ volts powers IC1 (on pin 5) through SW1. Pin 4 of IC1 is grounded. The output signal is taken from pin 2. The amount of audio inserted into the stereo channel is controlled by R4. C1 isolates and couples the controlled signal to the stereo input circuits. Use standard phono patch cords or homemade cords to connect the generator to the stereo inputs.

Stereo indicator circuit

The stereo indicator circuit consists of two sets of speaker jack input terminals, a DPDT switch, and a meter. Connect the generator to both stereo inputs so that the meter can be switched to either channel for balance tests. The other speaker is loaded with a 10-ohm, 10-watt resistor. Load higher-powered solid-state amplifiers when the speaker terminals are removed. R5 lowers the high wattage output to protect the meter.

Use regular phono-type inputs and single-wire speaker terminals for the input speaker jacks. Wire the red terminal as the hot lead and the

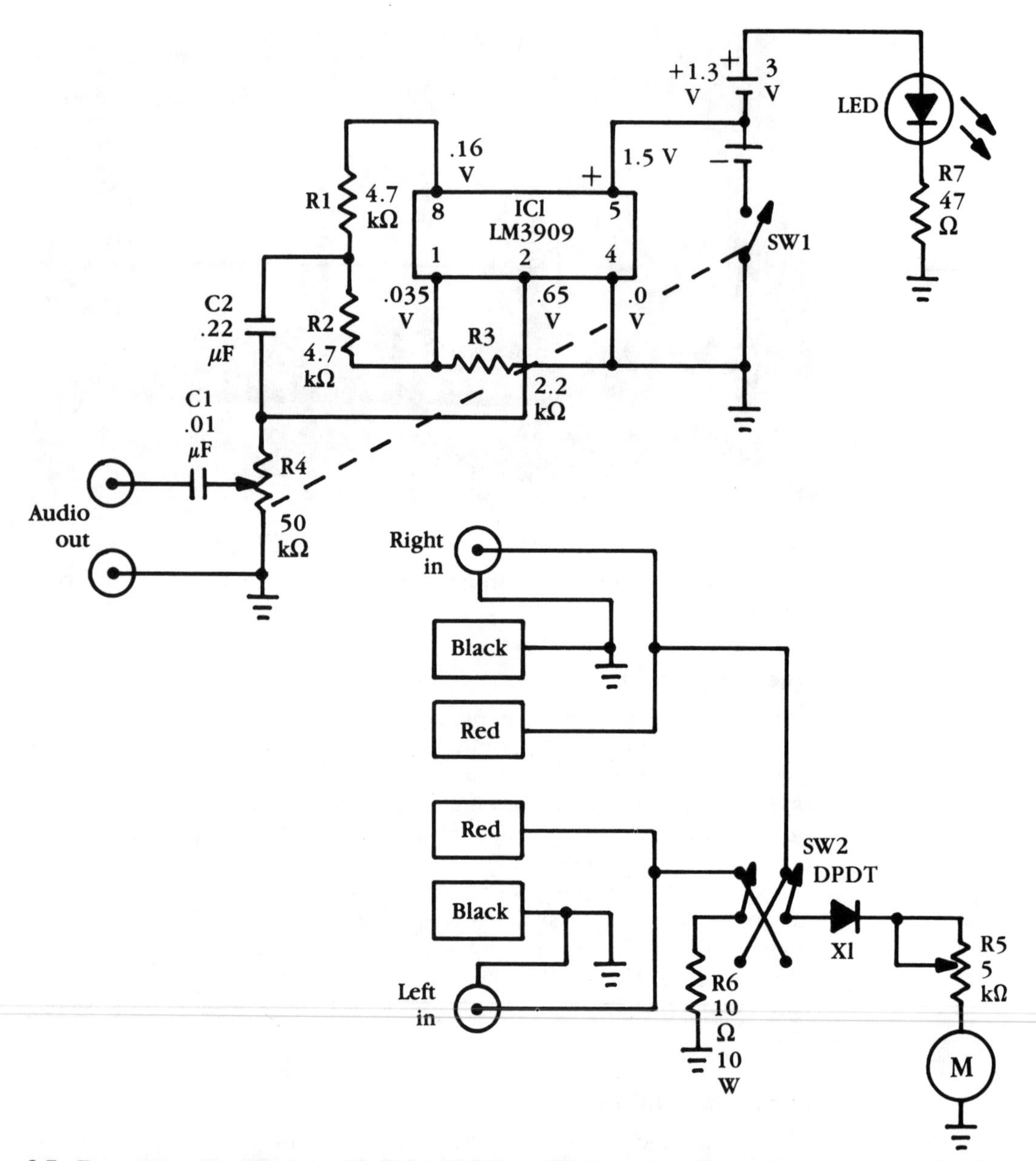

3-7 The audio oscillator circuit consists of IC1 and surrounding components. The speaker output circuit includes a meter, a 10-watt resistor, and speaker jack connections.

black terminal as the ground. Wire two different sets of speaker terminals in parallel so they can be switched by SW2. With this configuration, the speaker indictor will fit any type of speaker connection.

Construction

Cut out a 2-×-3 inch piece of pc board and layout the pc wiring (FIG. 3-8). Since only a few parts are required the pc layout is quite simple. Notice that the common ground runs around the outside of the pc board. Doublecheck all pc wiring before etching the board. Wash and wipe the etched board and brighten the pc wiring with steel wool. Now, drill all pin holes with a 1/16-inch or smaller bit. Drill two 1/8-inch mounting holes at opposite ends of the board.

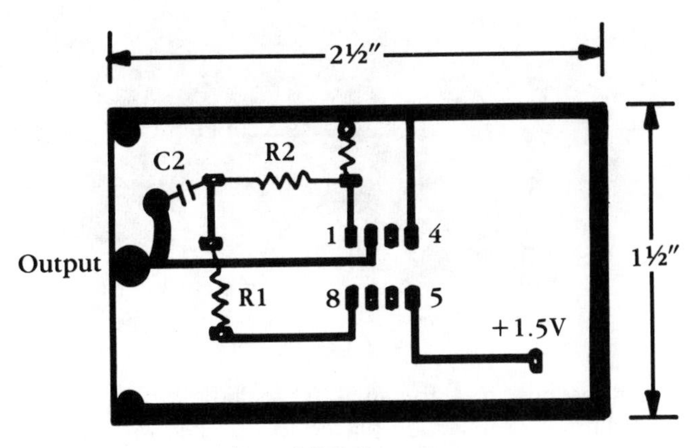

3-8 The small pc board is 2 1/2 × 1 1/2 inches and holds most components of the audio oscillator.

Component hookup

Solder each part in place and doublecheck the correct value of the resistors with an ohmmeter. Run a wire from pin 5 to SW1. Connect the positive terminal of the battery to the other end of the switch. Run a black lead from the battery to the common ground. Solder a 4-inch piece of hookup wire to the top side of the volume control and ground the other end of the control. Connect C1 to the center terminal of the volume control. Solder the other end of C1 to high-side phono input jack. Make sure that all IC socket terminals are soldered and not just shorted together (FIG. 3-9).

Testing

Check the signal generator by injecting an audio signal into an amplifier input. Tie a PM speaker to the output terminals and you should hear a weak sound with volume control wide open. If the signal generator is dead, suspect either that the battery polarity is wrong or that IC1 is backwards. Check the terminal of IC1.

3-9 The back view of the stereo-balance tester.

Notice if all prongs are in the small IC socket. Sometimes one of these prongs will fold underneath when placed in a socket. Make sure all pins or prongs are in line before inserting the IC. Bend or line up the row of pins with a pair of long-nose pliers. Often when the IC component is new, all pins stick outward—they can be bent inward slightly with the pliers.

Amp and speaker hookup

Both channels have two different speaker connections—one is an RCA plug-in and the other clips onto speaker wires. SW2 places the 10-ohm, 10-watt resistor as a speaker load when the other channel is tested. This load prevents amplifier output damage in solid-state amplifiers. The solid-state output amplifier should be loaded down at all times. R5 varies the meter into the circuit and prevents damage. All parts are mounted on the front panel.

Stereo balance tests

Connect the output of the audio oscillator to the stereo input jacks of the amplifier. Clip the two audio output amp connections to the respective speaker terminals on the tester (FIG. 3-10). Make sure that the inputs and outputs are connected to the proper channels. Rotate the volume control on the amplifier half way and set the balance control in the center.

Rotate R5 to the meter half scale. Switch between the two meter channels to check the balance of the amplifier. Both channels should have similar readings if they are properly set and balanced. If one reading is

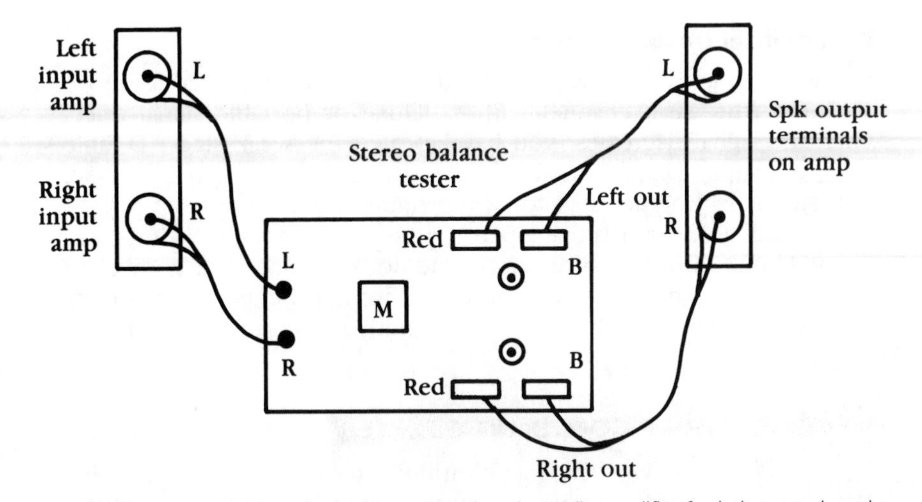

3-10 The stereo-balance meter is connected to the audio amplifier for balance and weak-channel tests.

much lower than the other, balance the channels with the balance meter on the amp. If this balance is far off, suspect that one channel is weaker than the other.

Parts list

IC1	LM 3909 LED flasher IC.
R1, R2	4.7-kΩ fixed ½-W resistor.
R3	2.2-kΩ fixed ½-W resistor.
R4	50-kΩ linear control.
R5	5-kΩ linear control.
R6	10Ω 10-W wire-wound resistor.
C1	.01-μF 450-V capacitor.
C2	.22-μF 100-V capacitor.
J1, J2	Banana jack.
J3, J4	RCA phono-type jack.
SPK1, SPK2	Pressure-type speaker lead test posts.
LED	RED 2.6-volt LED.
SW1	On rear of R4 (STSP).
SW2	DPDT toggle switch.
X1	1N60 fixed crystal detector or equiv.
Meter	Signal strength or 0-1 mA meter. (Can be surplus meters)
Case	Deluxe project case 7 7/16″ × 4 1/4″ × 2 3/8″, 270-224 or equiv.
Misc.	Solder, bolts, nuts, hookup wire, etc.

Weak channel troubleshooting

If either channel is very weak, leave the amplifier connected as above. Signaltrace the weak channel with a deluxe or portable signal tracer. When the signal decreases, you have located the weak stage. Double-check all coupling capacitors for weak sound on each terminal. An open or dried-up electrolytic capacitor will produce a weak output.

Suspect leaky transistors, open bypass electrolytic capacitors, and changed-bias resistors for weak amp conditions. Compare the good channel to the weak channel as you proceed through the amplifier circuits. The audio signal should be the same strength at the same point in both channels.

DIODE/SCR/TRIAC CONTINUITY TESTER

You can use this diode/SCR/triac continuity tester to check many low-resistance components for normal, leaky, open, or shorted conditions (FIG. 3-11). To avoid batteries, the small tester operates from ac power. Make a quick diode test with a fuse clip holder mounted on the top of the tester. Take regular continuity tests with the black and red test leads. Use all three test leads when testing SCR and triac components. Do not check small transistors with this tester.

3-11 Top view of the SCR, triac, and diode continuity tester checking out a defective SCR.

Circuit details

The power supply consists of a miniature 12.6-V ac stepdown 450-mA transformer (FIG. 3-12). D1 and C1 filter out the ac ripple. The negative return voltage is supplied from the black test leads. The B+ voltage feeds to one side of the quick diode test through R1 and the LED. Connect R2 and SW1 to the green gate terminal.

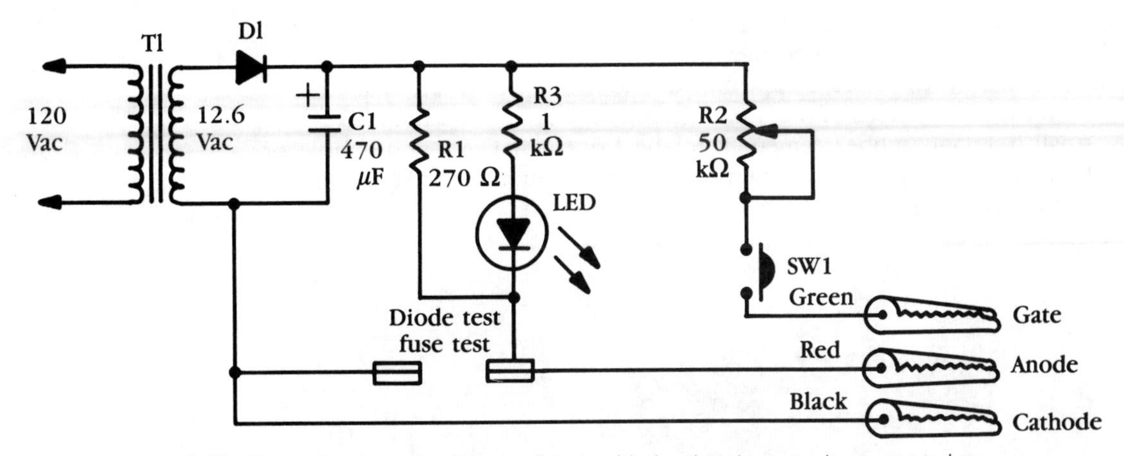

3-12 The main schematic of the small tester. Notice that the tester is ac-operated.

Make continuity tests with the black and red terminal leads. You can quickly check diodes and small fuses by placing them across the fuse clip holder. To test SCRs and triacs, clip the black lead to the cathode, the red lead to the anode, and the green lead to the gate terminal. Connect the black terminal lead to terminal 1 (MT1), the red lead to terminal 2 (MT2), and the green leads to the gate terminal (G) of the suspected triac (in the microwave oven in FIG. 3-13).

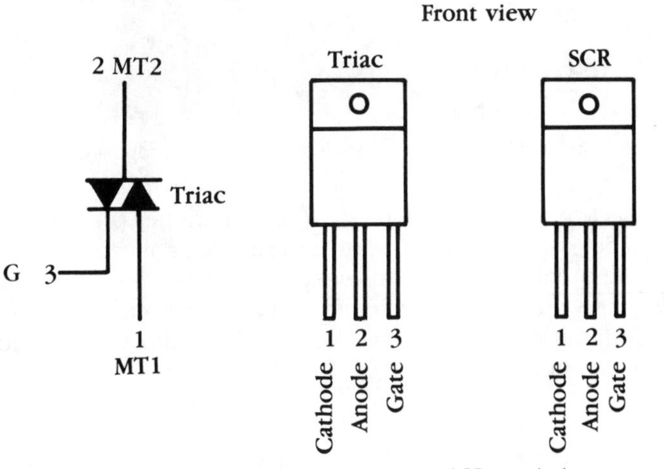

3-13 Test lead hookup of the triac and SCR terminals.

Mounting parts

You can mount most components as you solder them into the circuit. Connect two 1-inch-long bare solid hookup wires to each fuse clip terminal. Push the wire through the small holes of the top lid. Bolt the fuse clip to the front panel. Squeeze the clip points together so the suspected diode will lay in the groove area.

Underneath the fuse clip, bolt or mount a 3- or 4-terminal strip to hold the connections of the 3 test leads (FIG. 3-14). Mount the LED in the small hole and rubber cement it to the front panel. Allow a couple of hours for the cement to set, then connect one lead of the LED to one side of the fuse terminal.

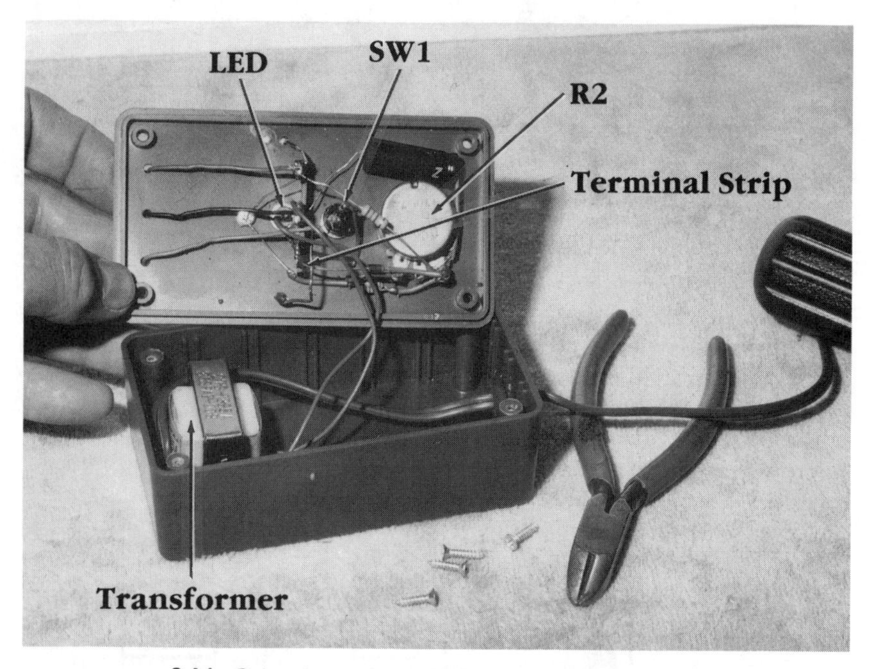

3-14 Only a few parts are located under the top panel.

Connect the cathode (K), diode (D), and emitter (E) test leads to the negative side of the fuse clip. Connect the anode, collector, and diode test leads to the other side of the fuse clip. If possible, use color-coded test leads and clips. Bolt the power transformer to the bottom side of the plastic case. Run the grounded lead to the fuse clip and black test lead. Solder the positive voltage to one side of C1, R1, and R2.

Diode tests

The fuse clip makes a quick handy test for the suspected diode rectifier. If you have a bunch of diodes to check, this is the ideal quick tester. Reverse the polarity of the diode if it lights when placed on the fuse clip terminal (FIG. 3-15). The normal diode only lights in one direction. If the LED lights with the diode in both directions, the diode is shorted. Discard it. If the LED does not light with the diode in any direction, the diode is open.

3-15 Quickly test the diode by placing it upon the fuse clip.

You can check any in-circuit LEDs or diodes with the red and black test leads. The normal diode or LED will light the tester's LED in only one direction. In fact, when checking several LEDs in series, the whole bank of LEDs might light. The normal LED will only light in one direction. Suspect that the LED is open if it does not light at all.

Normal SCR and triac tests

Connect the black lead to the cathode, the red lead to the anode, and the green lead to the gate terminals. The normal SCR will not light with these connections. If the light comes on, either the SCR or the triac is leaky. Push the momentary switch and rotate R2 until the LED lights at its brightest. With a normal SCR or triac, the red light will remain bright—no matter where the control has been set after being triggered on. Now remove the green gate lead; the red light should remain on with a normal SCR or triac. If the light goes out, suspect a defective component.

SCR and triac leakage tests

Use only the red and black terminal leads to check for leakage. The red light should be on when the leads are attached to the cathode and gate terminals. Does the LED light with the terminal leads reversed, indicating a normal junction between the gate and cathode terminals? If the LED lights when the red and black leads are connected to the gate and anode or to the cathode and anode, the SCR or triac is shorted or leaky. The results are similar with reversed test leads (FIG. 3-16). Most SCRs and triacs show leakage between cathode and anode terminals.

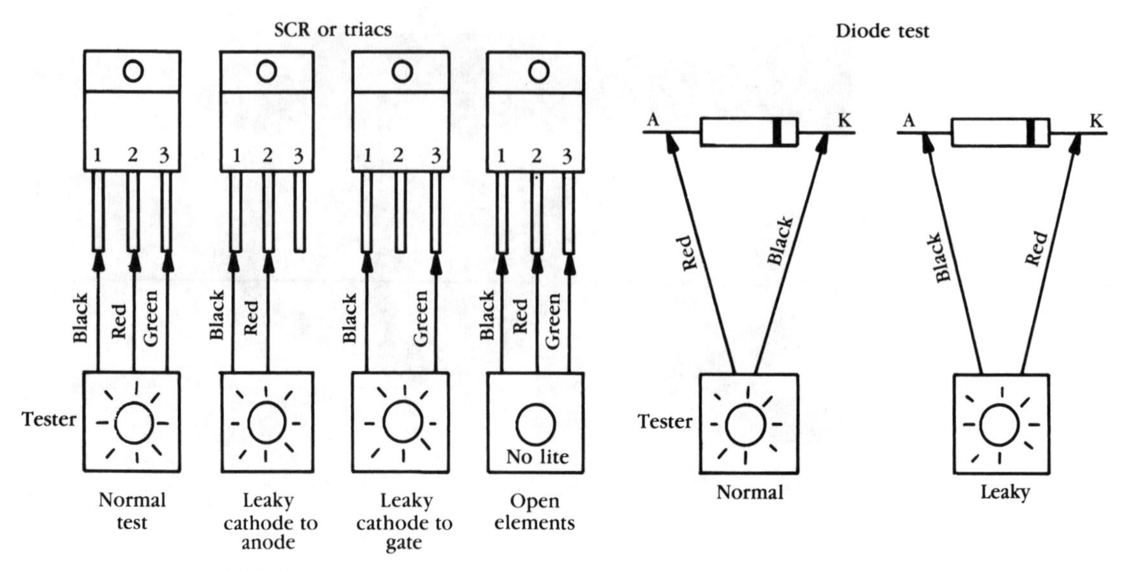

3-16 The various normal and leakage tests for SCRs, triacs, and diodes.

Continuity tests

Use the black and red leads to check switch controls or low continuity checks. Any continuity measurements up to 150 ohms will light the LED. Low-resistance coils and motor winds can be checked for continuity with this tester. Pilot lamps open fuses, or any low-resistance light bulbs can be checked with the continuity test. An unlit LED indicates an open component.

Parts list

T1	12.6-Vac CT 450-mA power transformer
D1	2.5-A 1000-V silicon diode
C1	470-μF electrolytic 35-V capacitor
R1	270-Ω ½-W carbon resistor
R2	50-kΩ linear control
R3	1-kΩ ½-W resistor
LED	Red high-intensity LED, Radio Shack #276-006A
SW1	Momentary SPST switch
Fuse Holder	Metal strip on plastic holder
Misc.	3 small alligator clips, a plastic box, 4- or 5-lug strip (Terminal), ac cord, etc.

CB INDICATOR

How many times have you known if the transmitter was actually operating within a small walkie-talkie? Does that citizen band transmitter really output the required rf? Besides indicating that the radio is operating, you can peak the CB and match its antenna with this indicator. Even the flea-power childrens' transceiver will register on the CB indicator, (FIG. 3-17).

The small walkie-talkie antenna can be slipped into the hole and through L1 for transmitting indication. Higher-powered CBs can be checked and tuned by clipping to the antenna. On high-powered transmitters, the indicator can merely be held close to the radiating antenna.

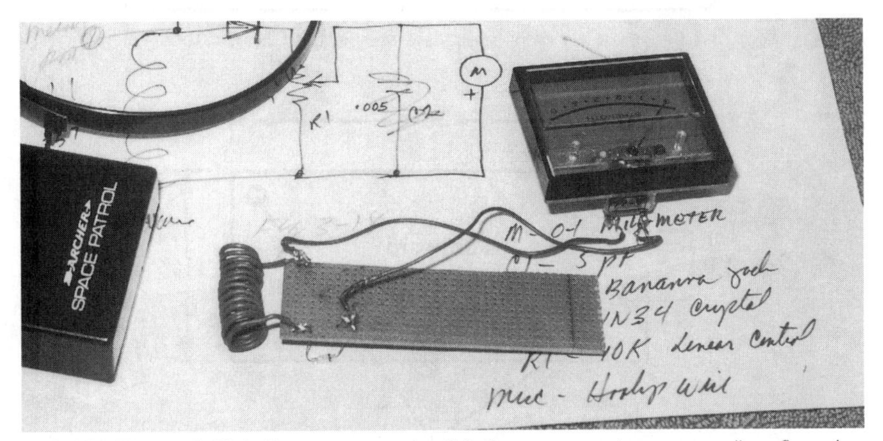

3-17 The small CB indicator parts and coil (L1) are mounted upon a small perfboard.

How it works

When the walkie-talkie antenna is inserted into the coil, L1 transfers the rf signal to the crystal detector (1N34). By clipping the meter directly to the output antenna circuit, a small portion of rf power is fed through C1 to L1 and the crystal. Here the signal is rectified and varied with R1. The signal from R1 is transferred directly to the meter (FIG. 3-18).

The flea-power walkie-talkie might only vary the meter 1/8 or 1/4 inch with R1 wide open. This measurement at least indicates that the transmitter is functioning. When connecting J1 to higher-powered CB units, R1 should be at its lowest rotation, so as to not damage the meter. A weak reading might indicate a weak battery in the walkie-talkie or a poor tune-up of the larger CB transmitters.

Construction

Wire L1 and all small components on a small perfboard. Cut the perfboard to 1 1/4 × 2 inches. Feed the two coil ends through the pin holes

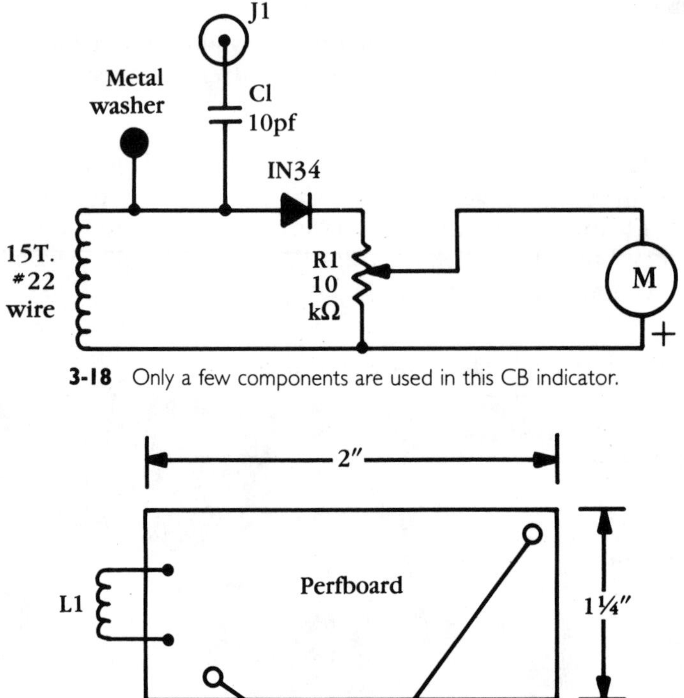

3-18 Only a few components are used in this CB indicator.

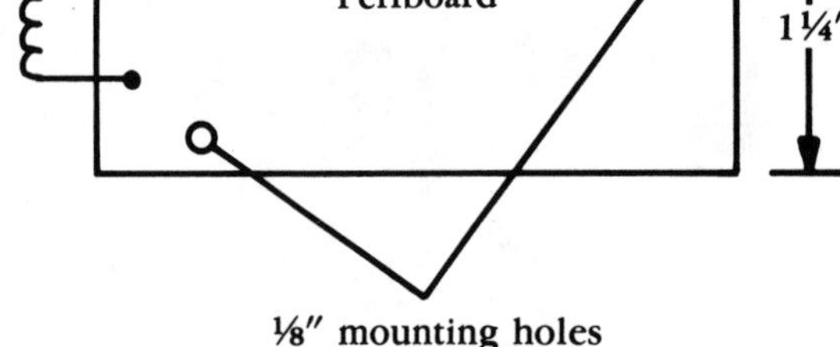

3-19 The coil layout and dimensions of the perfboard with mounting holes.

and bend them over to hold the coil into position. Mount the crystal (1N34) on one end of L1 and R1 (FIG. 3-19).

Place all components in a plastic economy case 4½ × 2½ × 1½ inches. The case should be large enough to safely mount the small meter. Place the meter to the left and the control to the right. Drill two ⅛-inch holes for the meter. Center the ⅜-inch hole for R1. Drill a ¼-inch hole for the banana jack and ⅛-inch holes for the mounting screws to the top left of the bottom case (FIG. 3-20).

Coil data

Wind L1 on a large fountain pen or any round object with ½-inch diameter. Since the coil is self-supporting, wind 15 turns of #22 solid hookup wire. Keep the turns close together. On each end, make sure that the leads do not block the center hole (FIG. 3-21), because the walkie-talkie antenna will be shoved through this hole when testing.

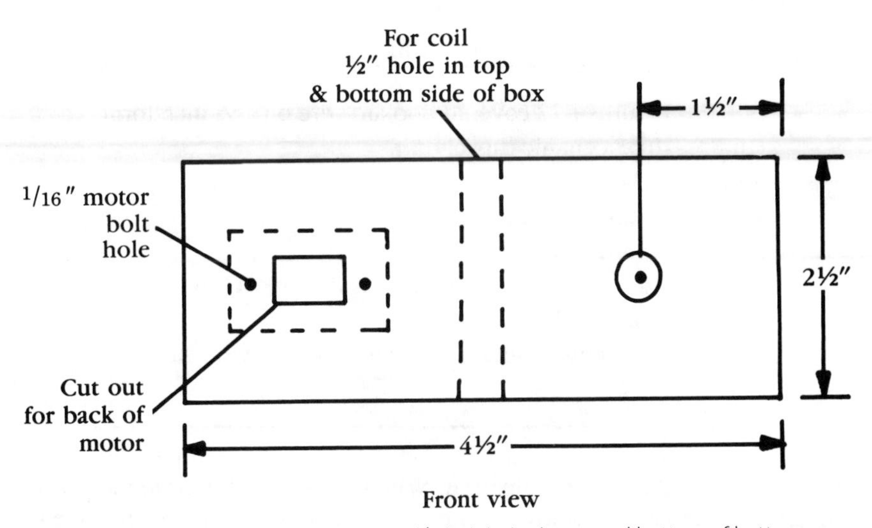

3-20 The top cover layout holes. Line up 1/2-inch holes in top and bottom of bottom case with L1.

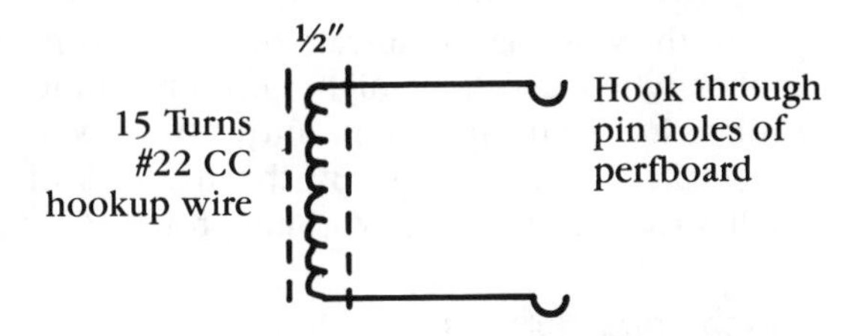

3-21 Wind 15 turns of #22 insulated hookup wire over a 1/2-inch form. Strip both leads 3/4 inch to fit in holes and tie them in.

Cement the coil in place with a dab of rubber cement, after the chassis has been mounted. Loop the bare ends of the coil through the end pin holes and back through the coil to help mount it.

Finishing

After the indicator is wired and tested, place it inside the plastic case. Drill two 1/8-inch holes in the case to match the holes in the perfboard chassis. Now drill a 1/2-inch hole in the center of the bottom case and one at the center of the top so that the antenna can stick through the coil for testing. Bend the coil to match the top and bottom holes. Place a 6/32-inch bolt with nut and metal washer near the banana jack for flea-powered transceivers.

Indicator tests

After the wiring has been completed, check the indicator by slipping the antenna of a walkie-talkie through the coil hole and press the transmit

Parts list

L1	15 turns #22 covered hookup wire on a ½-inch form.
C1	10 pF ceramic capacitor.
XTAL	1N34 type.
R1	10 k linear control.
M	0-1 mA dc meter.
Case	4½-X-2½-X-1½ economy plastic project box.
J1	Banana jack.
Misc.	Bolts, nuts, solder, perfboard, hookup wire, etc.

button. Notice the meter indication. When units are new, note the power readings. When power is low, you need to work on the unit or replace its batteries. Place the tip of the transmitting antenna on the washer/bolt and notice the signal measurement of a flea-power walkie-talkie.

To check larger-powered units, hold the backside coil toward the antenna or clip it to the antenna to indicate proper transmitting antenna tune up. Remember when checking higher-powered units, turn R1 to zero and then rotate it for an indication. Turn up R1 when testing the antenna gain of smaller walkie-talkies. By the readings of R1 and the meter, you can tell if the transmitter is working properly.

WHITE NOISE GENERATOR

This noise generator is quite effective in signal tracing rf, mixer, i-f, and audio circuits (FIG. 3-22), because most audio signal-injection oscillators will not work in the front end of any radio. You can start at the volume control and test your way towards the antenna. Work from stage to stage until the noise disappears, then you have located the defective circuit. Use the radio speaker as an indicator. Afterwards, take voltage, resistance, and transistor measurements to solve the problem.

The circuit

Just about any type of npn transistor will function as the noise generator. Of course, some npns require more B+ voltage before noise appears. In this circuit, Q1 operates nicely at 12 volts, but it will not produce noise at 9 volts (FIG. 3-23). The emitter terminal is tied to the 12-volt source with the collector terminal open. R1 and R2 provide bias for Q1.

C1 (1 μF) couples the noise through R4 to terminal 2 of IC1. The op amp operates in an inverting amplifier circuit with a single-polarity supply of 12 volts. The noisy signal is coupled from pin 6 to a 10 kΩ linear variable control. Choose C2 with higher voltage to isolate the noise generator from the higher voltages found in other circuits.

3-22 The front view of the white-noise generator with only one control.

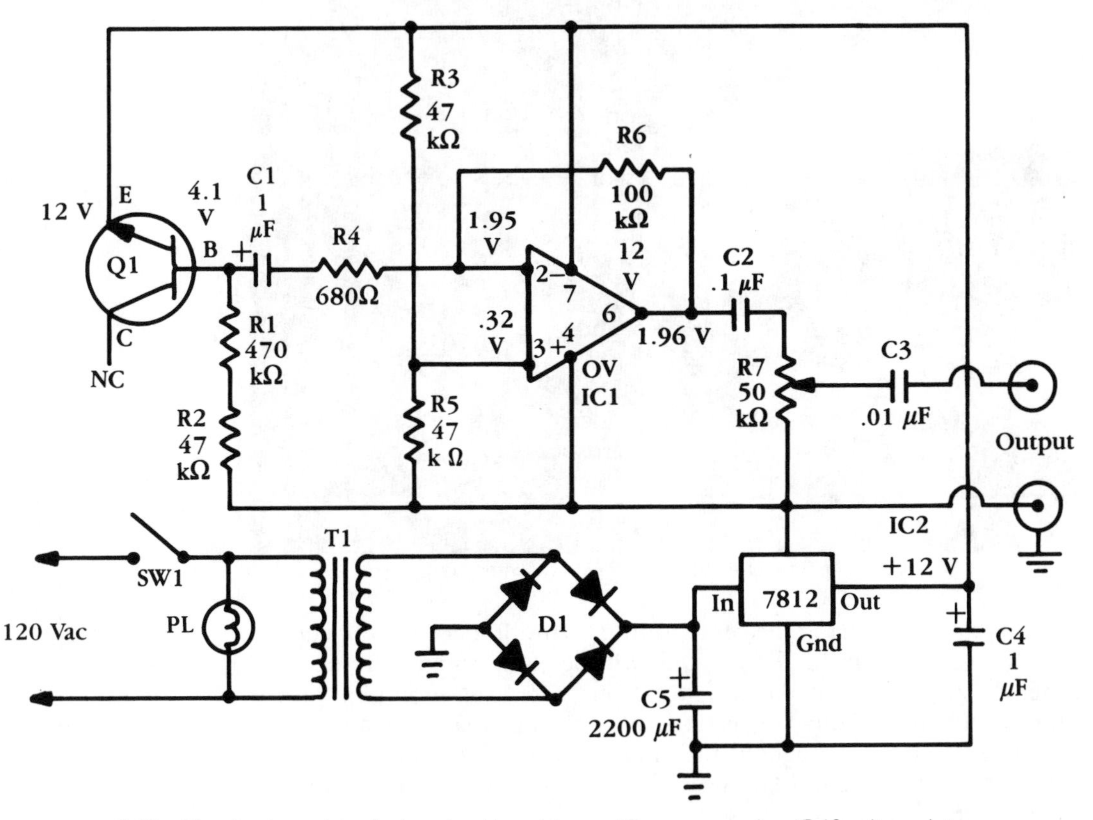

3-23 The circuit consists of a low-signal transistor, an IC op amp, and an IC 12-volt regulator.

The power supply consists of a small 12 Vac stepdown transformer with full-wave rectification. D1 and D2, 1-amp silicone diodes, rectify the ac voltage. C3 is the input filter with IC2 as the 12-volt regulator. N1 indicates when the unit is on and SW1 is located on the back side of R7.

Perfboard

Use a 2.83-×-1.95 inch IC perfboard to mount all small components, except the transformer, the neon light indicator, and the control (FIG. 3-24). The perfboard is ideal to mount the small components with IC and component solder-ringed holes. Use extreme care to avoid lopping solder into another hole in the circuit. Use a sharp-pointed low-wattage soldering iron.

3-24 A close-up view of the perfboard with mounted components.

Mount IC1 socket in the center of the board. Run a bare #22 copper wire across both outside edges. One will be the positive wire and the other the negative or common-ground bus wire. The mounting procedures for the rest of the parts is not critical. Notice that there are six 1/8-watt resistors used throughout the circuit; sometimes it's easier to take resistance measurements of these small resistors rather than trying to read the respective color codes.

Connect a four-inch lead from C2 to the variable control, R7 (FIG. 3-25). Solder a common-ground wire from the bare negative bus wire to the control and output terminals. Connect C3 between the center terminal of the control output jack. Connect the 12-volt secondary leads of T1 to the negative terminals of both diodes (D1 & D2) and mount the power transformer off the board, on plastic case. Rim the back blade edge of a pocket knife between each IC terminal of IC1 to remove any shorting solder points. It's wise to check each adjoining IC terminal with the ohmmeter on the low scale to indicate possible shorts.

3-25 Soldering wires to the controls and to the perfboard chassis.

Cabinet construction

The small noise generator board and transformer can be installed in any case that is large enough to hold all components. Drill a 9/32-inch mounting hole for the red-neon 120-Vac lamp. Drill a 3/16-inch hole for the ac cord in the bottom side of case. Two 1/8-inch holes are drilled to hold the small perfboard to the bottom chassis. Use 1/2-inch plastic spacers to hold the perfboard up from the case (FIG. 3-26).

Mount the two banana jacks on the top panel with 3/16-inch holes in the top case. Drill a 7/16-inch hole for R7. Mount the perfboard and then solder the extended leads to the top panel. Mount T1 last. Now, solder the secondary leads of T1 to the diodes on the perfboard. Tie a knot in the ac cord to prevent it from being yanked out.

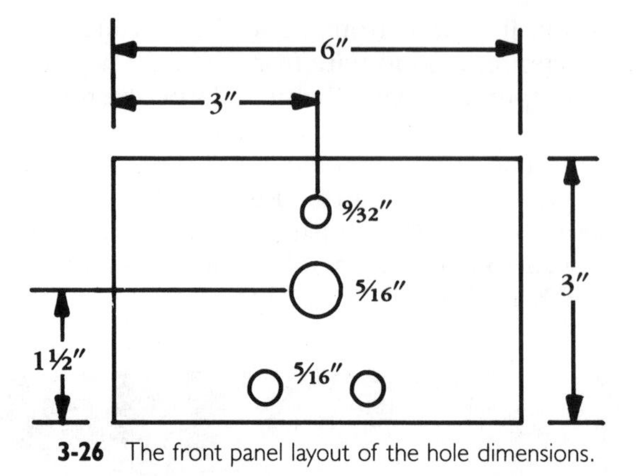

3-26 The front panel layout of the hole dimensions.

Testing

Clip the leads to the input of an audio amp. If you have constructed either the small hand-held or the larger audio tester, clip the output lead to the input terminals. Open R7 wide; you should hear a loud rushing of noise in the speaker (FIG. 3-27). If not, doublecheck all wiring. You can test the noise generator on any audio amplifier.

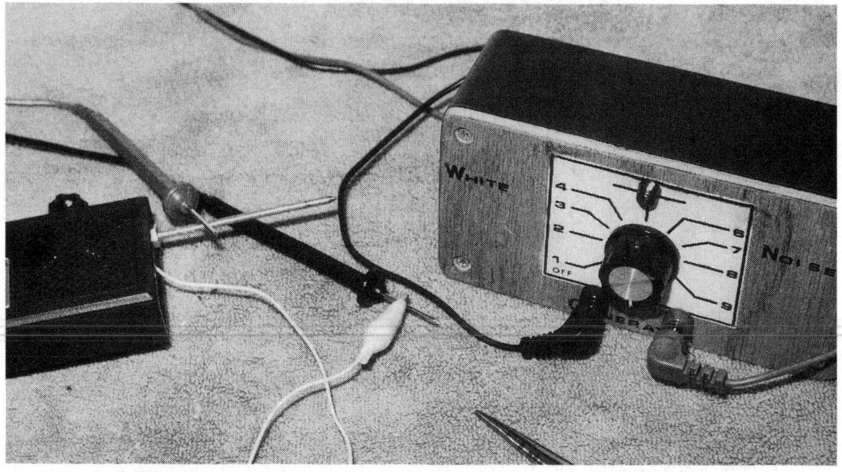

3-27 The small generator connected to the hand-held signal tracer.

Now, take voltage measurements of IC1 and compare them with the readings listed in the schematic. Make sure that the collector terminal of Q1 is open. Use the handheld audio signal tracer to check the noise from C1 to input terminal 2 of IC1. Check for output noise at pin 6 of IC1. Make sure that no IC1 terminals are accidentally soldered to each other.

Checking radio circuits

Inject white noise into the center terminal of the volume control, and if no noise is heard, inject it into the various audio stages. When you can hear noise at the volume control, start at the front end of the AM section and test toward the volume control (FIG. 3-28). Keep the noise control as low as possible. Sometimes you might hear a local radio station while signal tracing. Start at the AM rf coil and proceed to the base of the first i-f amplifier.

Next, test the collector terminal of i-f amp and proceed, by the numbers, through all i-f stages (FIG. 3-29). When you can hear the noise in the radio speaker, the dead stage is close by. You might notice a loss of volume in the i-f transformer and detector coils. The volume of noise should be the same on both sides of a coupling capacitor.

Tape player circuits

Start at the play/record head when the tape section is dead. Some players have two preamp stages and lower-priced units usually have only one (FIG. 3-30). Proceed to the base and collector of each audio amp transistor. The audio section can be checked from the volume control to the rf amps and power output transistors.

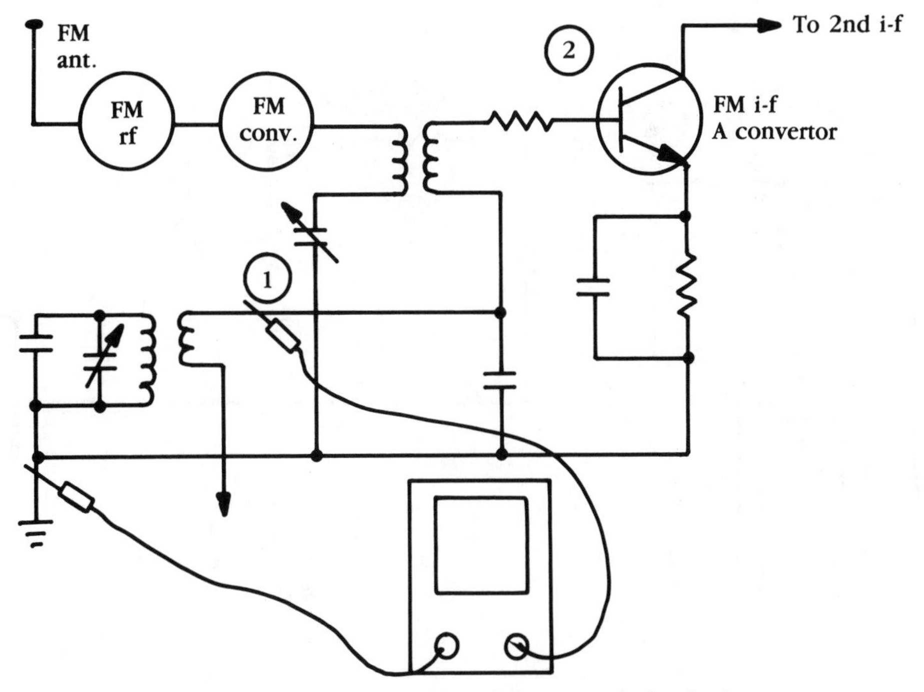

3-28 The hookup connections of the white-noise generator in the rf and converter stages.

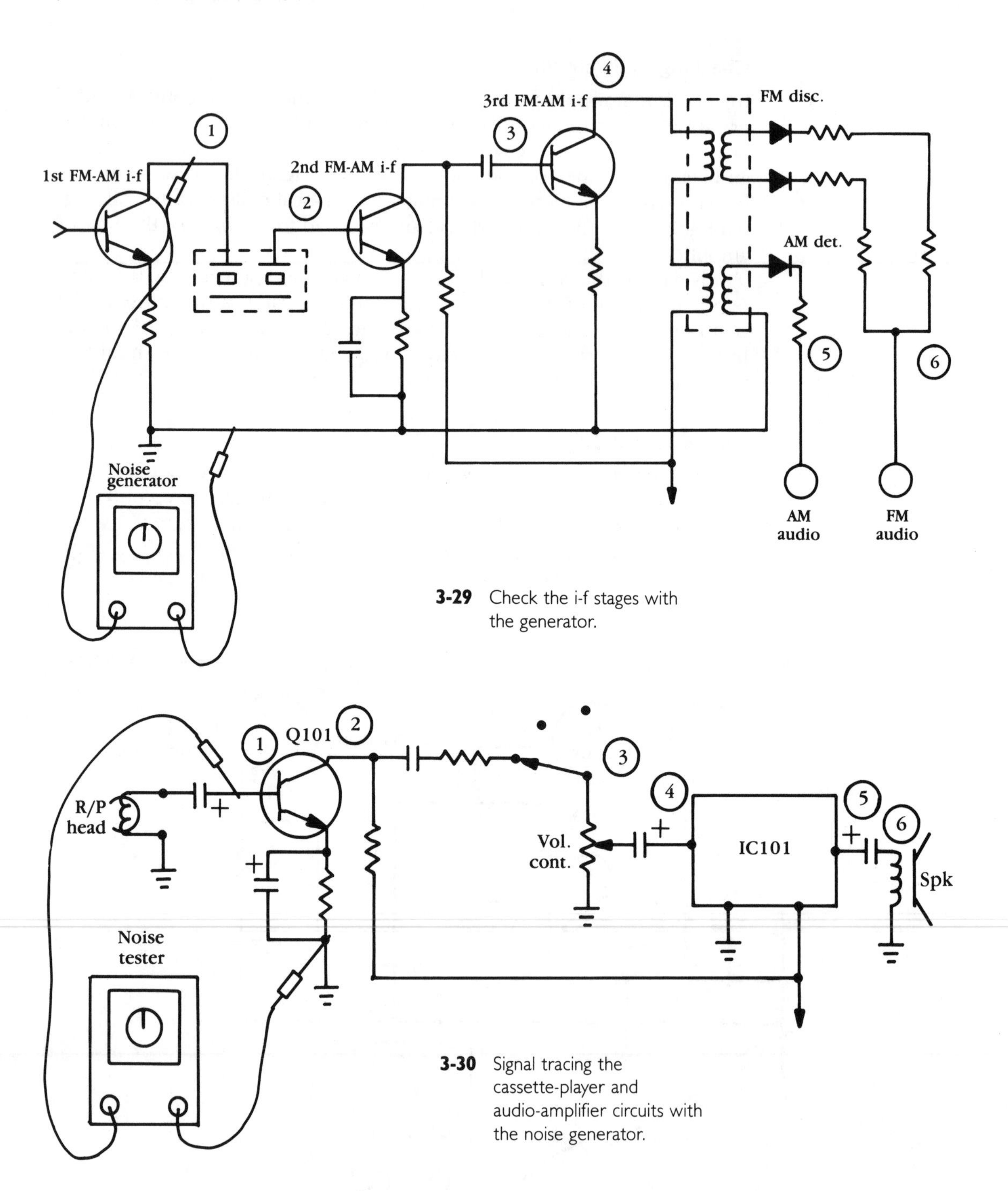

3-29 Check the i-f stages with the generator.

3-30 Signal tracing the cassette-player and audio-amplifier circuits with the noise generator.

Parts list

Q1	MPS 2222A or practically any npn. low voltage transistor or ECG 123.
IC1	741 op amp.
IC2	7812 12-V regulator.
C1, C4	1-μF 50-V electrolytic capacitor.
C2	0.1-μF 50-V ceramic capacitor.
C3	0.01-μF 500-V ceramic capacitor.
C5	2200-μF 35-V electrolytic capacitor.
R1	470-kΩ ⅛-W resistor.
R2, R3, R5	47-kΩ ⅛-W resistor.
R4	680-Ω ⅛-W resistor.
R6	100-kΩ ⅛-watt resistor.
R7	50-kΩ linear control with SPST switch.
D1	1-A bridge rectifier.
T1	300-mA 12-V secondary transformer, 273-1358 A or equiv.
Perfboard	Component 2.83 × 1.85, 276-149 or equiv.
SW1	SPST switch on rear of R7.
PL	120 Vac neon pilot lamp.
Case	Plastic economy project box, 3″ × 6″ × 2″.
Misc.	Ac cord, hookup wire, grommet, 8-pin IC socket, etc.

In IC component output stages, inject the noise signal at the input and then output terminals. Check each side of the output coupling capacitor. Check the noise at the speaker terminal; it should be quite weak. Follow the numbers when signal tracing the audio stages.

The white noise signal generator can be used in conjunction with the hand-held or audio signal tester (described in the book) to locate a dead or weak stage. If the audio output is dead, check the rf-i-f stages of the radio or tuner by connecting the audio signal tracer at the volume control. Proceed toward the front end of the radio or tuner with the noise generator. When the noise quits, you have located the defective circuit. In this case, the audio signal tracer is the audio indicator.

Advanced test instruments

This chapter starts with a simple fixed regulated power supply for 5, 9, and 12 volts. The audio amp checker is ideal for those who do a lot of stereo equipment servicing. The sine/square wave generator uses the reliable 8013 IC. The hand-held signal injector and hand-held signal tracer fit in the palm of your hand, but can locate those pesky, defective hard-to-locate parts.

The deluxe fixed and variable power supply can fit and operate on any service or workbench. The fixed voltages are −5, +5, +9, +12, +20 and +24 volts. A separate 1-amp variable power supply has a voltage supply from 1.2 to 27 volts. Both the current and voltage meters can be switched into any voltage source.

Last, but not least, this chapter features a valuable crystal checker with an old-type Pierce oscillator circuit. You can check any suspect crystals used in many commercial entertainment products—even saw filters and ceramic filters.

SIMPLE 1.5-AMP LOW-VOLTAGE POWER SUPPLY

This simple low-voltage power supply has voltages fixed at 5, 8, and 12 volts. Each voltage source is regulated with the IC regulator. This power supply can be used to supply voltage for any dc circuit up to 1.5 amps (FIG. 4-1). Just plug the positive (red) plug into the desired fixed voltage with a common ground plug.

4-1 Front view of the low-voltage power supply.

The circuit

The schematic consists of a 12.6-Vac stepdown transformer, T1. SW1 is a common off/on toggle or flip switch. N1 indicates that the power is turned on. The secondary ac voltage of T1 is applied to a bridge rectifier component, D1. T1 and D1 both have 2-amp ratings, although any 12.6-volt transformer rated from 1.5 to 2 amps will do (FIG. 4-2).

The rectified dc voltage is filtered with a fairly large 4700 μF filter capacitor rated at 35 working volts (C1). Be sure to choose the correct size of electrolytic capacitor to fit in the project case. This dc voltage is applied to the inputs of all three voltage regulators.

IC1 is a 12-volt regulator with a 1-μF and a 0.1-μF capacitor tied to the positive out terminal. Likewise, all three regulators have a bypass capacitor on the output terminal. Each dc voltage is connected to a separate banana jack. The center terminal of each regulator is grounded with the input terminal tied to C1 and output terminal to the voltage jack. Place a heatsink on each regulator if the power supply is to be operated at its peak output.

Experimenters perfboard

All components, except T1, N1, and SW1 are mounted on a small general-purpose IC pc board. This board has 417 indexed solder-ringed holes. Down the center of the board are two columns of holes, which are used for the common-ground terminal. Solder a piece of bare #22 hookup wire along one outside edge to connect the in terminals of the regulators and the positive terminal of C1 (FIG. 4-3).

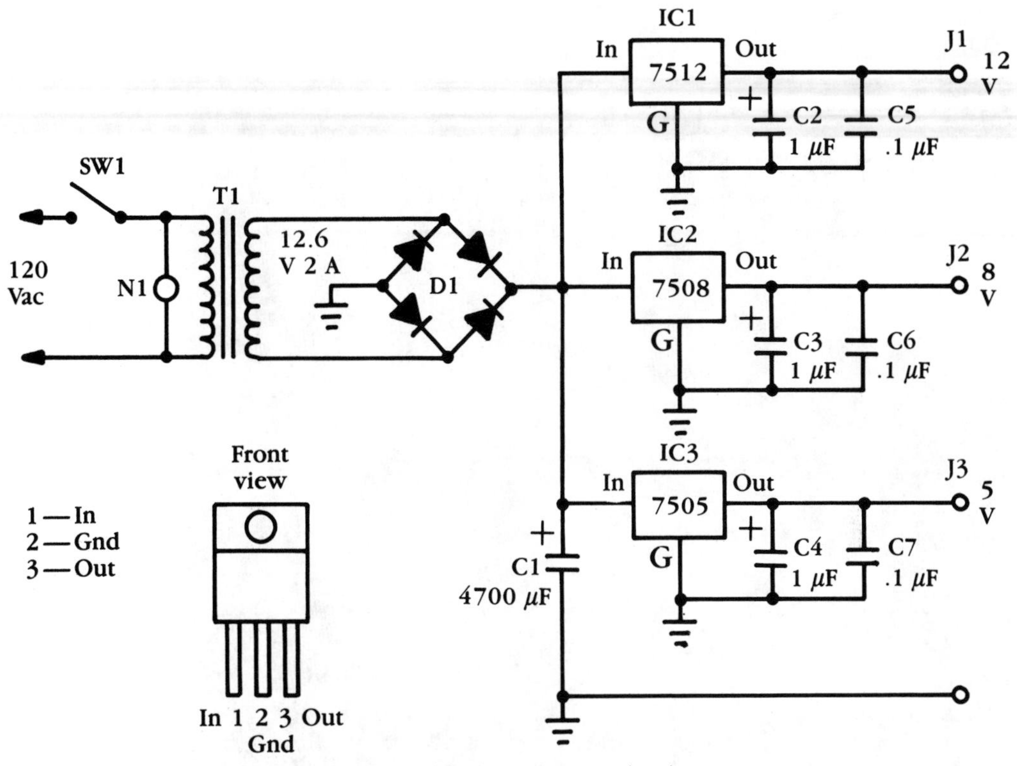

4-2 Circuit diagram of the low-voltage-regulated power source.

First, mount D1 at the extreme end of pc board. Next, place the large electrolytic capacitor just behind D1. Place all three regulators in line near the bare copper wire, starting with the 12-volt regulator. Now, solder the "in" terminal of each regulator directly to the positive bus wire.

Connect a short piece of wire to the center ground terminal of the regulators and solder it to the ground set of ringed holes. Keep the out terminal of all regulators free so as not to short it against a ground ringlet hole or bus bar. Solder C2 and C5, C3 and C6, C4 and C7 across each respective output terminal. Solder a 5-inch hookup wire lead to each output terminal.

Doublecheck all wiring. Make sure that no eyelets or ringed holes are touching. Use a DMM on the low-ohm scale and check for a measurement between each regulator terminal. Check each terminal and the associated circuit for continuity. Inspect the small wiring under a magnifying glass to determine if each connection is proper and not shorted.

Rotate the ohmmeter to the 20 kΩ-position and check from voltage output to ground. IC1 output should have a resistance of about 10 kΩ to ground. The IC2 and IC3 outputs should measure between 4 to 7 kΩ to the common ground. Suspect a leaky circuit or regulator if the resistance is below 3.5 kΩ.

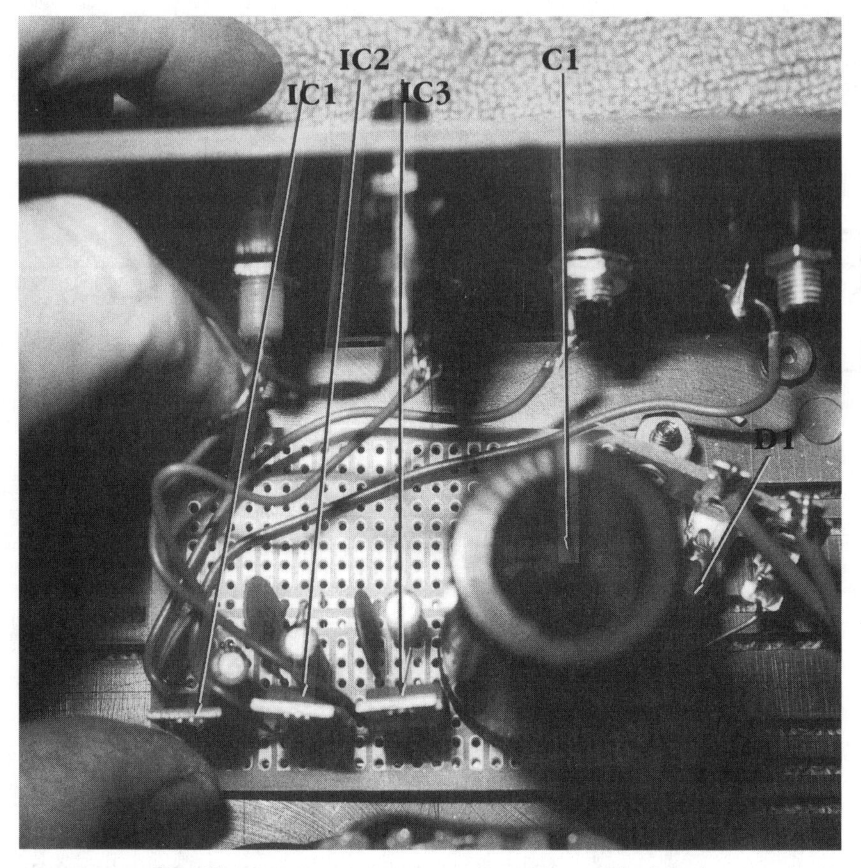

4-3 Close-up view of the mounted parts on small pc board.

Case layout

After the pc board has been completely wired, prepare the instrument case. In this particular project case, the plastic piece is for the front panel. The metal panel goes in the rear because it has a hole for the grommet and ac cord. Place front panel components well inside outside edge—both panels slip into the top and bottom grooves of the plastic case.

Draw a light pencil line lengthwise through the front panel, about 1 inch up from the bottom edge (FIG. 4-4). Equally space four 5/16-inch holes in line for the common and different voltage jacks. Make sure that the second jack from the right or switch is in the center of the panel. Near this hole, drill a 9/32-inch hole for the ac neon pilot light. Drill a round 3/8-inch hole for SW1. In this case, a fly-type switch is used with square-hole mounting. Drill two small 1/8-inch holes to mount SW1.

Clean all burr edges with a pocket knife. Before mounting the parts, letter each component on the front panel. Label the black front panel with white letters. Make sure all press-on letters and numbers are

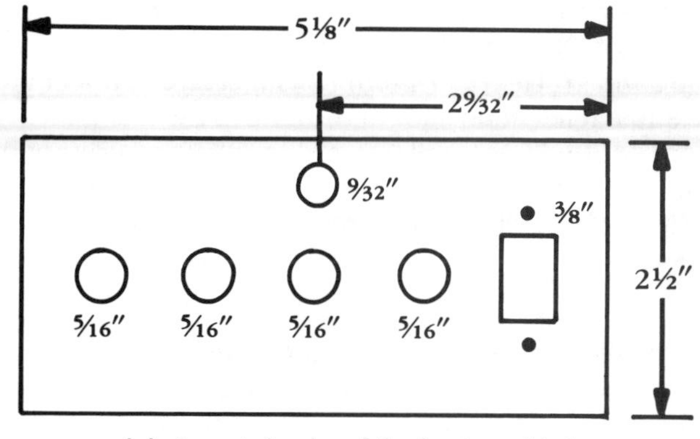

4-4 Layout drawing of the front panel holes.

smoothed down. Now, spray on a few coats of clear matte spray found in most art supply stores.

Connecting leads

Mount all parts on the front panel. Tighten each jack nut and connect all of the voltage jacks. Start with the 12-volt output wire, cut it to length and solder it to the 12-volt jack. Likewise, do each voltage source in the same manner (FIG. 4-5). If the regulators were put in line, just connect the wire coming from the correct regulator to corresponding voltage jack. Now, solder the common black wire to ground jack.

4-5 Soldering the connecting leads from pc board to front panel.

Solder the switch, transformer, and ac cord. Remember to solder N1 across one side of SW1 and the power cord. Bring the ac cord and primary winding wires (black) to a 4-terminal lug. Place a rubber grommet or ac line cord strain relief connector in the hole in the rear panel. Tie a knot in the ac cord so it will not pull out. Solder SW1 in series with the one primary lead of T1. Doublecheck the connected wiring. Place a dab of enameled paint upon each nut to prevent it from loosening.

Testing

Before attaching the top cover, test the power supply. Flip on the switch and the pilot light should come on. If not, doublecheck the ac wiring. If the power transformer beings to hum or groan, pull the ac cord—a direct short must exist in the rectifier or regulator circuits. Check the resistance across each voltage source. No resistance measurement should be under 4 kΩ.

Notice if the transformer or IC regulators are warm. All should be cool after several hours of operation. Doublecheck the location terminals of regulators. If a regulator runs warm, it is placed in backwards. Regulators are usually not damaged by having reversed terminals.

If C1 runs warm or red hot, the polarity is reversed or the capacitor has a low working voltage. If C1 is reversed, the transformer could smoke or make a loud noise. C1 should be a 35-volt dc electrolytic capacitor. Again, check the output voltage at all voltage jacks (FIG. 4-6). The tested voltage should at most vary only slightly from the required voltage.

4-6 Testing each voltage source with the DMM.

Parts list

IC1	7512 12-V regulator.
IC2	7508 8-V regulator.
IC3	7505 5-V regulator.
Z1	1.5-A zener diode.
C1	4700-μF 35-V electrolytic capacitor.
C2, C3, C4	1-μF 35-V electrolytic capacitor.
C5, C6, C7	0.1-μF 50-V ceramic capacitor.
J1, J2, J3	Banana jacks.
D1	2-A bridge rectifier or four 2.5-A single diodes.
T1	12.6-V 1.5- or 2-A stepdown power transformer, 273-1352 or equiv.
N1	Red neon 120-Vac indicator, 272-712 or equiv.
SW1	SPST toggle switch.
Case	2⅞"-X-5⅝"-X-5⅛" plastic instrument case, 270-250 or equiv.

AUDIO AMP CHECKER

The audio amp tester indicates proper channel balance, weak or defective channels, and provides output load while servicing. When troubleshooting direct-coupled transistor power output circuits, an unbalanced output can place excessive voltage across the voice coil of the speaker and burn it out. No dc voltage should be applied to speakers with a balanced output circuit. With this audio amp checker, you can first switch in the 10-ohm 10-watt resistors and check the speaker line voltage before applying the signal to the speakers (FIG. 4-7).

To check for the correct channel balance or if one channel is slightly weaker than the other, simply switch in the meter test. The internal speaker can be switched into the circuit for audio tests. Each channel input has another speaker jack for testing the amplifier speakers or for testing extra speakers after the amplifier tests. The amp checker can be used as an indicator when injecting an external audio signal to locate a weak, distorted, or defective stage.

The circuit

The speaker input circuit consists of a double set of jacks for phono-type and speaker wire clip-in terminals. A set of probe jacks are connected to each channel to measure the dc voltage on a possible leaky output circuit. Often, both speakers are damaged. The unbalanced circuit occurs when one speaker is subbed for the other to determine if the amp or speakers are defective.

The dc voltage test should be made first if you know that the amplifier contains balanced power output dc circuits. Simply flip the switch on

4-7 Front view of the audio amp checker.

both load transistors (Load) and take voltage measurements at the voltage terminals before opening the amplifier (FIG. 4-8). The speaker load can be switched on the speaker when tests are made on the other channel. In solid-state output circuits, the speakers should be loaded down at all times, especially if the gain or volume control is turned up. Otherwise, you might damage output transistors or IC components.

Each speaker channel can be checked for balance by switching to one channel, switching in the high (Hi) or low (L) resistors, and adjusting of R3. Always start with SW5 in the high position. Now, flip SW2 to the test position and SW1 to the other channel. If normal, both channels will be quite close when an audio signal is applied to the input terminal of both channels. When one channel is much lower than the other, suspect a defective channel or an improper adjustment of the balance control. The internal speaker can be switched in (SW2) for audio sound tests.

The defective channel can be signal-traced by injecting audio signal from a sine/square wave or audio signal generator at any point in the circuit. The test and meter circuit can be used as indicator. Simply switch to the internal speaker to check for clean audio. Keep a load resistor connected to the other output channel so as not to damage the output component. Another speaker can be plugged into the front test jack for the good channel.

Front panel construction

Construct the front panel from a double-sided pc board. The board was chosen for its metallic appearance, solid smooth construction, and work-

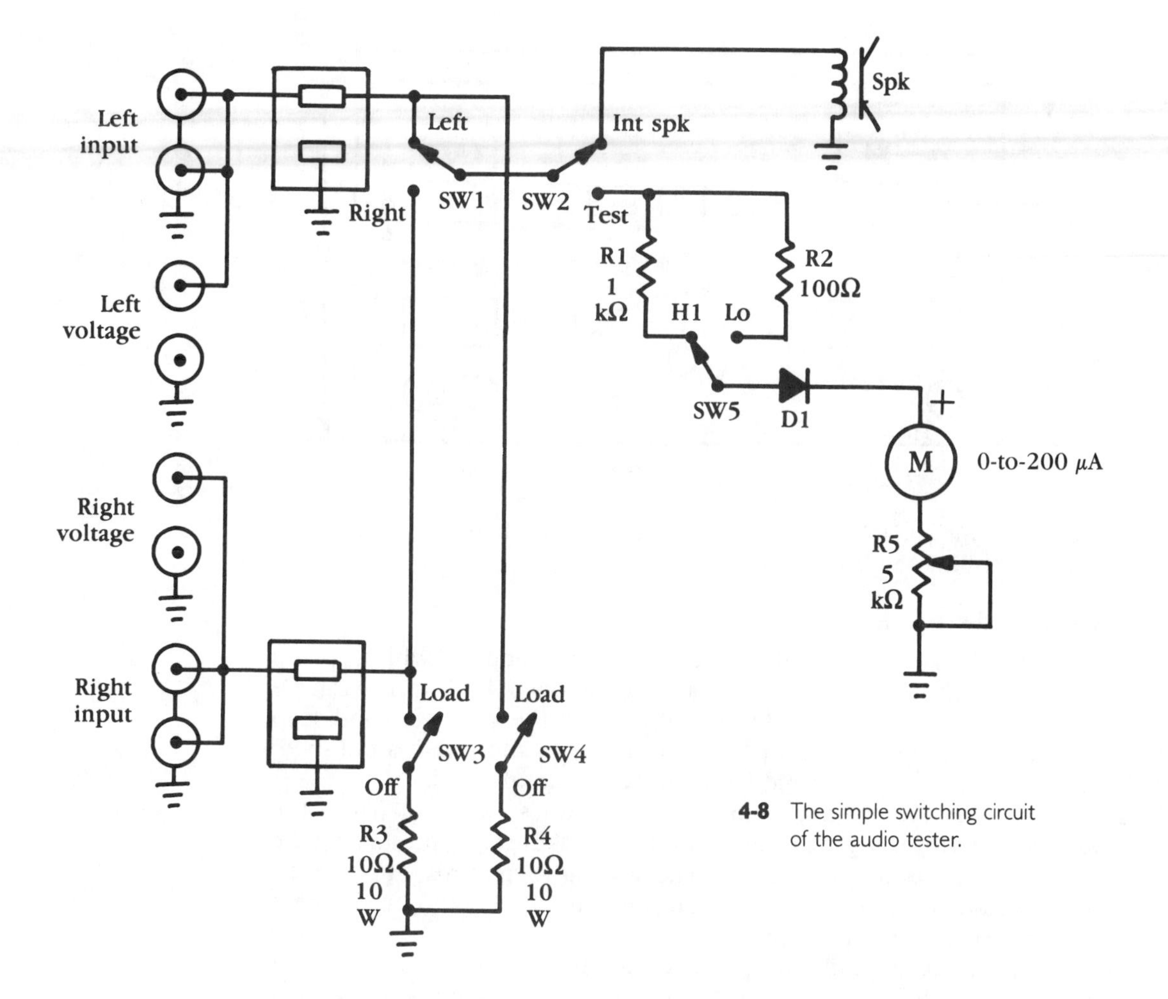

4-8 The simple switching circuit of the audio tester.

ing ease. Cut a 7¾-×-3-inch pc board. Lay out the dimensions and size the holes for the front panel controls (FIG. 4-9). Start with a small ⅛-inch bit and enlarge the hole with larger bits until the hole is the correct size. the panel to fit inside the front speaker sides.

Cut the holes for the input jacks wide enough that the jack terminals do not short against the metal copper side. Drill a hole for each jack terminal and enlarge each with a larger bit. Hold the copper-clad board down firmly when drilling holes so that it won't pop-up and whirl around with the bit. Clamp it down, if necessary.

Drill very small holes for the switch lever and break out the center area. Make sure the line holes are fairly even, for this opening can be seen from the front. The switches mount behind the metal front panel. Square the holes with a three-corner and a flat-type file and fashion them into a rectangular cutout. Likewise, enlarge and square the holes for the speaker

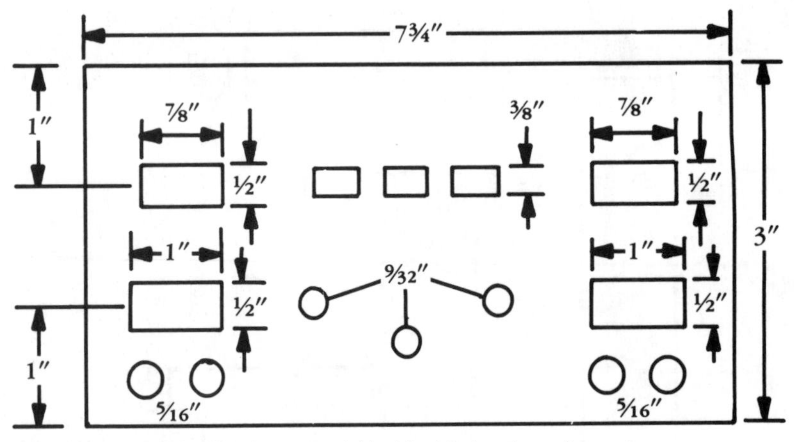

4-9 The front panel (double-sided pc board) layout.

input and output jacks. Make sure all parts fit into their respective holes before finishing the front plate.

Letter and symbols

After all holes are cut out and shaped, clean the front panel with soap and water to mark the various operation of components. All marks can be removed with steel wool and soapy water. Rub the steel wool and soap length-wise across the panel to remove scratches and give it a polished-copper appearance. Wipe and let the panel dry thoroughly.

Start lettering and give the front panel a professional look (FIG. 4-10). Keep all lettering straight by using scotch or masking tape and aligning the letters and numbers along the tape. The tape can be removed after the letters have been pressed on. Make sure the lettering is not covered by a switch or an input jack assembly.

Spray three coats of clear poster or art spray on the front panel. Let each coat dry before applying another coat. Now the panel is bright and copper-looking. You might want to use this panel as a template to drill holes in the speaker front panel. After the panel has completely dried, mount all parts on the front panel.

Preparing cabinet

Choose a speaker cabinet with fairly large speakers to house the audio amp checker. Lay the template on the front of the speaker grille cloth and drill the holes through the cabinet (FIG. 4-11). Make sure a portion of the speaker backside is open. Since the tester parts are mounted on the front metal panel, a portion of the speaker grille and cabinet can be sawed out with a saber saw. You might want to cut out the entire area so that the template can fit inside. Leave enough of the speaker board's edge to hold the mounting screws. Screw the panel to the wood front after you mount and solder the parts.

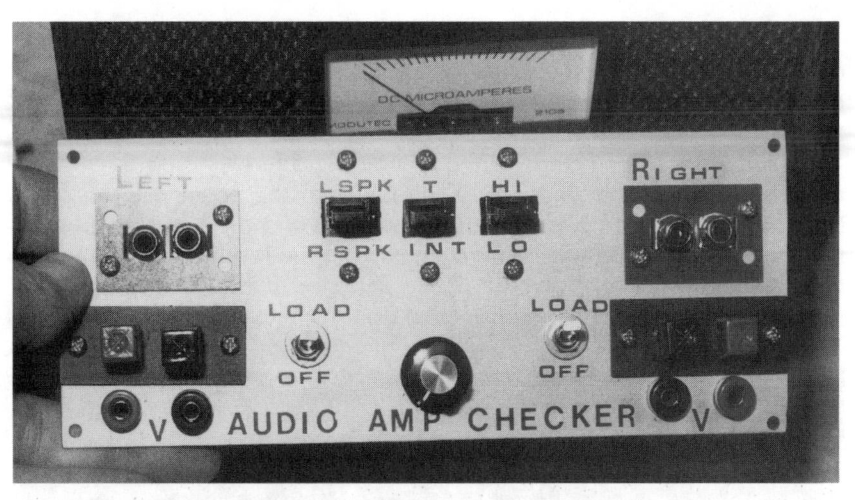

4-10 The front panel is lettered using three coats of clear spray.

4-11 The opening in the speaker cabinet.

Connecting components

Start with the left and right input jacks and solder the grounds together. Connect the two input jacks together and solder a piece of hookup wire to SW1. Likewise, connect both speaker input terminals to each side of SW1. Connect the voltage jacks to each side of the speaker input jacks. Run a piece of hookup wire to each side of SW3 from each respective input jack. Solder the 10-watt resistors from SW3 to ground. Each channel has its own 10-ohm load resistor.

Connect a piece of hookup wire from SW1 to the switching terminal of SW2. One side of the speaker will connect to SW2 and other lead to common ground. Insert R1 and R2 between the switched terminals of SW2 to SW5. Solder the crystal diode between the meter and SW5. Solder R3 in series with the meter and the common ground.

Place the positive terminal of the meter to the positive terminal of D1. If the meter has no terminal identification and the meter hand goes the opposite direction when audio signal is applied, simply switch terminals. Place the meter above the metal panel, on the speaker grille area. Test the meter after all wiring has been doublechecked (FIG. 4-12).

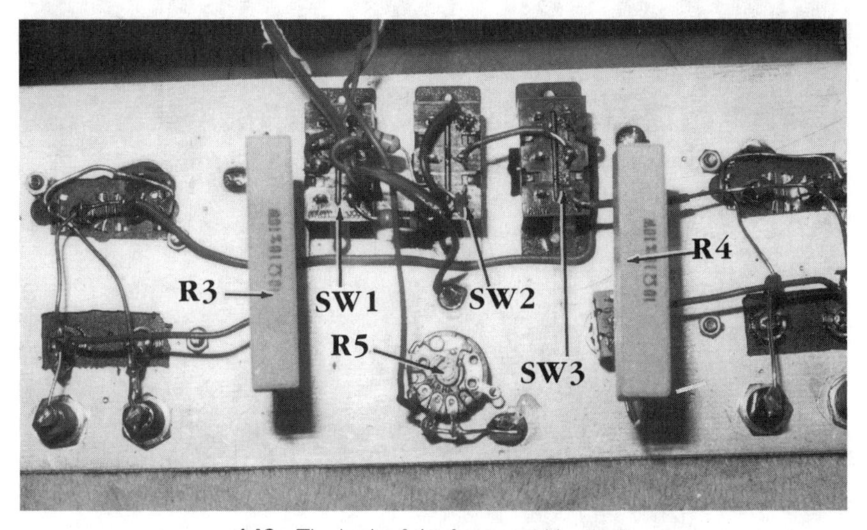

4-12 The back of the front panel is wired.

Testing

Connect the audio output from a radio, cassette player, or amplifier to the speaker inputs. Insert both right and left stereo speaker output terminals. Only one set of speaker input terminals are used when checking a manual amplifier. The stereo amplifier can be connected to either the phono receptacles or to the push-in wire terminals. Check chapter 6 for the correct input cable connectors.

Switch on the two load resistors at SW3 if the output stages might have dc audio-balance circuits. Suspect balanced outputs if one or both of the channels is dead or if there is no speaker indication. When the speakers show either a dead short or an abnormal resistance, suspect an unbalanced audio channel. If one or both original speakers are damaged, switch in the load resistors.

The load resistors might run quite warm when a dc voltage is applied to the speaker output. Mount the resistors away from the metal panel to dissipate heat. After identifying the leaky channel, let the load resistors

Parts list

J5, J6	2 dual phono jacks, 274-332 or equiv.
TR1, TR2	2 push-button terminal strips, 274-315 or equiv.
J1, J2, J3, J4	4 banana jacks.
S3, S4	2 SPDT miniature toggle switches.
S1, S2, S5	3 SPDT lever switches.
R1	5kΩ linear control.
D1	1N60 or 1N34 crystal diode.
PC	7¾"-X-3" double-sided pc board.
M	0-200, 0-500 microamp meter or surplus-type, All Electronics Corp.
R1	1000-Ω 1-W resistor.
R2	100-Ω 1-W resistor.
Cabinet	Spk. cabinet with 5- or 6-inch PM speaker.
R3, R4	10-Ω 10-W resistors.
Misc.	Bolts, nuts, solder, hookup wire, etc.

remain switched to that channel while troubleshooting. Do not connect the internal test speaker or any other external speaker to the shorted channel or you will quickly burn out or damage its voice coil. Measure the dc voltage on the defective channel at the front voltage jacks. It's best to switch in the load resistors and take low-voltage dc measurements before making any other tests.

To check for channel balance, switch SW2 to the test position, adjust the amp volume and R3 for a half-scale reading. Now, compare the other channel. If both channel levels are fairly close, the amplifier balance control can balance them. If one channel is lower than the other and will not balance, suspect a weak audio channel. Always use an audio generator signal at both inputs, not music, for an accurate channel balance.

The weak channel

A small loss of audio gain in the amplifier can be adjusted with the balance meter. When no balance meter is in the circuit and the gain in one channel is noticeably greater than the other, suspect a defective component. Notice the gain of the good channel on the meter volume, the volume and gain control settings.

Inject the audio signal from a sine- or square-wave oscillator, audio signal generator, or hand-held injector at the input of both channels. Notice the gain settings in the good channel compared to the weak channel. Check from stage to stage to see where the loss occurs with the generator. A small audio loss is very difficult to find. Check each point in the good stage and compare it to the defective one.

4-13 Checking the output channels in a "boom-box" cassette player with the audio amp checker.

Weak audio can be caused by transistors and IC components. Often, small audio loss results from dried-up electrolytic coupling or bypass capacitors. Although transistors will not weaken, a component in the transistor circuit might cause the transistor stage to appear weak. The defective transistor might become leaky or open.

Distorted sound

Suspect leaky or open audio output transistors or IC components if one channel is distorted. Extreme distortion and weak audio might occur in the audio output stages. Slight distortion can appear in any audio circuit.

Weak distortion should be signal-traced with a scope and a square- or sine-wave generator. Extreme distortion can be located with an audio signal generator using a speaker as an indicator. Simply start at the volume control and inject the signal toward the speaker. Usually, a leak, a short, or an open transistor will cause extreme distortion. If both audio stages are distorted, suspect a leaky power IC. Low or improper voltages applied to the transistor or IC output circuits might cause weak sound or distortion. Remember to check the bias resistance and bias controls for distortion.

SINE/SQUARE/TRIANGLE GENERATOR

This little generator comes in handy when troubleshooting and repairing amplifiers and high-fidelity sound systems. The output signals are in sine, square, and triangular waveforms. The waveforms can be injected at the input of the audio amplifier or at defective stages that are distorted or lose

4-14 The front view of the sine-, square-, and rectangular-waveform generator.

signal. The ideal sine wave to pass through the amplifier stages should be perfect, with no distortion (FIG. 4-14).

If the sine waveform has its top section flattened, there is positive peak clipping, usually caused by a decrease in gain of a transistor stage. Negative peak clipping can be attributed to excessive transistor gain, collector current, and overdriven transistors. Overloading can flatten the top of the sinewave.

Flattened positive and negative peaks are caused by excessive overloading and distortion. Poor low-frequency response with a square waveform makes dips in the top and bottom waveforms. Poor high-frequency response can cause the square wave to look like a positive and negative sawtooth waveform. A lagging or leading phase shift of the square wave at low frequencies will make the square wave tilt backward and forward. You can quickly locate the defective stage by injecting a sine or square wave from the sine/square/rectangular generator. The generator delivers a signal from 20 Hz to 20 kHz.

The circuit

The waveform generator is built around the popular 8038 function IC. The voltage developed across R4 and R6 produces current to charge and discharge the timing capacitor, C6. When the voltage across these two resistors change, the output frequency of the oscillator varies greatly. The frequency of the generator is controlled by R1. The oscillator frequency of IC8038 is determined by C1 (0.0047 μF) and the voltage applied to pin 8 (FIG. 4-15).

R8 and R9 are vertical-mounted screwdriver or thumb-operated parts

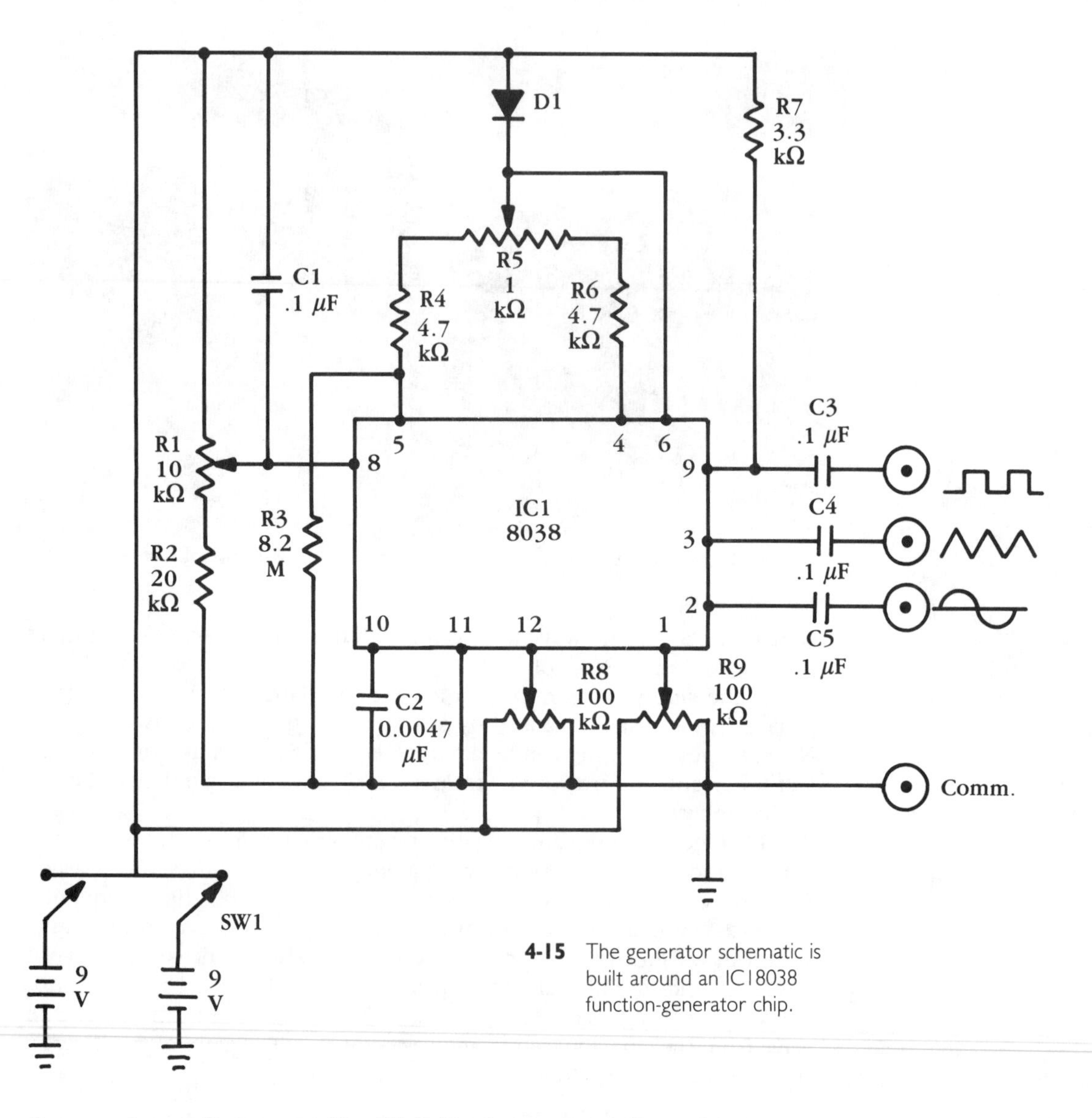

4-15 The generator schematic is built around an IC18038 function-generator chip.

for waveform adjustments. The R5 (1 k) trimmer pot adjusts the symmetry of the triangle waveform. R8 and R9 are alternately adjusted, while checking the sine wave on the scope. Touching up all three controls might be necessary for correct sine, square, and triangle waveforms.

PC board construction

Lay out the small pc board on a 3-×-4 inch piece of one-sided copper-clad board. Mount the 14-pin IC socket in the center of the board. The three small vertical pots are found at the bottom of the board. Draw the three waveform outputs at the top of the board. Run a common-ground pc connection around the outside of the pc board (FIG. 4-16). After all component mounting positions have been checked, drop the pc board into etching solution. It should take about 35 minutes to etch the pc wiring.

Drill holes with the smallest bit possible to mount all the generator components, except the power supply. Drill through the board for clean holes. Drill two 1/8-inch holes at each end for mounting.

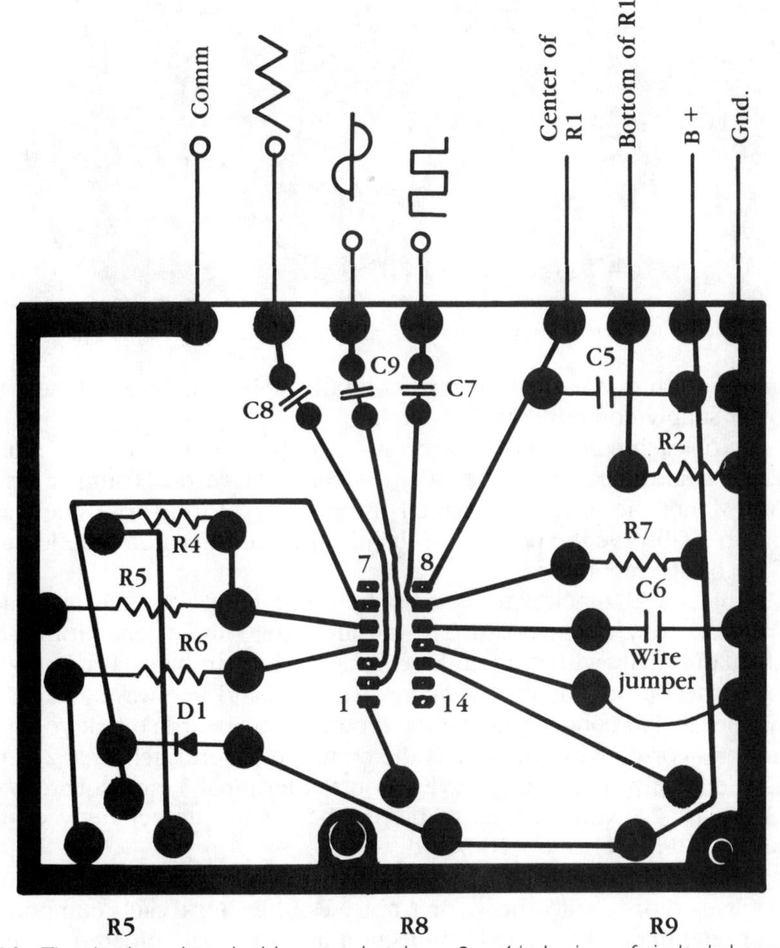

4-16 The simple pc board wiring was placed on a 3- x -4-inch piece of single-clad copper board.

Mounting parts

You will notice that only a few components are needed besides the IC generator. Fasten the IC socket to the pc board. Do not solder IC1 directly

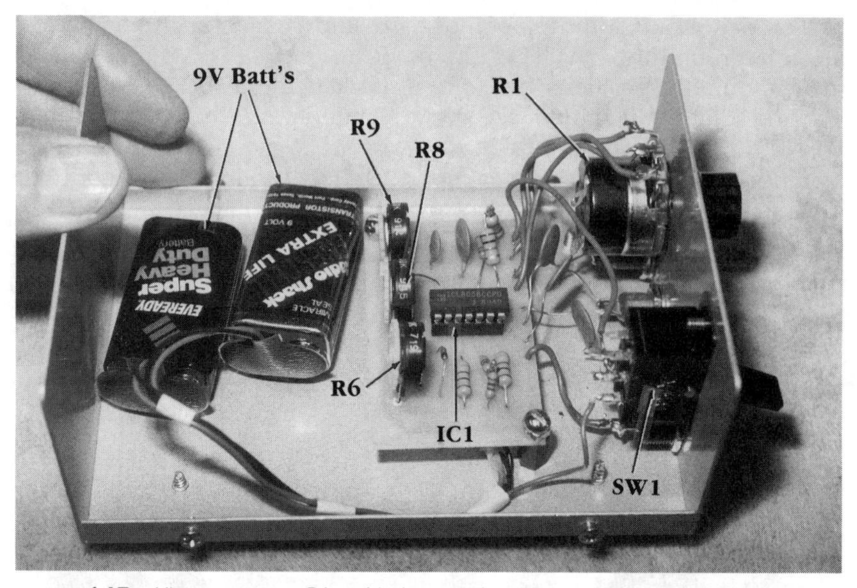

4-17 All parts, except R1 and jacks, are placed upon the small pc board.

into the circuit. Mount all small parts on the pc board, except R1 and the power supply components (FIG. 4-17).

Drop each part into the correct holes and solder a good joint. Mount the resistors and capacitors first, then solder in the vertical trimmer capacitors. Mount the IC last. Solder D1 using a pair of long-nose pliers as a heatsink. Observe the polarity of the electrolytic capacitors, diodes, and of the bridge rectifier.

Solder a 2½-inch extension hookup wire for R1 and all four waveform jacks from the pc board. The output of this simple generator is controlled externally within the audio channels. The four banana jacks on the front panel output common-, square-, triangle-, and sine-wave forms. Follow the labeled connections on the top of the pc board so that you wire terminals correctly: terminal 1 is the common ground, terminal 2 is the triangle waveform, the sine wave outputs at terminal 3, and square wave at terminal 4. Do not forget to place a bare wire jumper between the ground and pin 11 of the IC board.

Very little trouble should result when finished wiring, unless a component is dipped into the wrong hole. Doublecheck each component mounting, internal soldered connection and external soldered connection. Use two separate 9-volt batteries to power the generator. Snap the batteries in place and test it (FIG. 4-18).

The oscilloscope must be used to correctly adjust the linearity and waveform. Either borrow an oscilloscope or take your generator to a place that has a scope—a TV shop or elsewhere. Connect the scope leads to the sine-wave jack and the common ground. Adjust R5, R8, and R9, to obtain a perfect sine wave.

4-18 Checking the distorted amplifier section of a radio/cassette player.

Adjustment

Connect the scope probe to the triangle waveform and adjust R5 for minimum distortion and a clear waveform (FIG. 4-19). Next, place the probe on the sine-wave jack and adjust R8 and R9 for minimum distortion. You might have to touch up each one of these controls for a good clean sine wave. The sine wave should be rounded at the top and not too high a peak (FIG. 4-20). Now, check the square-wave form. Re-adjust R5, if needed, for a clear square waveform. Recheck the sine and triangle waveform if they are still normal. It only takes a few minutes to acquire the correct waveforms.

Testing

If the generator has no output sine or square waveforms, check the current across the switch terminals with the switch off. The dc current mea-

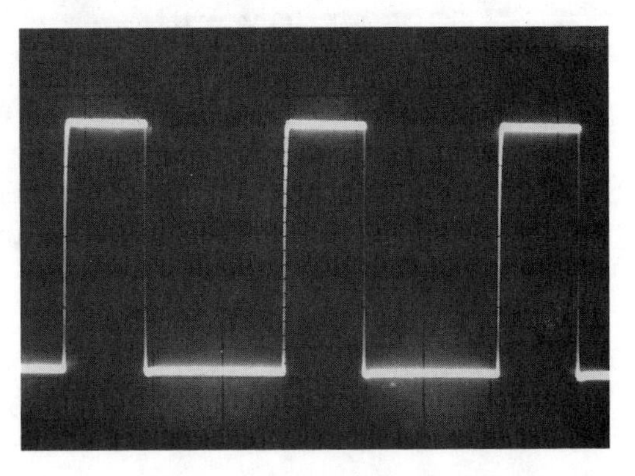

4-19 A true square waveform at the output terminals.

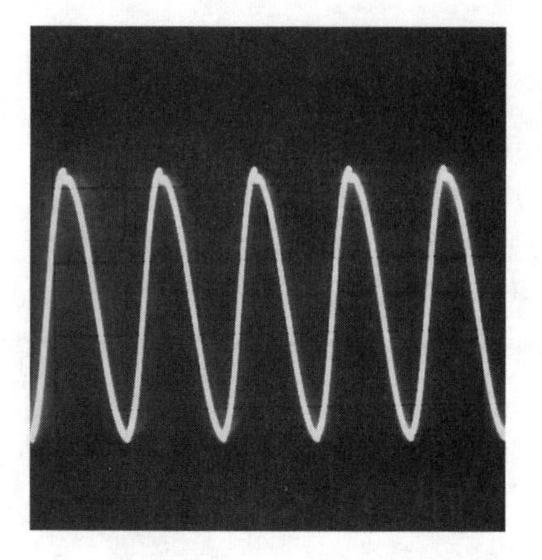

4-20 The correct sine waveform after proper adjustments were made.

surement should be around 17.3 milliamperes. Suspect a poor battery connection, switch contacts, or improper wiring if there is a no-current reading. Inspect the wiring connections and battery leads. If the current measurement is over 20 milliamperes, suspect a leaky IC or a defective component.

Take a few voltage measurements at pins 8 and 9. Compare these readings to the schematic. In fact, check all voltage measurements at IC1. If the voltages appear normal, suspect a broken output lead or coupling capacitor. When two waveforms are normal and the other is not, inspect

Parts list

SW1	DPDT toggle or paddle switch.
D1	1N914 silicon switching diode, 276-1122 or equiv.
Jack	4 banana jacks.
C1, C3, C4, C5	0.1-μF 100-V ceramic capacitors.
C2	0.0047-μF 100-V mylar capacitor.
R1	10-kΩ linear control.
R2	20-kΩ ½-W fixed resistor.
R3	8.2-mΩ ½ W resistor.
R4, R6	4.7-kΩ ½-W resistor.
R5	1-KΩ trimmer screwdriver or thumb vertical control.
R7	3.3-kΩ ½-W resistor.
R8, R9	100-kΩ trimmer screwdriver or thumb vertical control.
IC1	ICL8038 waveform generator, from Circuit Specialties.
Batt.	Two 9-volt batteries.
Cabinet	Metal cabinet with aluminum front, back, and bottom, 270-253 or equiv.
Misc.	Hookup wire, nuts, bolts, solder, etc.

the coupling capacitor and the connections on the banana jack. The common jack terminal connects directly to the signal and battery ground circuits.

HAND-HELD SIGNAL INJECTOR

Signal tracing is a quick method to locate a defective audio or rf stage. Start at the volume control and isolate the defective section before or after the volume control (FIG. 4-21). Use either speakers or earphones as an indicating device. Simply connect the ground clip to the common ground and probe away.

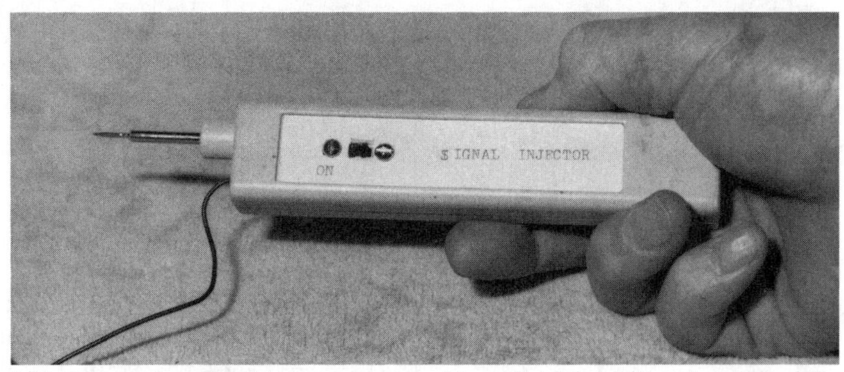

4-21 The small commercial case was obtained from Global Specialities. A probe and clip lead have been added.

Special case

This little signal injector is housed in a commercial blank CTP-2 Global Specialties case. The probe case can be used for pulser, probe, or injector enclosures, including tips and predrilled top plates. The gray high-impact styrene case is only 5.8 × 1.0 × 0.7 inches and it costs $7.95. A white stick-on label can be typed or marked before applying over the various controls (FIG. 4-22).

The circuit

The single 1 kHz audio injector is constructed around an LM 3909 LED flasher-oscillator IC. The circuit is powered by one AA flashlight battery (FIG. 4-23). Only five terminals of the 8-pin IC socket are used. The 1½-volt battery is turned on with a miniature sliding switch (SW1).

The output audio signal is taken from pin 2 of IC1 through a 0.01-μF 500-volt capacitor. A higher voltage coupling capacitor (C1) should be used if the injector is to be used on tube circuits. Usually, tube circuits have higher dc voltages. The alligator ground clip is fastened to the com-

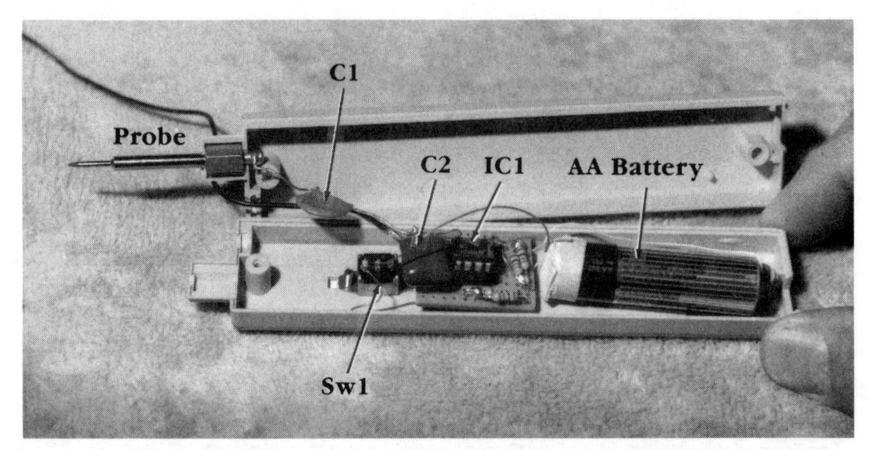

4-22 Top view of the small hand-held signal injector. Notice that only one AA cell operates the circuit.

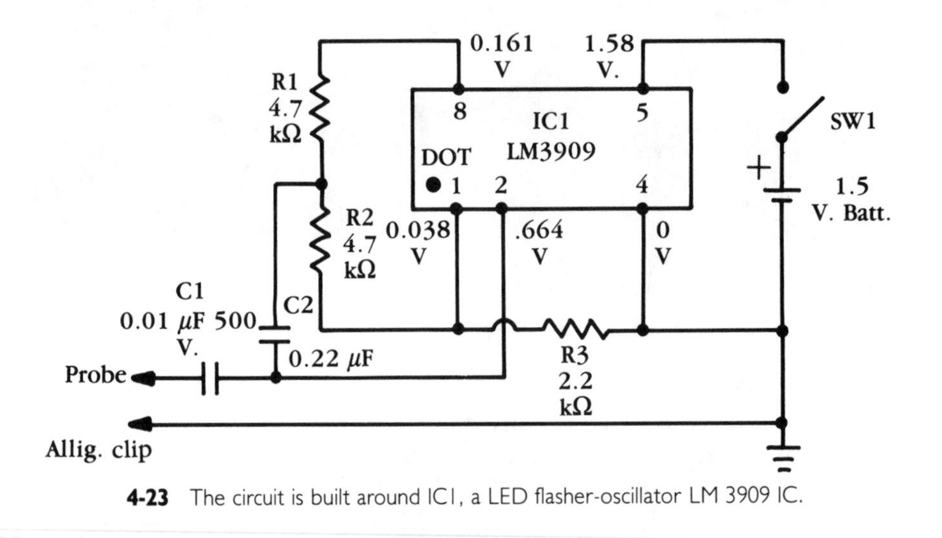

4-23 The circuit is built around IC1, a LED flasher-oscillator LM 3909 IC.

mon ground. Pin terminal 1 can be marked with a white dot on each side of the perfboard for easy wiring.

Wiring

Cut a 1-×-7/8-inch piece of perfboard with hacksaw (FIG. 4-24). File or grind the cut edges. Mount the 8-pin dip IC socket in the center of the perfboard chassis. Mount the small parts in the order that they should be soldered into the circuit. Start by connecting a stranded hookup wire to pin 5 and SW1.

Since there is not enough room to include a battery holder, solder a flexible hookup wire to the positive and negative terminal of the battery.

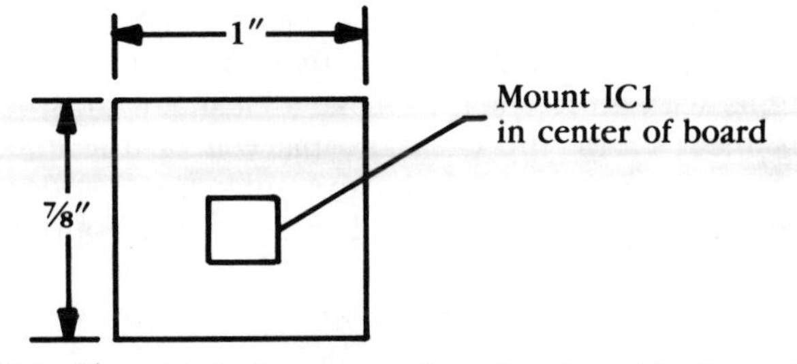

4-24 Cut a 7/8-×-1-inch piece from a piece of perfboard for the small chassis.

Solder an area on each end of the battery. Do not leave the iron on too long and destroy the positive connection. Place a piece of masking tape over the positive end. Keep these leads as short as possible.

Part mounting is not critical. Keep all component leads as short as possible. Solder all parts on the perfboard chassis before connecting the outside leads. Layout the chassis, switch, and battery in the probe case. Now, solder C1 to the probe terminal. Connect a 10-inch black flexible ground lead with alligator clip to the common ground. Tie a knot in the ground lead so it cannot be jerked out of the probe assembly.

Testing

The injector should be tested before enclosing it in the commercial case. Clip the probe tip to the high side of a volume control in a radio or to the input of an audio amplifier. Use a longer clip lead for this test. Now clip the ground connector to the common ground of the amplifier. When the switch is turned on, you should hear a loud 1-kHz tone in the speaker—if not, take voltage, current, and resistance measurements.

With SW1 on, you should measure 1.5 volts at pin 5. If the voltage is very low, IC1 might be leaky or inserted backwards. Check the wiring of the entire injector circuit. Remember, voltages at each pin of the LM 3909 IC are very low and should be checked with the low-ohm scale of the DMM. Check the low voltages listed on each pin of the schematic diagram.

Before replacing IC1, measure the current across SW1. Leave the slide switch turned off. A high current measurement (over 10 mills) indicates a leaky IC or an improper wire connection. The normal current should be approximately 1.23 milliamperes. Check for broken or improper connections with the low-ohm scale of ohmmeter.

Prepare probe case

Drill two 1/8-inch holes to mount the small sub-mine slide switch in the ready-made slot. Fasten SW1 to the top panel with 4/40 bolts and nuts, then solder a flexible wire to it. Next, prepare the white stick-on label (provided with the probe case) to go over switch area.

Cut a small rectangular hole for the slide switch into the stick-on label with a sharp razor blade. Punch two small holes for the switch mounting bolts with an ice pick to make a clean-cut finish. Type "on" and "signal injector" on the label and place it on the injector before mounting the mini-switch with bolts and nuts. Now, solder the switch wires into the circuit. Cement the small chassis to the bottom case with a dab of rubber cement.

How to use

This little signal injector can be used on car radios, AM-FM receivers, audio amps, cassette player pre-amps, phono amps, intercoms, and IC amplifiers to locate defective components (FIG. 4-25). Rotate the volume control half way and inject the signal at the center top of the control. If the audio amp signal is normal, you should hear a loud tone in the speakers. If you can't hear the tone, the audio stages are defective.

Now, start at the speaker and the last output stage. Place the probe on the base terminal of the last transistor or on the input terminal of the IC. If a signal is heard, proceed to the driver af amp. When no signal is heard, measure the voltage and resistance at the output transistor or IC.

Proceed toward the audio input stages when the signal is loud and clear on the volume control. Inject the signal right up to the tape head in the cassette player. Remember, as you work towards the input stages, the volume at the speaker should become louder. Lower the volume control for adequate indication.

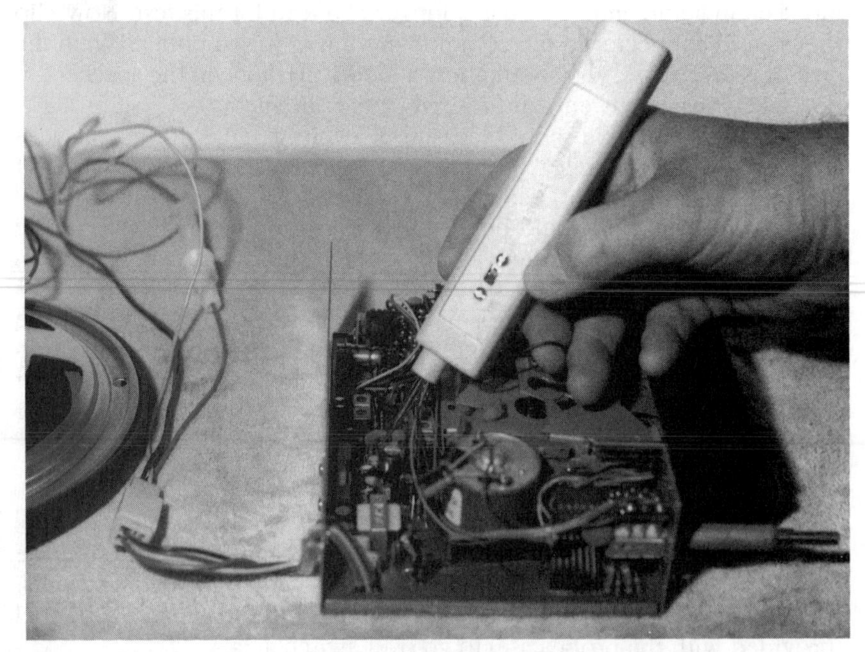

4-25 The signal injector is used to troubleshoot the different sections inside a car radio.

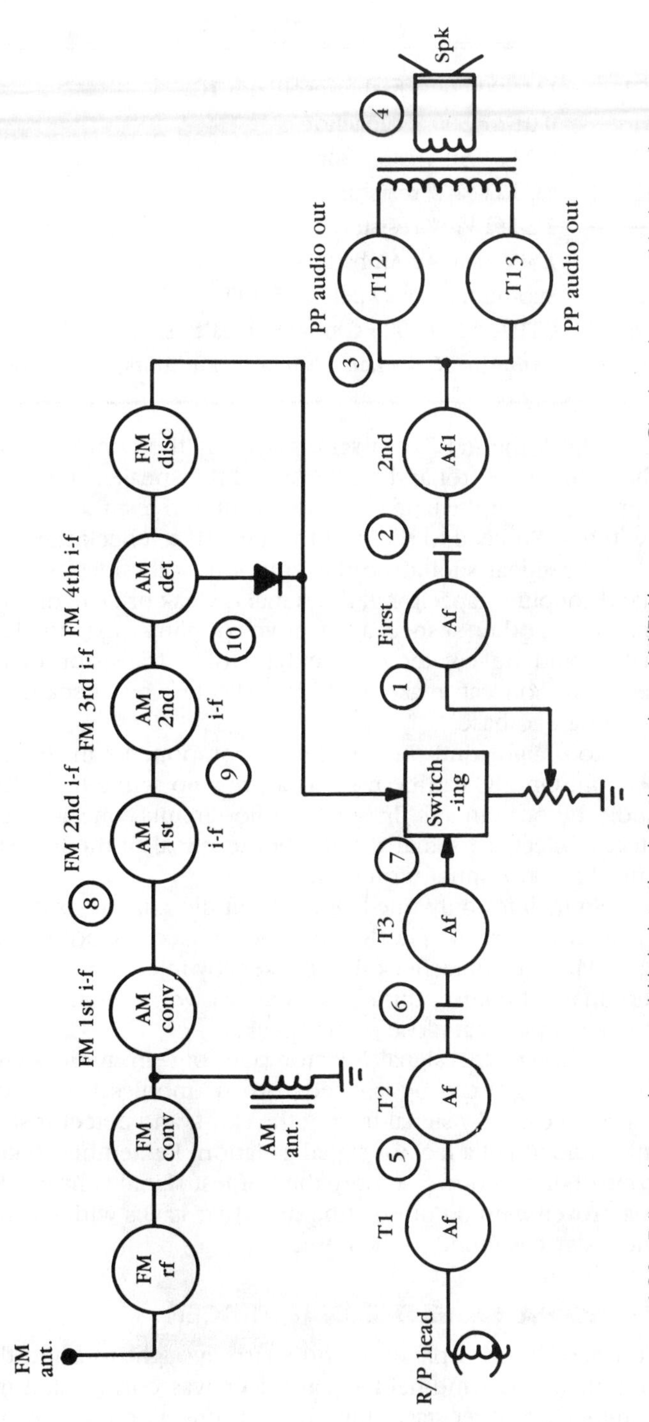

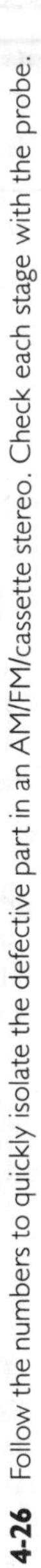

4-26 Follow the numbers to quickly isolate the defective part in an AM/FM/cassette stereo. Check each stage with the probe.

Parts list

IC1	LM 3909 LED flasher-oscillator IC.
C1	0.01-μF 500-V capacitor.
C2	0.22-μF 50-V capacitor.
R1, R2	4.7-kΩ ¼-W resistor.
R3	2.2-kΩ ¼-W resistor.
Batt.	1.5-V alkaline AA battery.
S1	Sub-mini slide switch, 275-409.
Case	CTP2, #110-0006 Global Specialties.
Misc.	8-pin dip IC socket, 4/40 bolts and nuts, solder, hookup wire, etc.

Check the AM-FM cassette player by the numbers (FIG. 4-26). Start at the volume control and work toward the speaker until you can hear the signal. Begin at the tape head and point 5 to test the cassette pre-amp and switches. Proceed to points 8 through 10 to check the AM stages.

The signal should be the same on both sides of an electrolytic or fixed coupling capacitor. If the signal is weak or dead on one side and normal on the other, suspect a defective coupling capacitor. Likewise, the signal should be greater on the base of a transistor than the collector terminal. You can make a quick audio test by checking each transistor from base to base.

You might find one or two IC components in the integrated audio section. One IC can be used as a pre-amp while the other contains the audio output circuits. In some audio circuits, one IC includes all audio stages. Inject the signal at the input terminal of the IC—stereo audio circuits have two input terminals.

Sometimes only one-half of the audio channel is dead and the other channel is normal. The IC must be replaced if both channels are contained in it. Doublecheck the speakers by injecting a signal at each output terminal. Of course, this signal is very low, but you should still hear it if you keep your ear close to the speaker.

Although the signal injector puts out an audio signal, the rf, and detector stages can be isolated with harmonics from the audio squarewave injector. Try signal-tracing the rf, i-f, and detector stages with a normal radio to get a feel for signal isolation. Remember to keep the volume control turned down so only the loudest signal is heard. Signal injection is a proven method of isolating defective stages with electronic entertainment devices found in the home.

AUDIO HAND-HELD SIGNAL TRACER

You need a lot of patience and sharp eyesight to build this small signal amplifier. The hand-held signal tracer was constructed inside a surplus commercial pager case, but the versatile two-piece project case from

Radio Shack will do the job (270-257). A piece of experimenter's IC perfboard was used for the pc chassis.

Only seven small working parts are in this signal tracer. The small 4-pin IC socket was placed in the middle of the cut board. Select a 3-inch long bolt as test probe. All parts, including the speaker, fit inside the slender case (FIG. 4-27).

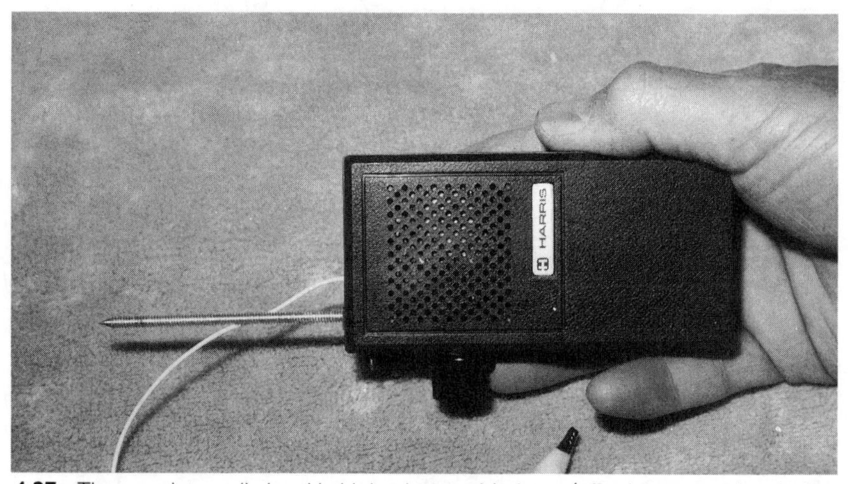

4-27 The complete audio hand-held signal tracer. Notice on/off volume control at the bottom.

The circuit

The simple audio circuit consists of a single LM 386 audio power amplifier plugged into a 4-pin socket. R1 controls the input signal, picked up by the test probe. C2 couples the audio signal from the volume control and feeds to socket pin 3. The audio output signal is coupled from pin 5 with C4 to a 1-inch PM speaker. The pickup volume, while signal tracing any audio circuit, is adequate (FIG. 4-28).

Construction

Cut a 7/8-inch strip from the IC perfboard and sand or grind the edges. You should have at least 7 rows of solder-ringed holes left to mount parts in. Place the 4-pin socket in the middle with the holes and tabs to the side for connections. Solder all 8-pin socket terminals. Start with C2 and connect to pin 3. Ground pins 2 and 4.

Be careful when soldering each pin so that the solder does not lop over. Use a small-wattage iron with a sharp point with only a little bit of solder. Solder C4 and C6 to pin 5 and solder a 10-ohm resistor (R2) to C6 and ground. Connect a flexible wire to the other end of C4 (220 μF) for one speaker wire and solder C3 across pins 1 and 8. Observe the polarity of C2, C3, C4, and C5. Connect the positive end of C5 to pin 6 and

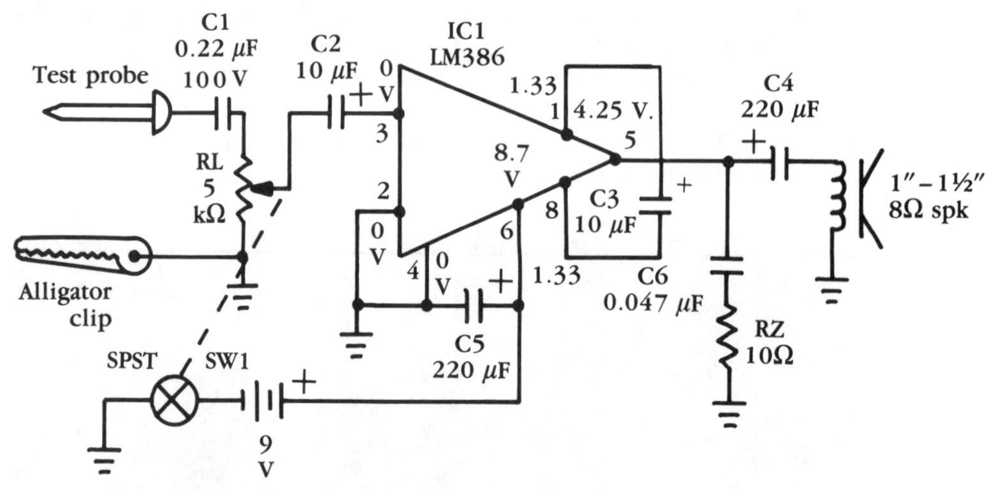

4-28 The simple wiring diagram of the hand-held signal tracer with enclosed speaker.

ground. Solder the red battery cable to pin 6 and connect the black cable wire to ground (FIG. 4-29).

Place the back side of a pocket knife or screwdriver blade between each IC terminal to remove any points of solder. Check each connection with a hand-sized magnifying glass to make sure that no wires or connections are touching or soldered together. Examine the wiring at least twice before connecting the volume control and battery.

Preparing case

Select a 1- or 1 1/2-inch PM speaker that will fit inside the plastic case. Drill 1/16-inch holes in a square or circle for the speaker grille cloth. Glue the

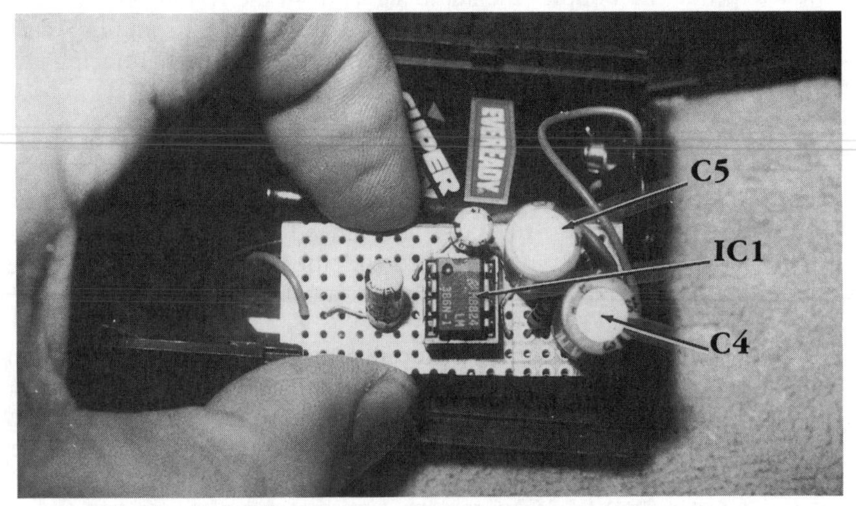

4-29 The small 7/8″ audio pc board. All parts are mounted close together.

small speaker to the plastic case. Drill a 11/32-inch hole for the small volume control (R1) in the bottom side of case. Drill a 5/32-inch hole, 1/2-inch up from the bottom, in the front end of the case for the metal probe. Finish the cabinet with a 1/8-inch hole at bottom front for the alligator clip lead.

Metal probe

The metal probe was constructed from a 3-inch long 5/32-inch bolt and nut. Place the nut on the bolt and rotate it down a couple of inches. Grind or sand (with an electric sander) a round fine-pointed tip at the end of the metal bolt. Make the point as sharp and long as possible (FIG. 4-30) so the point can connect to the transistor or IC terminals. Attach a star washer to one end and connect the hookup wire before placing it in the hole. Tighten the small nut on the outside of case to fit snugly.

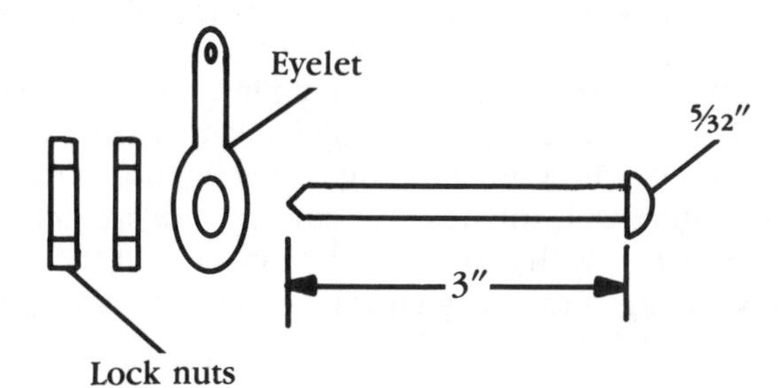

4-30 Drawing of the metal probe made from a 3″ long 5/32″ bolt. Grind the end to a sharp point.

Connecting

Connect all wiring before you place the small chassis in the case. Solder C1 to the top of the volume control and leave the other end long enough to solder it to the test probe. Connect the black battery lead to the on/off switch to the center lead of R1. Ground the other switch terminal. The 9-volt battery should be in series with switch, ground, and terminal 6 of IC1. Solder a hookup wire from the end of C4 to the one side of the PM speaker (FIG. 4-31). Ground the other speaker terminal to the ground side of the volume control. Glue the small speaker behind the grill with model glue or rubber cement. Let the cement set up before connecting the wires. Test the signal tracer before cementing the chassis in place.

Testing

Turn on SW1 and hold the test probe. You should hear a low hum in the speaker. Touch the probe to a piece of metal, and a hum and a scratchy noise is heard. If not, shut the unit off and place the milliammeter across

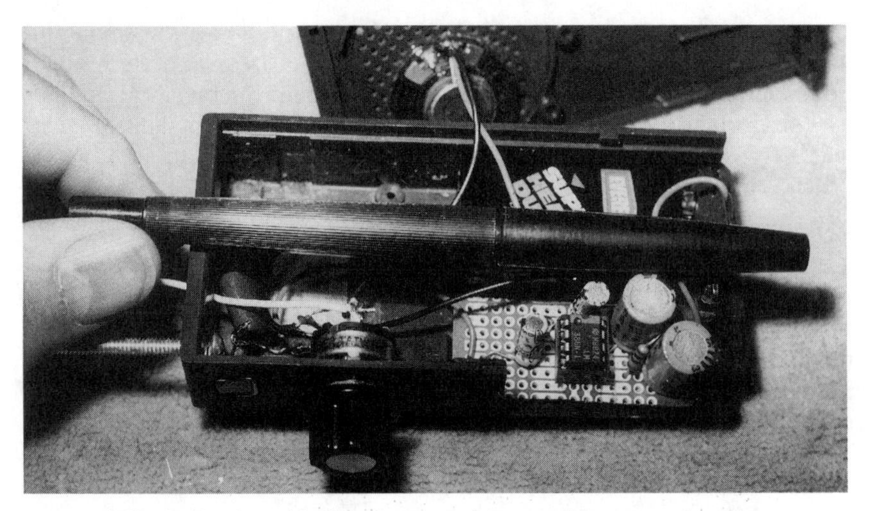

4-31 Inside view with all parts mounted, except the cover and speaker.

the switch terminal. Suspect a leaky IC1 or an improper connection if the current is over 10 mills. The normal idling current is around 8.5 mA's.

Doublecheck the wiring for binding or connections that touch near the switch and volume-control area. Measure the voltage across each pin of the IC and compare it with those on the schematic. If the wiring is normal, but the low voltage has heavy current, replace the LM 386.

Signal tracing

Although the small signal tracer is not as powerful as some amplifiers, you can signaltrace audio from the detector stage of the radio to the speaker terminals. Always keep the volume low in case you accidentally place the probe against a pin with a greater volume. As you go from stage to stage, the volume should increase. The alligator clip should be fastened to the common ground.

When signal tracing any amplifier, start with the signal on the center terminal of the volume control. This procedure cuts the circuit in half and saves service time. Test the next stage if the audio is present. If not, start from the volume control and work toward the front of the audio circuits. Remember, with this method the volume will decrease as you proceed toward the tape head in a cassette player (FIG. 4-32).

Start at the tape head terminals of a small cassette player and signaltrace the audio to the speaker terminals. Then go to the base of the first audio amp and collector terminal. Keep checking, from base to collector, turning the volume down each time. The volume will increase as you work toward the speaker. When the audio stops, you have located the defective stage. Check both sides of the coupling capacitor as you would for both sides of a transistor.

When locating a defective transistor with no sound at the collector

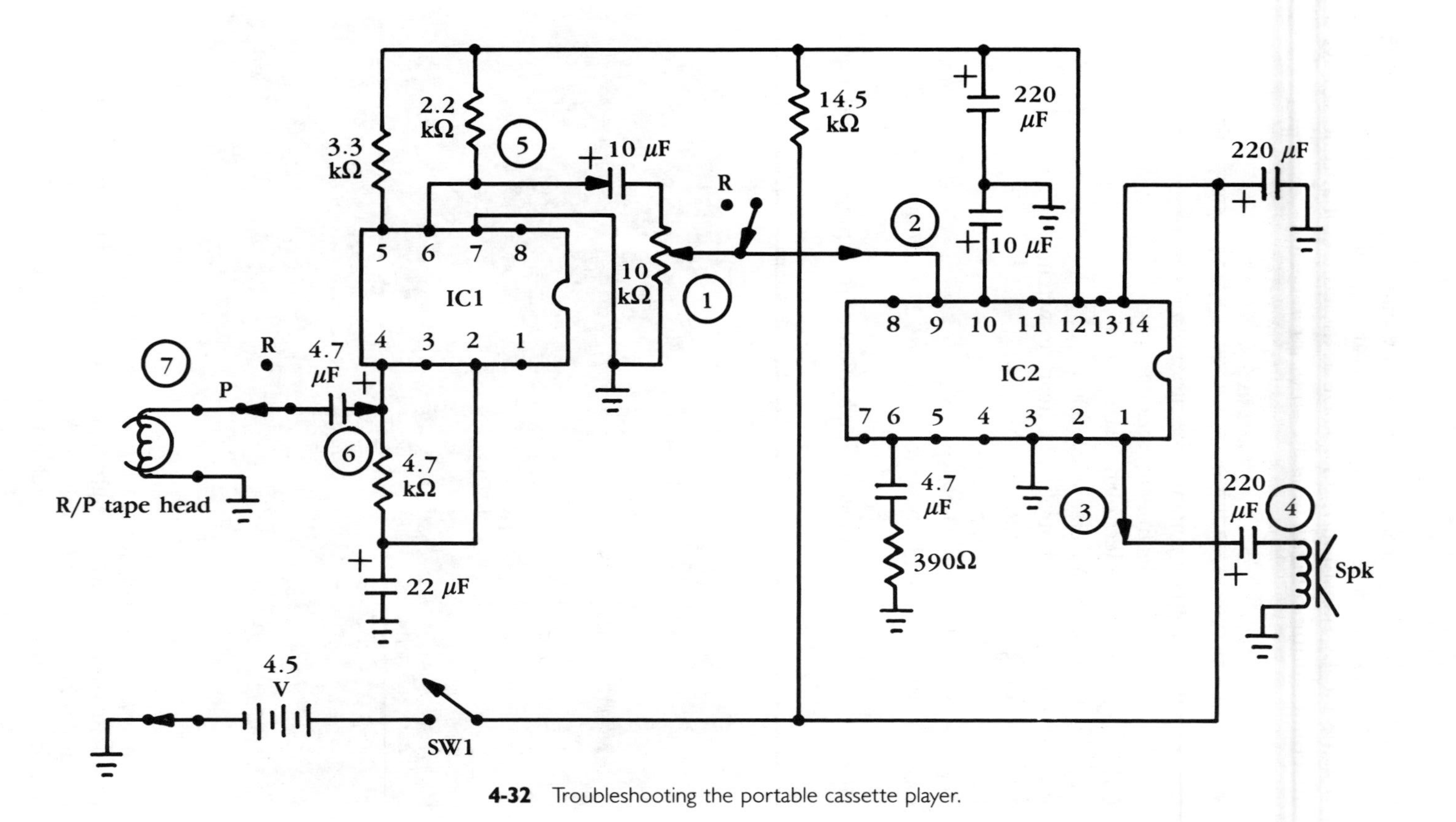

4-32 Troubleshooting the portable cassette player.

terminal take voltage measurement. A leaky transistor might have comparable voltages on each pin. Low voltage at the collector terminal might indicate a leaky transistor. High voltage at the collector and no voltage at the emitter terminal might indicate an open transistor.

Parts list

C1	0.22-μF 100-V ceramic capacitor.
C2, C3	10-μF 35-V electrolytic capacitor.
C4, C5	220-μF 35-V electrolytic capacitor.
C6	0.047-μF 50-V ceramic capacitor.
R1	5kΩ volume control and miniature switch.
R2	10-Ω ½-W resistor.
IC1	LM 386 audio amp.
SPK	1- or 1½-inch 8-Ω PM speaker.
Case	Two-piece high-style pro-enclosure, 270-257 or equiv.
Batt.	9-V battery.
Misc.	Battery cable, alligator clip, hookup wire, 4-pin IC socket, nuts, and bolts.

4-33 Always turn down the volume when troubleshooting large power amplifiers.

A few years ago, audio stages were all transistors. Today, the small radio and cassette players might just be ICs. Signaltrace the ICs starting at the audio input terminal and check with a stronger signal at the output terminal. Do not forget to check for signal loss at the speaker terminals. In many cases, the small electrolytic capacitor that couples the speaker to the IC output terminal was open. Be careful of high volumes when working on large amplifier components (FIG. 4-33). Keep the volume down at all times.

DELUXE FIXED AND VARIABLE REGULATED POWER SUPPLY

This heavy-duty power supply can be used to troubleshoot circuits in the TV or audio amp, where two different voltages are required. Of course, each power source can be used for supplying voltage to any project or circuit. Each voltage source is regulated in a 1-amp circuit (FIG. 4-34). The variable voltage source can vary from 1.2 to 26.5 volts dc. The fixed regulated sources are 24, 20, 15, 10, and +5 volts.

The circuit

Two different power supply circuits are provided with two different power transformers. Each circuit is fused with a 1-amp fuse and line indicator (N1) showing that the power supply is on. In power supply 1, T1 is

4-34 Front view of the dual 1-amp power supply.

an 18-volt 2-amp power transformer (273-1512). A 2-amp bridge circuit (D1) provides fullwave rectification. One large filter capacitor (6800 μF) provides input capacity filtering to the variable regulator (FIG. 4-35).

The variable regulator output voltage is controlled by a 5-kΩ control (R2). A 220-ohm 1-watt resistor ties the control into the "out" voltage circuit. C3 and C4 provide rf and loading capacity. Since the voltage can vary between 1.2 and 26.5 volts, a 0 to 50 volt meter is included in this circuit. Notice that the common ground for the variable voltage supply is separated from the fixed voltage source.

The second fixed voltage source has 1-amp fuse protection and a 120-volts ac light indicator. Both ac circuits are tied into one fuse block.

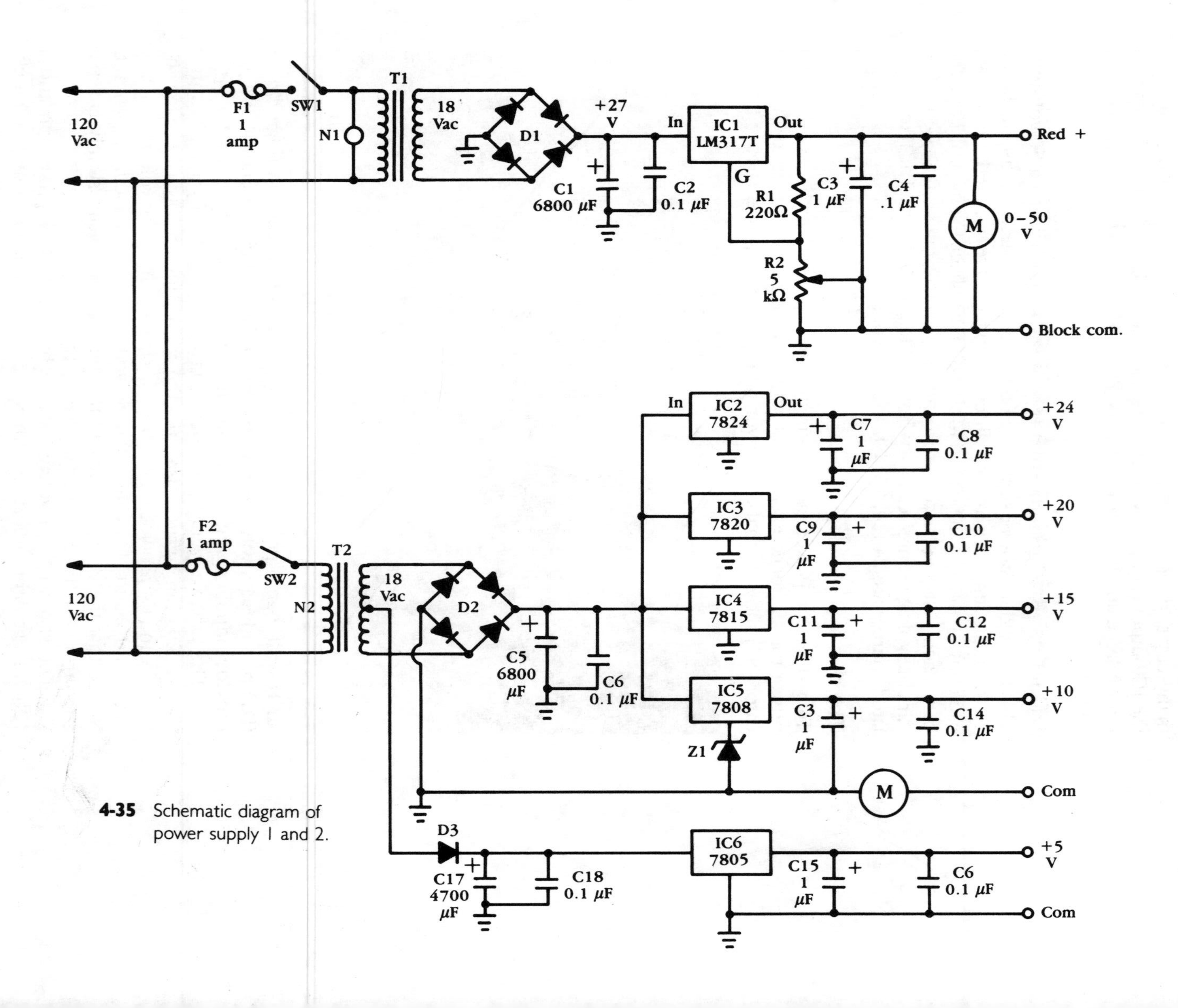

4-35 Schematic diagram of power supply 1 and 2.

T1 provides 18 volts ac at 2 amps and is applied across a 2-amp bridge rectifier (D2). Notice that the bridge rectifier negative terminal is at common ground, and the center top of the power transformer provides ac voltage to the 5-volt circuit.

Capacitor (C5) provides input filtering with a 63-volts dc rating. Both C1 and C5 are large audio amp-type capacitors; however, a 50-volt-type will work nicely in this circuit. IC2 through IC5 supplies fixed voltage sources of 24, 20, 15, and 9 volts. Although IC5 is an 8-volt fixed voltage regulator, Z1 in the ground lead provides a 10-volt output. Each fixed voltage has its own output jack. Notice that the ammeter is inserted in the common lead of the fixed voltages, feeding the positive 5-volt sources from the center top winding of T2. A 3-amp diode provides dc rectification. The half-wave rectification circuit is tied to two different filter capacitors: C17 (4700 μF 35 volts) and C18 (.1 μF 35 volts). IC6 is the positive voltage regulator (7805). Notice that the 5-volt source has a common banana jack.

Cabinet preparation

The front panel layout consists of the dc voltmeter and variable power source to the right and the ammeter with fixed voltage sources to the left. Place both meters evenly, centered down from the top panel. Layout the 5-volt source jacks even with the bottom edge of the two meters (FIG. 4-36).

Drill four 1/8-inch holes for meter holes and cut a round hole for the back side of both meters. The large meter holes can be cut with a circle

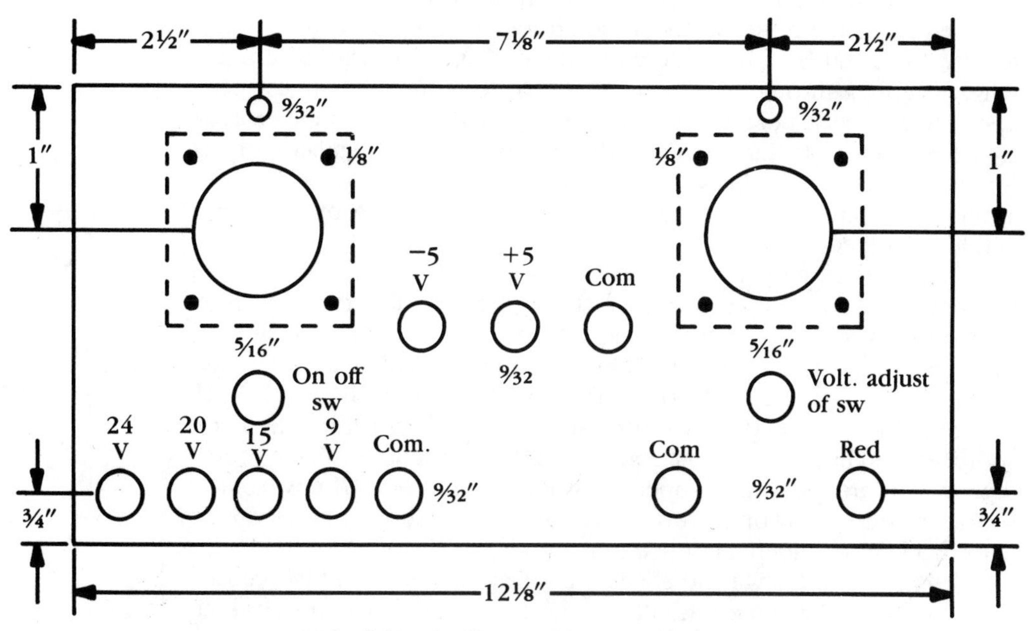

4-36 Cabinet and layout of front-panel holes.

cutter or you can drill many small holes around the perimeter and break out the circled piece.

Drill three 9/32-inch holes for the 5-volt power source in the middle of the panel. Likewise, drill seven 9/32-inch holes at the bottom for fixed and variable voltage jacks. Try to keep all jacks evenly spaced in one section. Drill a 5/16-inch hole for the 5 kΩ voltage control (of the variable power supply) below the voltage meter. Be very careful not to scratch the front panel while drilling the front panel holes.

Laying out pc board

Since several circuits and components must be mounted on the pc board, take time to layout the pc wiring (FIG. 4-37). Leave room in each corner for a mounting hole. In this case, the voltage and ground wiring were made larger than normal with pc tape. Mount the fixed voltage regulators along one outside edge. Leave enough room for the regulators to attach to heatsinks. Fit the 9-volt regulator sideways, next to the large filter capacitor (C1). Now etch the board.

Mounting parts

Solder a bare wire on the top of the board from the outside ground to the inside ground. This is the only ground wire loop used on the board. If desired, you can tin the copper wiring with a large soldering iron. Mount all components on power supply 1, then do the same with power supply 2. Be sure to make good soldering joints (FIG. 4-38).

Bend over the center wires of each bridge rectifier unit so the transformer wires can later be soldered to them. Cut off all excess terminal wires that stick through the board. It's best to mount the largest parts first. Always check polarity of diodes, bridge rectifiers, and electrolytic capacitors before soldering them. After mounting each regulator, check for the in and out terminals. Clean excess rosin solder flux between or around connections. Brush the wiring area after scraping off any rosin and loose solder.

Doublecheck all parts mounted and make sure that the polarity of capacitors and diodes is correct. Now, take a measurement from the common ground to each power supply voltage source. The resistance between ground and the 24- and 20-volt sources should be above 25 kΩ. The resistance of the 15-volt source to ground should be above 12.5 kΩ. The resistance of the 10-volt source should be above 4 kΩ. The 5-volt source should have a resistance above 4.5 kΩ. The variable voltage source to common ground will be open until R2 is connected. If any voltage source measurements on the board are much lower, suspect a leaky regulator or a sloppy-soldered connection.

Start with 2-kΩ range and check the leakage across each filter capacitor with a DMM. The meter should start near zero and slowly rise through the 2 kΩ reading and meter will go to O L. Check all four electrolytic

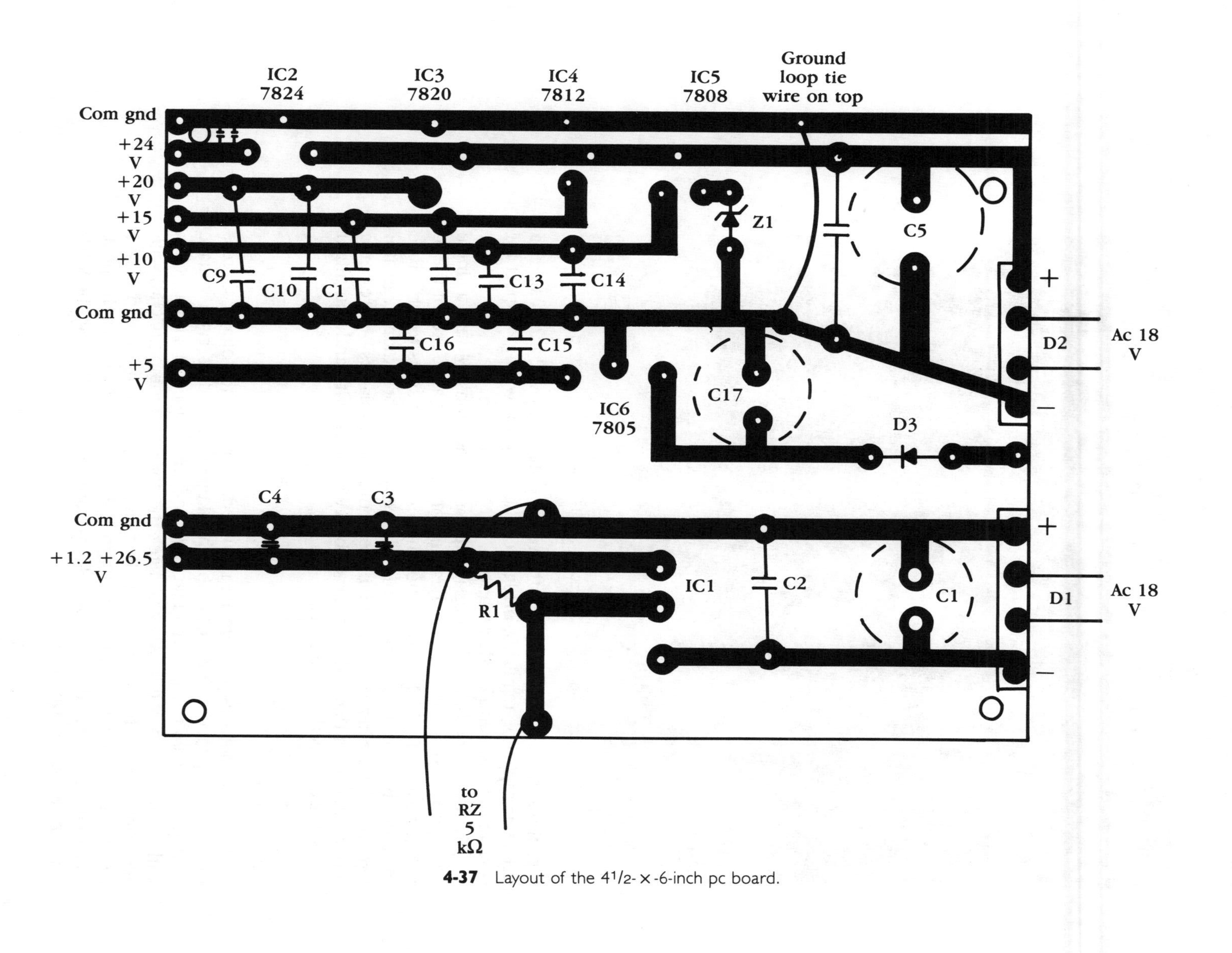

4-37 Layout of the 4½- × -6-inch pc board.

4-38 Close-up view of the components mounted on pc board.

capacitors in the same manner. If the reading stays below 2-kΩ, suspect a defective capacitor, regulator, bridge rectifier, or poor wiring. Inspect for correct capacitor and diode polarity.

Check the pc board for possible poor connections. Mount a single black anodized aluminum-alloy heatsink on each regulator, except the 5-volt regulator (TO-220). It does not matter if the heatsinks touch, since the metal body is ground potential. Be careful not to break any regulator terminals. These heatsinks can be bolted onto the regulators before being mounted, except that they are rather bulky to work with and you might break the connection terminals.

Connecting

After the cabinet components are mounted, connect the voltage sources to the front panel. Solder a flexible hookup wire lead to each voltage source and tie it into the correct jack on the front panel. Run all wires in a path as direct as possible. Always check the correct common (COM) voltage source of each power supply. Remember, the common lead of the fixed voltage supply goes in series with the ammeter. Connect the positive ammeter lead to the banana jack common terminal. Reverse the leads if the meter reads backwards.

The variable dc power supply has only two output jacks. Connect the common jack to the variable power supply. The variable and fixed voltage common terminals are connected together to provide two different voltage sources at one time. Solder the 0 to 50 voltmeter to the variable output jack (FIG. 4-39).

4-39 The pc board connected to the front panel.

Solder the secondary power transformer leads to the respective full-wave bridge circuits. Connect the ground terminal of the bridge rectifier (D2) to the common terminal of the fixed dc source jack. Connect the center tap lead of T2 to D3. Recheck the polarity of each diode and bridge circuit before wiring. Now, solder the correct 5 volt positive leads to the correct front panel terminal jacks.

Place a rubber grommet in the ac line hole and tie a knot in the cord. Run one ac line to both sides of the ac fuse block. Connect hookup wire to the other fuse terminals and solder the transformer primary windings through the separate on and off switches. The switch for T1 is on backside of R5. A toggle switch is used to turn on power supply 2.

Testing

Turn on control R2 to apply power to power supply 1. The voltage should register on the voltmeter. If nothing happens, check if IC1 is inserted backwards. If the power supply indicator light dims and the fuse blows, suspect a defective diode in the bridge rectifier or a leaky C1.

Check if C1's polarity is reversed. The capacitor will appear quite warm if the connections are reversed. Measure the voltage across C1 for

+27 volts. If you measure voltage before IC1, suspect that IC1 is defective. Make sure that the connections at R1 and R2 are correct.

Test power supply 2 by flipping the line switch. If the fuse blows at once, suspect C5 and D2. Check D2 like any diode—on the diode test of the DMM.

When one or more voltage output sources are missing and the others are normal, check the IC regulator with the defective voltage source. Either the IC voltage regulator is in backwards or there is a poorly-soldered connection. Suspect that the wrong wire is connected if the improper voltage source is at the output jack. If the ammeter reads backwards with one of the voltage sources connected, reverse the meter leads.

Checking horizontal oscillator circuits

The internal power supply circuits can be eliminated from a TV chassis when it shuts down or if you suspect that a flyback transformer is defective. In most TV sets with flyback-derived voltages, the horizontal oscillator and output circuits must function to keep the chassis operating. Just locate the supply voltage to the horizontal oscillator on the schematic. This voltage usually ranges from 9 to 24 volts. Check the horizontal oscillator by applying an external voltage to that circuit (FIG. 4-40).

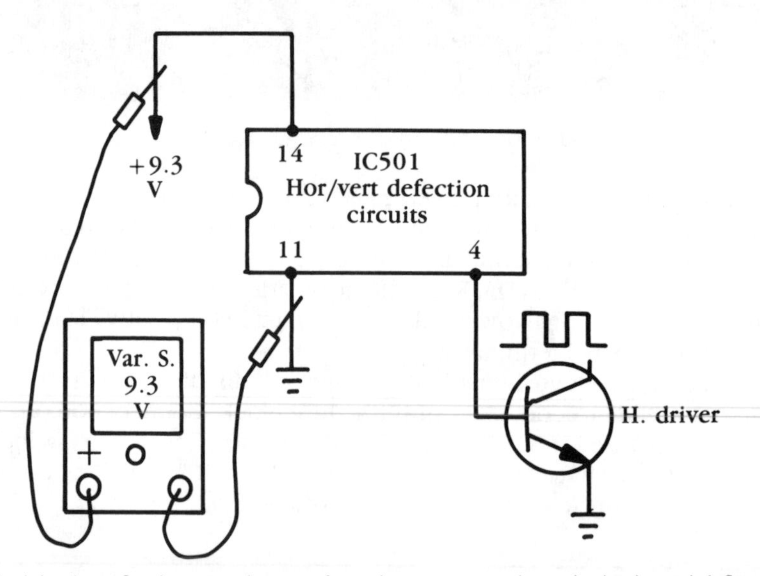

4-40 Injecting a 9-volt external source from the power supply to the horizontal-deflection IC component and checking output waveform with the scope.

Unplug the TV power cord and clip power supply 1 to the supply source on the oscillator IC and common ground. Monitor the horizontal waveform at the IC deflection pin (FIG. 4-41). Rotate the voltage upward with R2. Apply the correct voltage on the IC supply terminal. Check if the

4-41 The horizontal-IC output waveform applied to the horizontal driver transistor.

horizontal oscillator and deflection circuit (IC or transistorized) is working with the correct waveform. If the horizontal oscillator and drive circuits are functioning, go on to the horizontal driver and output stages.

Checking vertical circuits

The vertical circuits can be checked in the very same manner as the horizontal circuits—with an external voltage. You might find that two different voltage sources feed the vertical stages, so use both power supplies on the vertical section. In this case, use the 24-volt supply for the fixed voltage, and power supply 1 for the lower voltage. Reverse this procedure if more than 24 volts are used in the vertical circuits.

In chassis, where one IC component powers the deflection circuits and the output vertical and horizontal circuits have a separate voltage, place both power supplies into the circuit (FIG. 4-42). For instance, if the supply voltage to the IC deflection is +9 volts, connect the 9-volt fixed source to this supply connection. It's best to solder a piece of hookup wire to the IC pin connection instead of trying to connect the meter clip onto the circuit.

The vertical output transistor or IC operates at 44 volts. Clip the adjustable power supply to this voltage source and slowly raise the voltage to 26.5 volts. With the scope lead connected to the vertical output terminal (going to the yoke), you should have the correct waveform. Although the variable voltage source is not as high as that specified in the schematic, the vertical circuits are working (FIG. 4-43).

Checking audio TV circuits

The audio output stages in the present-day TV set can be powered from 9 to 30 volts dc. Some of the early sets had +120 volts applied to the output stages. Of course, the input stages operated from 9 to 24 volts. If the audio circuits are suspected of shutting down, the TV chassis might have a defective power supply. You can check the audio circuits by applying external voltage.

Some power supplies might have two different voltage sources and others have only one (FIG. 4-44). In this case, the transistor output and audio input circuits are powered with 9 and 26 volts. Simply connect the +9-volt supply to the input audio IC circuits and the 24-volt fixed power source to the transistor output stages. Now, the sound stages can be repaired by testing the voltage, resistance, and external audio. Inject the

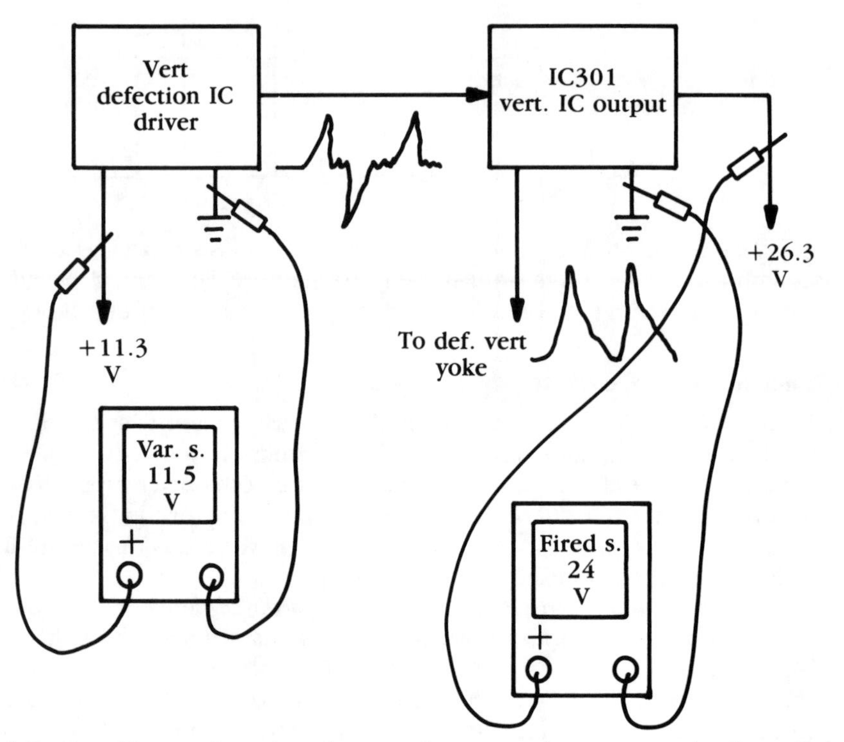

4-42 Two different voltages from the external power supply are connected to the vertical oscillator/driver IC and the IC output.

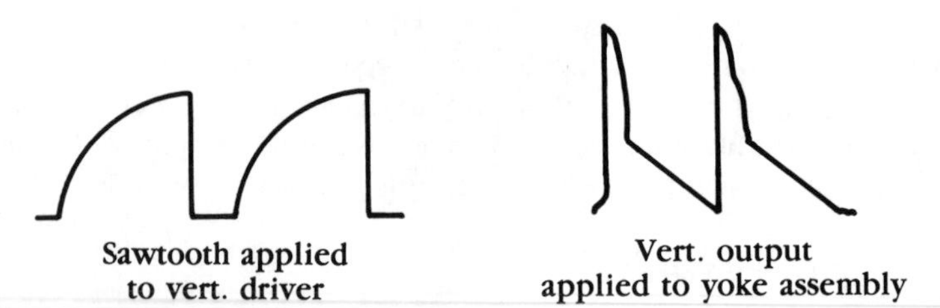

4-43 The sawtooth and vertical output waveforms found in most vertical circuits of a TV chassis.

audio signal from the sine- or square-wave generator and hand-held audio oscillator to check the audio output circuits.

Checking the audio amp

Amplifiers not pulling over 1 to 1.5 amp current can be serviced with the external power supply. A larger power amplifier might have voltage from 30 to 60 volts, and a smaller amp might operate on 18 to 30 volts. One stereo channel can be repaired by applying an external dc voltage and the

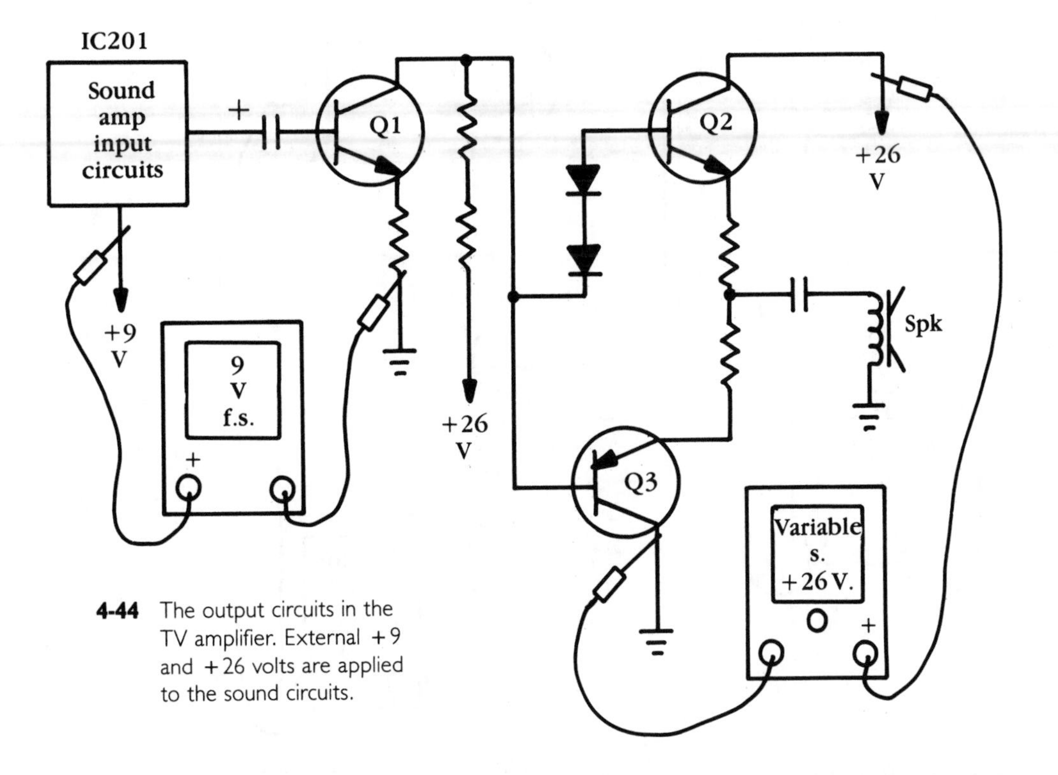

4-44 The output circuits in the TV amplifier. External +9 and +26 volts are applied to the sound circuits.

power output circuits of the smaller wattage amp can be checked with the external universal variable power supply.

Simply leave the power off of the defective amp and insert the correct voltage to the output circuits (FIG. 4-45). Apply the two different voltages to the driver and the audio output circuits. Apply the 20-volt fixed source to the driver stage and the 26.5 volts of the variable power supply. By injecting external voltages, you can service the audio amplifier stages of various electronic circuits.

Conclusion

The external voltage supply is handy when a defective part in the TV flyback circuits shut down the whole chassis. With this power source, one voltage source can be connected to one section while another supply is tied to other voltage sources. Each voltage source from the power supply might shutdown when overloaded with a defective component in the attached circuit—this is normal and no harm is done to the power supply. At least you know that the defective section is causing the entire chassis shutdown. Besides applying voltage to radio, TV, or amplifier circuits, the power sources can charge nickel-cadmium batteries and supply voltage for electronic projects and battery-operated equipment.

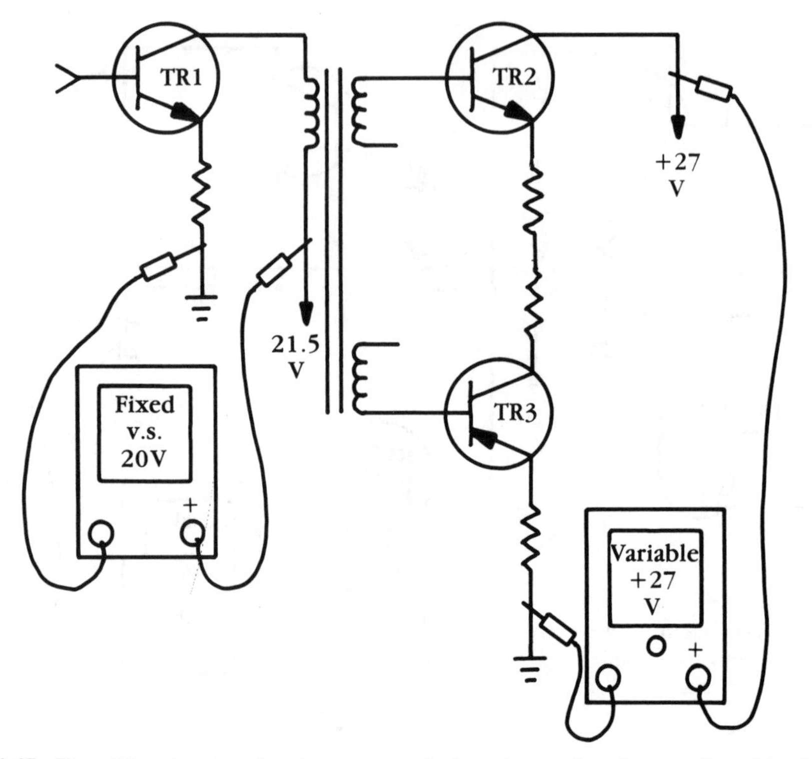

4-45 Two different external voltages are applied to the small audio amp (found in the radio, cassette player, or phono-player combination).

CRYSTAL CHECKER

This crystal checker will test most any crystal—those in ham receivers, two-way radios, scanners, or TV sets. The defective crystal can become weak from old age, have a broken internal connection, or maybe it just will not oscillate. The crystal can be inserted into the banana jack or be clipped with miniature alligator clips (FIG. 4-46).

Connect the suspected crystal into the jacks or clips. The polarity makes no difference. Turn on SW1 and rotate R3 until the meter hand begins to move upward. A defective crystal will not oscillate or make the meter hand move. You can also check ceramic filters and TV saw filters with this little crystal checker.

Simple circuit

The Pierce-oscillator circuit is built around a low-priced 2N2222 transistor. Of course, just about any low-signal npn transistor can be used in the tester (FIG. 4-47). CD1 is a variable-voltage capacitance diode used in place of those hard-to-find variable capacitors. Just an increase in voltage changes the capacity of the capacitance diode. C1 provides the ground return of CD1. Voltage applied to CD1 provides a greater frequency range.

Parts list for power supply 1

F1	1-A fuseholder and fuse.
N1	120-Vac red neon indicator.
T1	18-V 2-amp power transformer.
D1	2-Abridge circuit rectifier.
C1	6800-μF 50-V electrolytic capacitor, Hosefelt Electronics, Inc.
C2, C4	0.1-μF 100-V ceramic capacitor.
C3	1-μF 50-V electrolytic capacitor.
R1	220-Ω 1-W resistor.
R2	5-KΩ linear control with SPST switch.
SW1	On back of R2.
IC1	Variable regulator, LM 317T or equiv.
Jacks	2 banana jacks.
Meter	0–50 Vdc meter, 20-1124 Circuit Specialists.

Parts list for power supply 2

F1	1-A fuseholder and 1-Amp fuse or double fuseholder.
N1	120-Vac red neon indicator.
T2	18-V 2-A power transformer, 273-1512 or equiv.
D2	2-A bridge rectifier.
C5	6800 μF 50-V electrolytic capacitor.
C6, C8, C10, C12, C14, C6	0.1 μF 100-V ceramic capacitor.
C7, C9, C11, C13, C15, C18	1 μF 35-V electrolytic capacitor.
C17	4700-μF 35-V electrolytic capacitor.
IC2	AN 7824, 24-V fixed voltage regulator.
IC3	AN 7820, 20-V fixed voltage regulator.
IC4	7812, 12-V fixed voltage regulator.
IC5	7808, 8-V fixed voltage regulator.
IC6	7805, 5-V fixed voltage regulator.
D3	2.5-A 1000-V silicon diode.
M2	0–1 A circuit meter, 20-1115 from Circuit Specialists
Z1	3.3-V zener diode.
J1, J2	2 banana jacks.
SW2	On/off SPST toggle or rotary switch.
Case	4 7/16" × 12 3/8" × 6 9/16", MG 12 from Hosefelt Electronics, Inc.
Misc.	Line ac cord, nuts, bolts, pc board, hookup wire, etc.

4-46 Front view of the crystal checker.

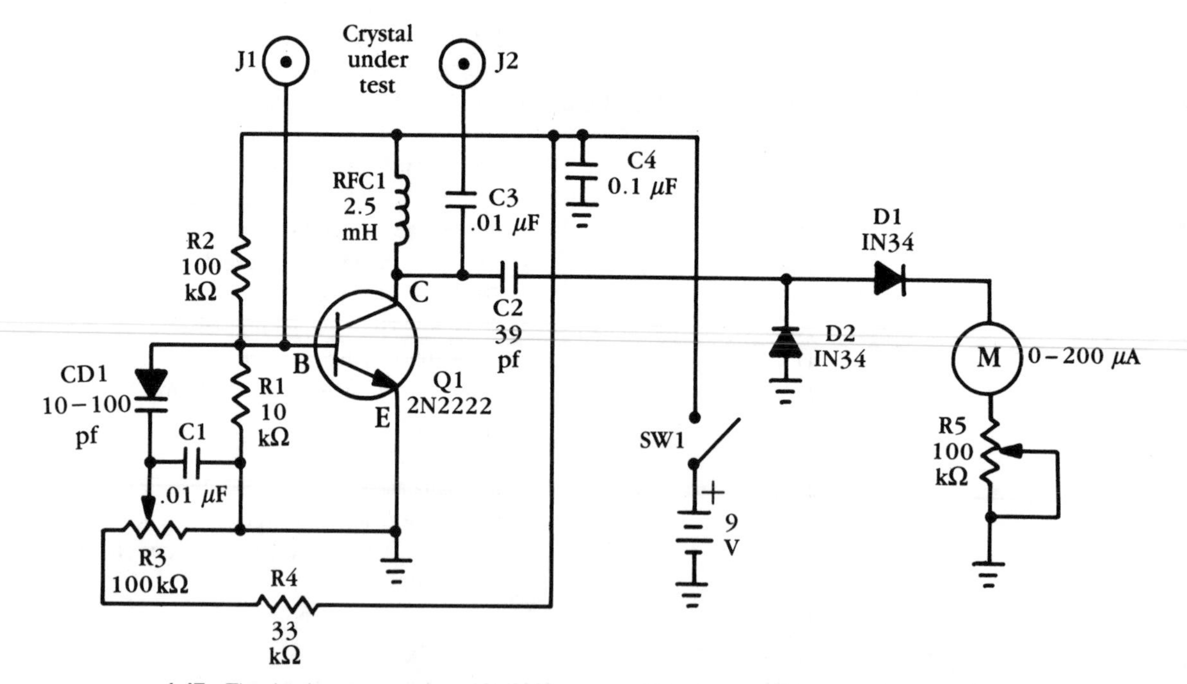

4-47 The simple tester consists of 2N2222 low-signal transistor in a Pierce oscillator circuit.

The crystal jacks connect the suspected crystal across the base and collector of Q1 in a feedback circuit. C2 couples the oscillator to full-wave rectifiers D1 and D2 (either 1N34 or 1N60). The 0 to 200 μA meter registers the oscillating signal. R5 prevents and adjusts the meter movement from possible damage. The crystal circuit is powered with a 9-volt battery.

PC board construction

Cut a 2-×-2 inch piece of copper-clad board and layout the actual-size pc wiring (FIG. 4-48). After etching, drill all holes with the smallest bit possible. Make sure the bit goes through to clear out any small holes. Drill a 1/8-inch hole to bolt the pc board to the bottom meter case. Use a low-wattage (35 W) soldering iron when wiring the pc board.

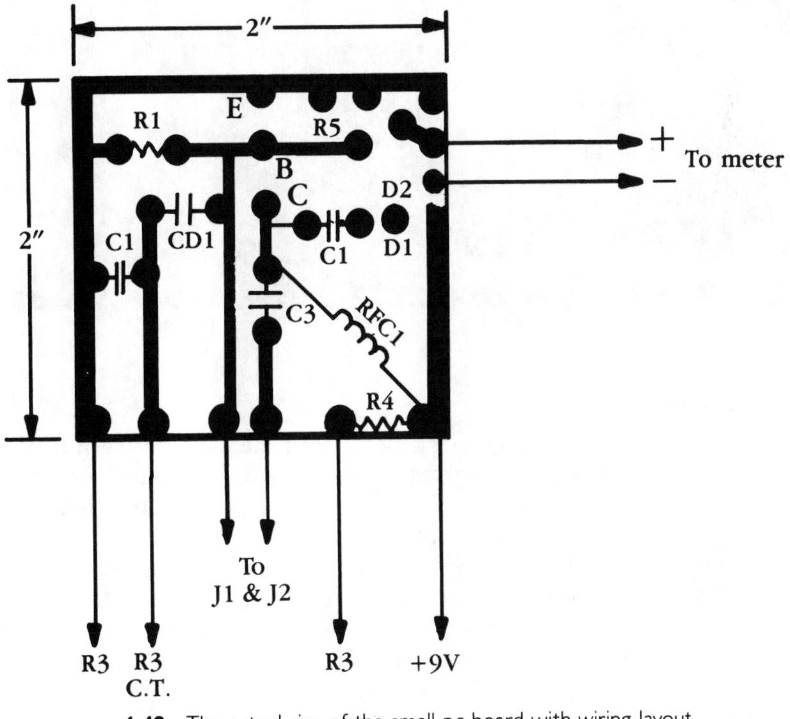

4-48 The actual size of the small pc board with wiring layout.

Mounting components

Simply drop the small parts in the correct holes and solder. Do not leave long leads on any of the small parts. Be careful not to heat a connection too long or the pc wiring will pop. Solder the transistor leads when all other connections are made. Observe the correct polarity on CD1, D1 and D2, and the meter. Connect the positive terminal of the meter to ground. If the meter reads backwards, reverse the meter wires.

Connect a four-inch piece of flexible hookup wire to the ground and the positive terminal. Solder two 3-inch flexible wires to the crystal jacks. Connect a ground wire to A and a 3-inch long wire to B and E for the variable control, R3. Solder two 4-inch wires to the end of R3 and D2 for the meter. Doublecheck all wiring after you solder in the parts (FIG. 4-49). Always check the condition of transistors and diodes with the DMM transistor tester after you solder the circuit. Make sure that the heat does not damage the semiconductors. Cut off each wire after you solder the units together.

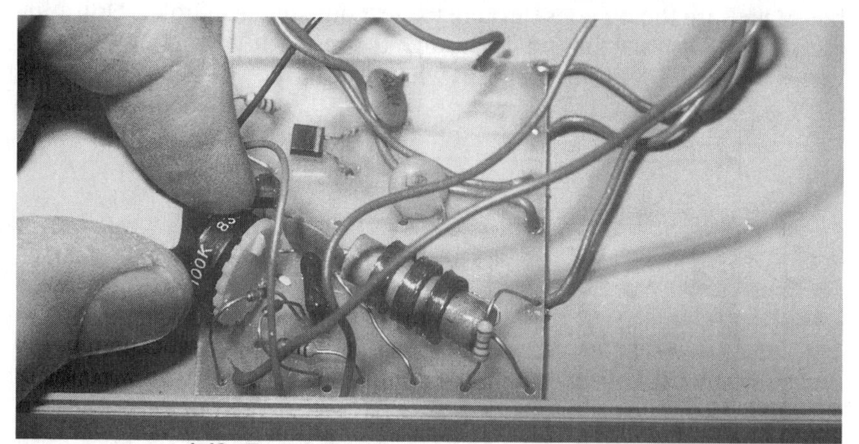

4-49 The small pc board and mounted components.

Case preparation

You must prepare the small meter case to hold the meter and controls. Fit the meter on the slanted surface and mark the back side mounting holes. This $3^1/_4$-x-$2^3/_8$-inch meter base covers the large hole in the meter case. Drill four $^3/_{16}$-inch holes to mount the meter (FIG. 4-50).

Center the variable control and jacks inside the case. Drill a $^{11}/_{32}$-inch hole for R3 and two $^1/_4$-inch holes for the banana jacks. Keep the two jack holes close together so that the small crystal can plug snugly into them. Separate phone jacks can be used since the small crystals have even smaller mounting terminals. Keep the crystal holes away from the battery compartment.

Place a dab of enameled paint on the control lock nuts and on the jacks, so that they will not loosen.

Testing

Before firing up the tester, turn R5 wide open. Just turn on SW1 when R3 is rotated, the hand should not hit the peg. Notice that the meter will not register if the crystal is out or if the circuit is not oscillating. Now, adjust R5 until the meter reads full scale. If the meter does not move from the

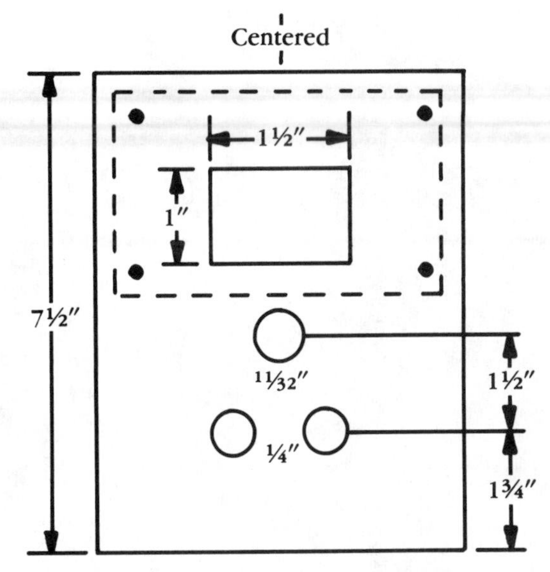

4-50 Drill the meter, R3 control, and banana jack holes in the top case.

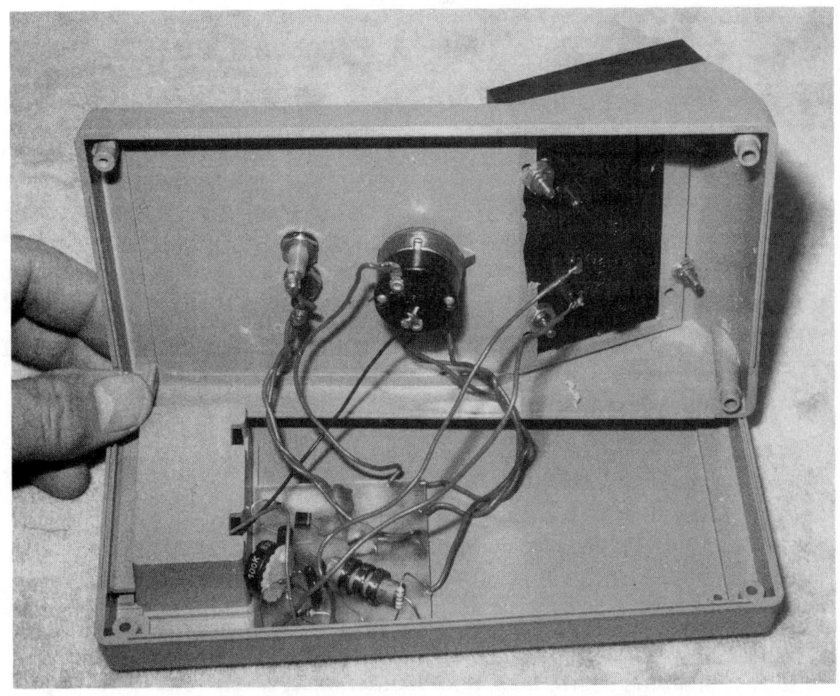

4-51 Backside view of pc board with meter and controls.

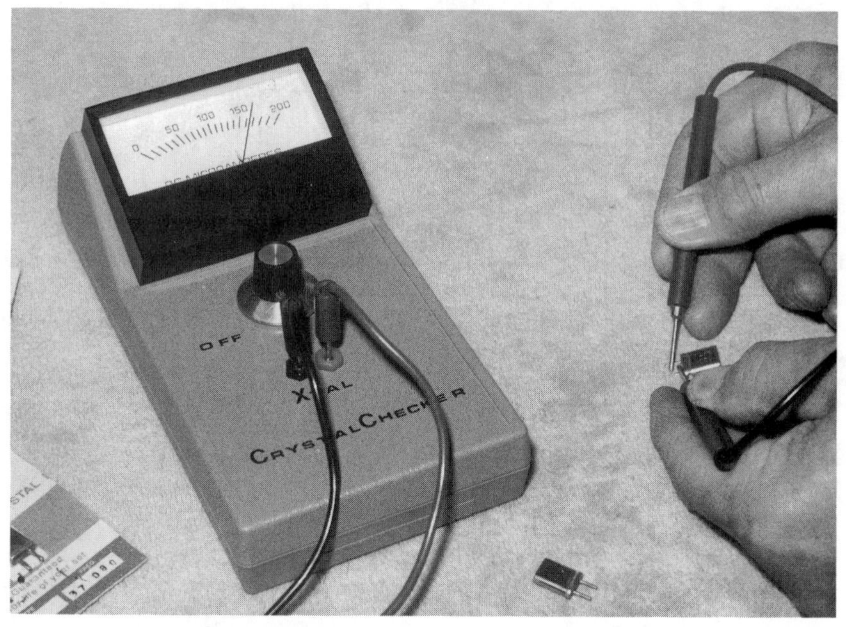

4-52 Testing out a suspect police scanner crystal.

Parts list

Q1	2N2222 npn transistor, or MPS 3904, ECG 123, 2N3904, or equiv.
R1	10-kΩ ½-W resistor.
R2	100-kΩ ½-W resistor.
R3	100-kΩ linear control with switch.
R4	33-kΩ ½-W resistor.
R5	100-kΩ variable pot pc mount, 271-220 or equiv.
C1, C3	0.01-μF 100-V ceramic capacitor.
C2	39-PF silver mica 100-V capacitor.
C4	0.1-μF 100-V ceramic capacitor.
CD1	Variable capacitance diode, 10–275 PF MV1662 Hosefelt Electronic Inc. or 0–100 PF MV 2115 Circuit Specialists.
Batt.	9-V battery.
M	0–200 μA Modutec panel meter MET10, All Electronics Corp. or equiv.
RFC1	2.5 mH rf choke coil.
D1, D2	1N34 or 1N60 diodes.
J1, J2	Banana jacks.
Case	Sloping meter case, HSP 14-109 Hosefelt Electronics or equiv.
Misc.	Small pc board, nuts, bolts, solder, battery clip, etc.

zero position with crystal being tested, check the circuit once more (FIG. 4-51).

Rotate R3 when the meter is at the full scale. The voltage placed on CD1 should make the meter hand move over a large range of the dial. On some crystals, when more voltage is applied to CD1, R3 will make the meter read full scale; then when rotated slowly, the meter hand will slowly drop to zero. Test the checker by shorting crystal test leads and the meter should go up at least halfway if circuits are normal. No damage results from this test.

If the meter hand does not move at all, reverse CD1. Sometimes these diodes are marked with the voltage on the right when the flat side of the diode is up, or vice versa. A brief test with reversed leads will not damage CD1. Simply reverse CD1 if the hand does not move when rotating R3. If the meter hand goes backward when connected, reverse the meter leads.

Soon, testing the unknown crystal becomes a snap. If a crystal will not oscillate or meter hand does not move, suspect a defective crystal. If the meter hand hits 50 on the dial, but will not move when R3 is rotated, replace the crystal. If the meter hand does not move from zero with R3 rotated, the crystal is defective.

Clip the small crystal into the jack sockets or insert the short extension leads (FIG. 4-52). Keep the leads as short as possible; longer leads can be added to test the color crystal and the saw filter in the TV receiver. Instead of waiting to order another crystal for substitution, test it on the small crystal checker.

Chapter 5

Technician test instruments

This chapter contains thirteen different test instruments that the TV technician, telephone lineman, and sound repairman can use most every day. You can check for broken or intermittent cables and wires in the wall or in telephone lines with the tone tracker/speaker aid. The sound technician can use the deluxe signaltracing amp, deluxe sine/square-wave generator, variable regulated power supply, and the sound level meter when troubleshooting and repairing sound systems.

The electronic or TV technician can use the laser/infrared CD diode checker to test CD players, the infrared remote-control tester to check those pesky remote-control transmitters, the deluxe signaltracing amp and the deluxe sine/square-wave generator to test the TV monaural and stereo audio channels, the variable regulated power supply to inject voltages into the various TV circuits, the TV power supply to eliminate the power source while checking other units in the TV chassis, and a tuner subber to determine if either the tuner or the TV chassis is defective.

The motor box voltage source can be used to check the various motors in the tape players, CD players, or camcorders. The 5-amp auto radio power source can furnish dc power while you repair car radios. Besides the DMM and the scope, these test instruments will come in handy every day on your service bench.

Most TV shops have an isolation and a variable step-up transformer to service today's ac/dc TV chassis. The TV chassis to be serviced must be plugged into the ac isolation transformer. Not only does this action provide safety for you, but it prevents damage to the TV chassis. When you attach test instruments to the TV chassis that are not plugged into the isolation transformer, a fuse and several critical components might blow. So, play it safe when working on present-day TV chassis.

TONE TRACKER/SPEAKER AID

The tone tracker with the speaker aid can be used to trace broken, intermittent, open or shorted wires in the wall or in buried cables. The tracker sends out a 1-kHz tone at one end and the aid unit receives it at the other. Both units use alligator clips to easily connect to the suspected wires (FIG. 5-1). A sharp probe can be connected to the aid input. The speaker aid can receive the tone if a probe is placed near the wire connected to the tone tracker.

The tone tracker sends out a tone that is used for continuity tests. Besides checking wires, the tracker can be used to make many different continuity tests. The speaker aid uses an internal speaker, eliminating the need for a handset or to use your hands. Just clip the tracker to or touch the end probe to the suspected wire and locate it with the speaker aid. The speaker sounds when the correct wire is located.

Tone tracker circuit

The tone tracker consists of a small 8-pin LED flasher IC in a 1-kHz resistor-capacitance circuit (FIG. 5-2). Only seven small electronic parts are needed. The audio tone frequency is selected with R1, R2, and C1. The tracker's single 1.5-volt battery should last for a long time since the circuit only draws 1.5 milliamperes. SW1 can be a toggle or slide SPST switch since no pilot light is needed. The small AA battery is placed in a battery socket for easy removal. C2 isolates the tracker from the line to be tested.

IC pc board

Build the tracker on a regular general-purpose-IC pc board with indexed holes. Round the corners so that the pc board will fit inside the case (FIG. 5-3). Straddle the 8-pin IC socket over the two connecting lines of holes

Tone tracker parts list

IC1	LM 3909 LED IC, 276-1705 or equiv.
R1, R2	4.7-kΩ ½-watt carbon resistor.
C1	0.22-μF 100-V capacitor.
C2	0.1-μF 100-V ceramic capacitor.
V1	1.5-V battery.
Case	3¼″-X-2⅛-X-1⅛″ experimenter's box.
Battery holder	AA battery holder, 270-401 or equiv.
Pc board	General-purpose IC pc board, 276-150 or equiv.
SW1	SPST off/on miniature toggle switch.
Misc.	8-pin IC socket, nuts, bolts, solder, alligator clips, hookup wire, test wire, etc.

5-1 Front view of the signal tone tracker and the chassis of the speaker aid.

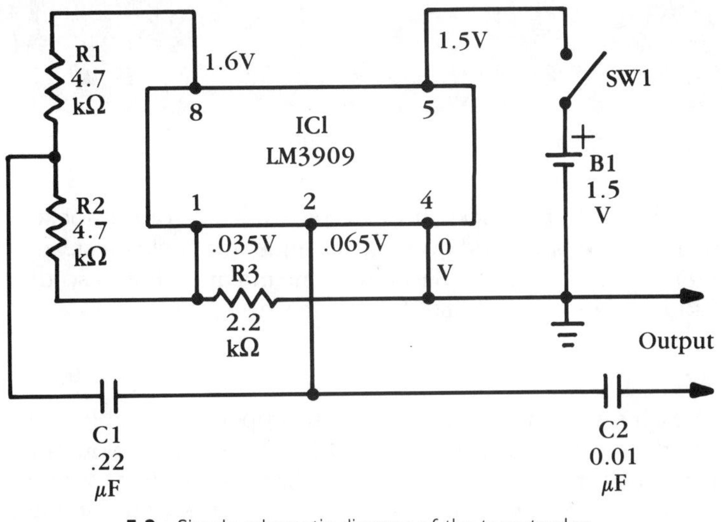

5-2 Simple schematic diagram of the tone tracker.

between the IC socket. Use one line of holes for the B+ and the other common ground terminal. Place the IC socket at the very end so the battery holder can be bolted to one side. Bunch the rest of the small parts together down the other side.

It's best to bolt the battery terminal in place with two 2/56 bolts and nuts before mounting any other components. Then you can mount the small components as you solder them into the circuit. Next, mount the IC socket at the far end so the switch leads can be brought out the other.

5-3 Tone tracker layout on a general-purpose IC pc board.

Make sure each IC pin sticks through the other side of the socket. Sometimes one or two pins bend over and lay under the socket instead of going through the indexed holes. Bend over one pin at each end so the socket will hold in place until you solder it.

Wiring

Solder each pin of the IC socket to each copper-ringed hole. Place a dot on both the top and bottom of pin 1 with a felt pen. Now you can identify each pin number instead of turning the board over while mounting or soldering the parts. Use a 35-watt pointed or battery-powered soldering iron so solder doesn't lop onto the adjacent wiring. Use extreme care—all pins are quite close together. Each pin only needs a dab of solder.

Mount the small resistors and capacitors as you complete the circuit. Cut the battery terminal wires and solder them directly into the board. Use a ringlet hole for the on/off switch terminal wires. Connect two 3-inch flexible hookup wires to these two switch holes. Solder the common ground and the output leads directly to the board when board is mounted. Tie a knot in each test lead so they can't be yanked out. Doublecheck all wiring and connections with a hand-held magnifying glass.

Testing

Test the tone tracker before placing it inside the small case. Connect the two output test leads to a speaker or pair of earphones. Of course, the tone heard in the speaker is quite weak but sound in the earphones is much stronger. Once the unit is performing, bolt the board to the bottom of the small box.

If, by chance, the tracker does not make a sound, check for voltage at each pin of IC1 and make sure the battery is placed in correctly; the positive terminal should go to the on/off switch. Next, check the current across SW1; if the current is over 5 mills, suspect a shorted IC or incorrect wiring. Again, check the wiring. Inspect the parts for poor or broken connections. Make sure that IC1 is inserted in the socket with the white dot (or line) at pin 1. Some of these LM3909 ICs have a white dot or indented dot area and others have a white line indicating where pin 1 is located.

Case preparation

Place the small wired pc board in the bottom of the plastic case and mark one mounting hole. Often two small 1/8-inch holes will hold the board in place. Drill a 9/32-inch hole at the top for the toggle switch. If a slide switch is used, cut a slot into the plastic. Drill several small 1/8-inch holes in a line, so that the remaining piece can be broken out. Level the slotted hole with a 3-corner or a flat file. Drill two 1/8-inch holes at one end for the test leads. Just solder on two small alligator test clips (FIG. 5-4) and the tone tracker is now ready.

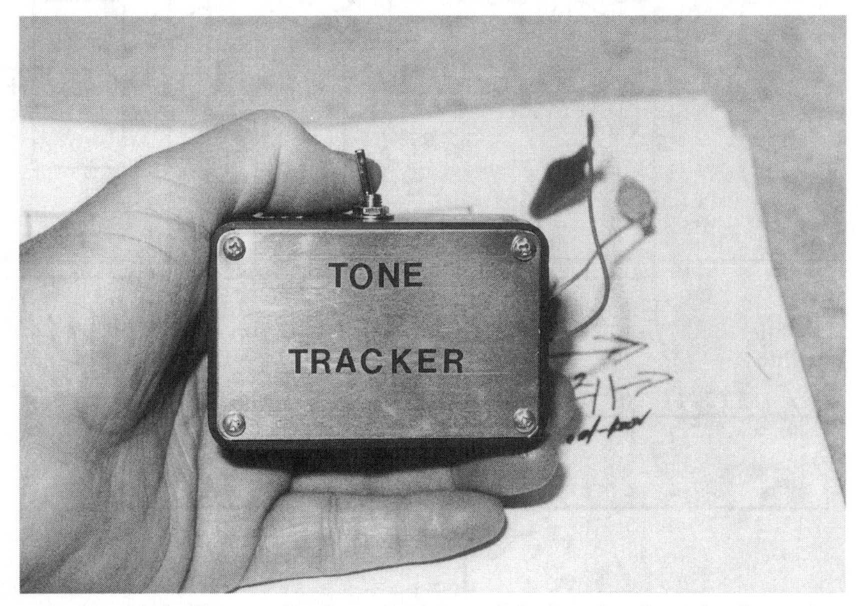

5-4 The completed tone tracker can fit in the palm of your hand.

Speaker aid circuit

The speaker aid consists of two op amp ICs. IC1 is a quad amp 14-pin component that will use only one-fourth of the amplifier section. The first four pins and the common-ground pin (11) operate as the IC amplifier. The alligator clips connect to the suspected wires and C1 couples the picked-up tone to IC1 (FIG. 5-5). IC1 amplifies the signal and R8 controls it. R8 is a regular miniature audio control, but can be subbed for a prefixed trimmer resistor with a screwdriver adjustment. If R8 is a trimmer or a thumb variable resistor, a toggle or slide switch must be added for the on/off switch (SW2).

The controlled sound is coupled by C5 to pin 3 of the audio IC amp. The amplified audio at terminal 5 is capacity-coupled to a small-pin speaker. The size of the speaker is not critical to hear a tone. Choose the correct-sized speaker (1 to 2 inches in diameter) to fit in your case. C9 and R9 provide greater volume and correct frequency response. You will need

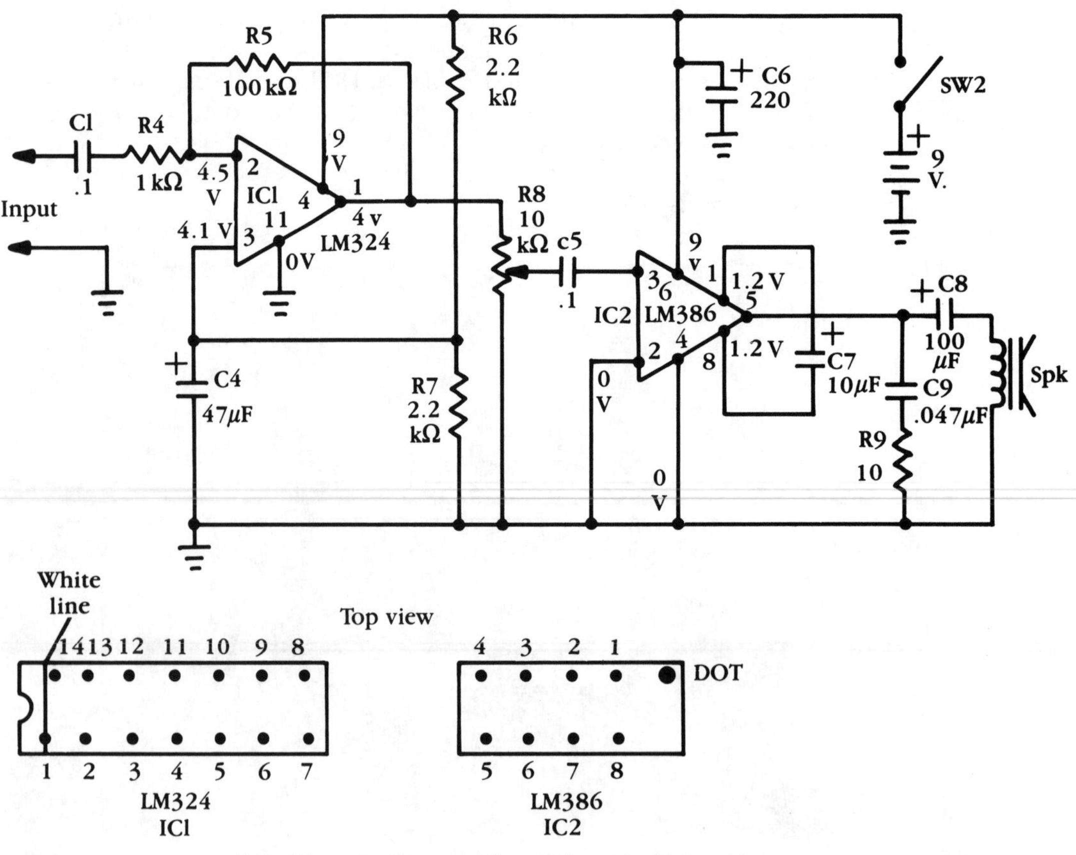

5-5 Schematic diagram of the speaker aid with two IC op amps.

Parts list

C3, C5	0.1-μF 100-V ceramic capacitor.
C4	47-μF 15-V electrolytic capacitor.
C6	220-μF 15-V electrolytic capacitor.
C7	10-μF 15-V electrolytic capacitor.
C8	100-μF 15-V electrolytic capacitor.
C9	0.047-μF 100-V ceramic capacitor.
R4	1-kΩ ½-W carbon resistor.
R5	100-kΩ ½-W carbon resistor.
R6, R7	2.2-kΩ ½-W carbon resistor.
R8	10-kΩ audio control and switch.
R9	10-Ω ½-W carbon resistor.
IC1	LM 324 IC quad op amp, 276-1711 or equiv.
IC2	LM 386 IC audio amp, 276-1731 or equiv.
SPK	1- to 2-inch PM speaker.
Pc board	General-purpose IC pc board, 276-150 or equiv.
Case	Deluxe project case 4⅝″ × 2⁹⁄₁₆″ × 1⁹⁄₁₆″, 270-222 or equiv.
Batt.	9-V battery.
SW2	On back of R8.
Misc.	8-pin and 14-pin IC socket, battery clip terminals, solder, hookup wire, alligator clips, etc.

to solder C6 right on pin 6 of IC2 to prevent "motor-boating" and oscillations. A 9-volt battery powers the speaker aid.

Board layout

Mount the 14-pin IC at one end and the 8-pin IC at the other end of the pc board. Mount the small parts as you wire them into the circuit. Of course, IC1 has more parts clustered around it than IC2 does. You can use ⅛-watt carbon resistors and 15-volt electrolytic capacitors that require less room on the board. Mount both ICs into separate sockets after the board is completely wired.

Again, use a general-purpose IC pc board with 417 indexed holes. The two ICs will straddle the common ground and B+ buss-ringed pc wiring. Some of the components mounted away from the IC sockets might need longer leads. If you do not want to use center-ringed wiring, run a bare common ground wire down one side and a bare common B+ wire down the other side of the board. Be extremely careful when soldering the IC pins so that you do not solder more than one terminal together. Mark pin one on both sides of the board (FIG. 5-6).

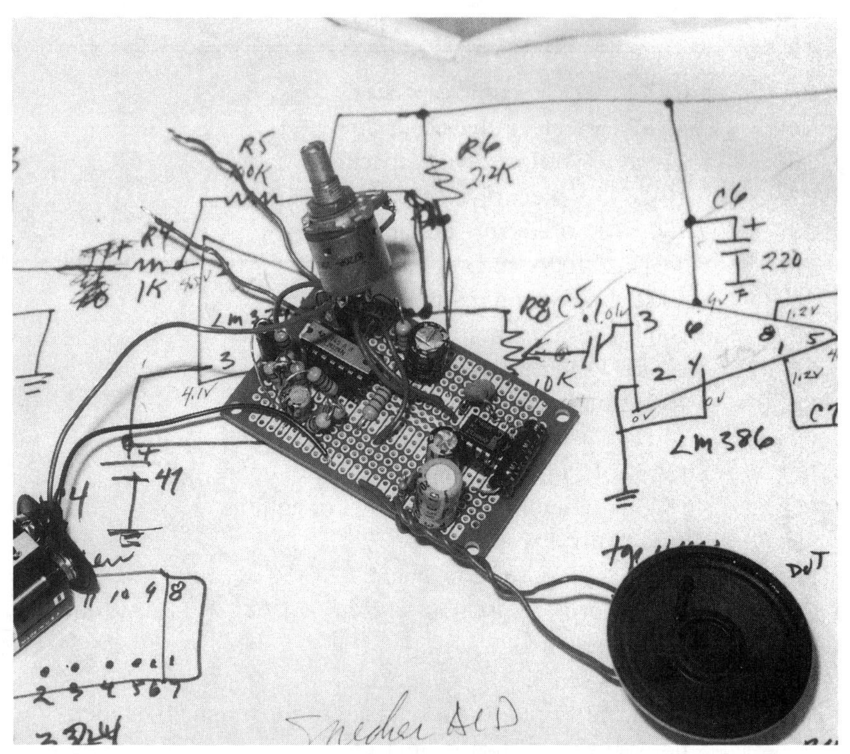

5-6 Close-up view of the soldered components on the speaker aid pc board.

Wiring

Cross off the parts on the schematic as you wire them into the circuit. Observe the correct polarity of the electrolytic capacitors. Cut the leads of each component as short as possible. Solder in three 4-inch flexible leads for R8. Cut two 3-inch leads for the speaker wires. Solder the 9-volt battery cable directly into the pc board circuit. Cut two flexible 10-inch leads for the input wires and solder them to C3 and ground. Tie a knot in both wires so they cannot be pulled out. The speaker and volume control leads can be cut off after you mount the chassis in the plastic case.

Testing the speaker aid

Before mounting the pc board in the plastic case, test it. Plug in the 9-volt battery and check the current across SW2 with switch turned off. A normal 9.5-milliampere reading is okay (FIG. 5-7). If the meter draws current over 12 milliamperes, suspect a leaky IC or improper wiring. Double-check the wiring. Make sure that both ICs are plugged in correctly. The front side of each IC should point toward the outside of the pc board. Notice if any of the small pins are bent over.

When touching the ungrounded lead of C1, you should hear some

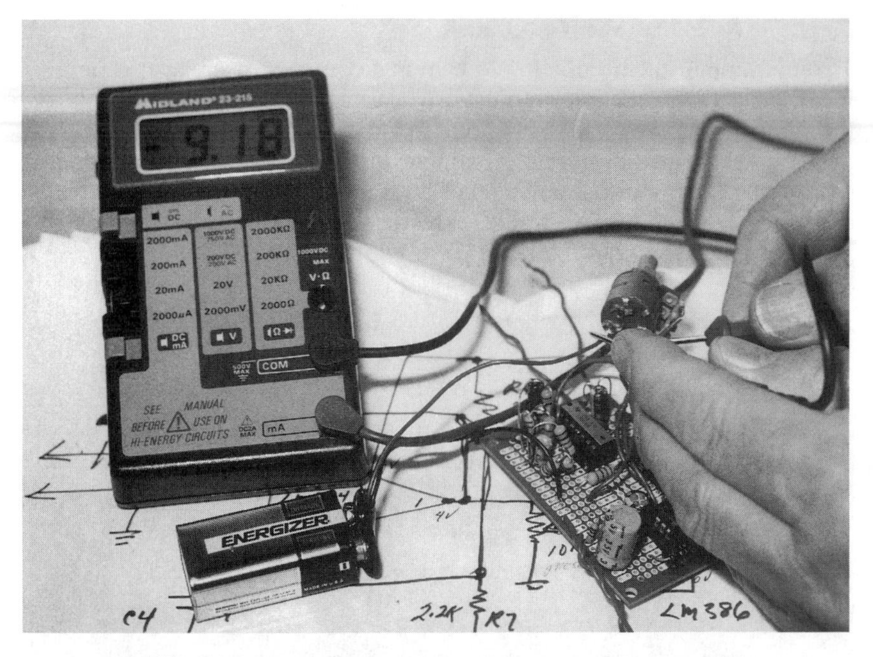

5-7 Take a current measurement of the speaker aid across SW2.

hum. If not, place a screwdriver blade against the center terminal of R8. Each time it's touched, a clicking noise should be heard in the speaker. Measure the voltages at IC2 and compare them to the schematic. If you find no voltage at pin 6, check the battery and SW2 connections. Check IC1 if a noise or low hum is heard at the volume control.

Check the low voltages on IC1 when no hum is noticed at C1. If pin 4 is low, suspect that IC1 is leaky or has been installed backwards. The white line at the "U" symbol of IC1 should be pointed at the outside of the pc board with pin 1 at the left bottom side of IC1. Inspect the wiring for poor or improper soldering connections. Check each connection if hum is intermittent when touching the pc board. You can use the speaker aid as the external amplifier by connecting the input alligator clips to the circuit to be amplified.

Mounting

After the tests, mount the pc board and parts inside a deluxe project case. The $4\frac{5}{8}$-x-$2\frac{15}{16}$-x-$1\frac{9}{16}$-inch plastic case is large enough to house both pc board and 9-volt battery. Drill four $\frac{8}{32}$-inch holes to mount the pc board. Drill a $\frac{1}{8}$-inch hole to mount the battery clamp. At the top center of board, drill a $\frac{5}{16}$-inch hole for R8. Drill two $\frac{1}{8}$-inch holes in one end of the enclosure for the input cable. Solder the input alligator wires directly to the pc board. Attach two small alligator clips to the end wires. These will clip on the external wires to be tested with the tone tracker.

CD INFRARED LASER CHECKER

To determine if the infrared laser beam in a compact disc player is operating, the electronic technician must use a laser power meter or some infrared light device (FIG. 5-8). This homemade device can be used to check the laser beam in the portable or table model compact disc player. Although the CD infrared laser checker does not measure the power level of the CD laser beam, it does indicate that the laser beam is performing.

5-8 The CD laser checker operates from a 9-volt battery.

The laser pickup assembly in the CD player is critical and delicate. Never stare directly at the laser optical lens assembly. It can quickly destroy your eyes even though the beam is infrared and cannot be seen. The laser diode optical assembly usually has a wavelength of 765 to 805 mm with a continuous wave and an output power from 0.26 to 0.7 mW.

With a focus-controlled process, the laser beam strikes the bottom area of the compact disc. The compact disc contains small pits and landings that is digitally encoded music. The laser beam is pin-pointed on each pit and landing as the disc rotates. If the laser light beam is missing the pits, then the compact disc player is silent. In this case, the small CD infrared laser checker is handy to indicate if the laser beam is present.

How it works

The infrared phototransistor (Q1) is placed over the CD laser beam and if working properly, the small buzzer (PB1) will sound. Q1 is directly coupled to PB1 to detect the infrared rays. All components are wired in series. SW1 turns the checker off and on. LED1 indicates and pulsates the tone of the piezo buzzer or element. A 9-volt battery powers the checker.

The circuit

This laser checker is comprised of only 6 parts, including the battery and case. Connect the collector terminal of Q1 (pin 2) to the negative terminal of PB1. Notice that all components will be wired in series (FIG. 5-9). Connect the positive terminal of the battery to the cathode of LED1. Connect the positive terminal of PB1 directly to the anode terminal of LED1. Wire the black battery lead to SW1.

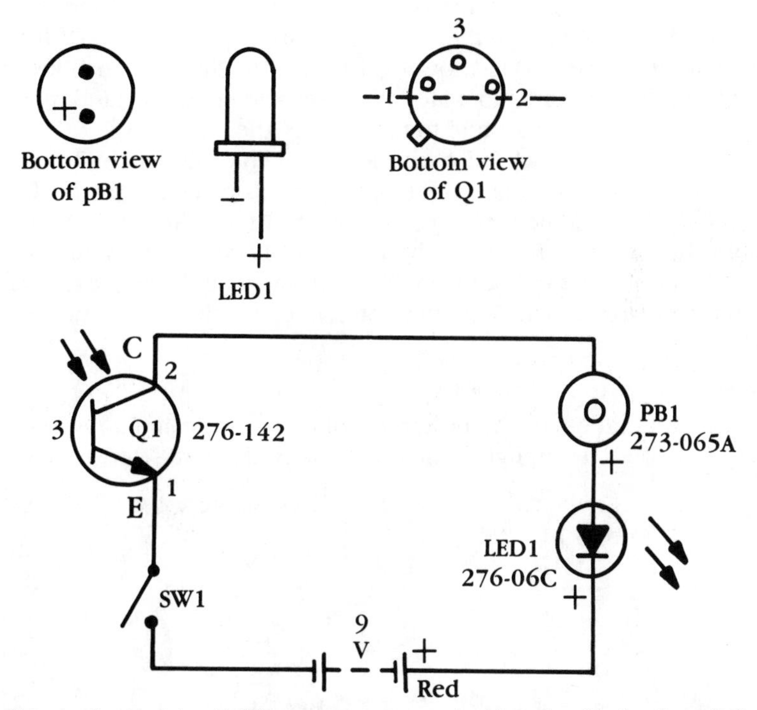

5-9 This simple circuit has 5 different components wired in series. Notice the polarity of the 9-volt battery, LED1, PB1, and Q1.

The piezo buzzer was chosen because of its size and pleasant tone. The buzzer operates on 3 to 20 volts dc and has an operating frequency of 2.8 kHz. You can use any buzzer frequency from 1 kHz to 6.5 Hz. The 273-065, 273-060, or 273-064 from Radio Shack or piezo buzzers from various electronic stores and catalogs can also be used. The 273-065 and 273-065A have a positive sign (+) marked on the case and either must be connected to the positive side of the battery in series with LED1.

The blinking red LED combines a MOS IC driver and red LED within a plastic LED housing. Because the resistances of the MOS driver transistors limit the current through the LED, no external current-limiting resistor is needed. The typical supply voltage is 2.5 to 3 volts dc with a blinking rate of 2.0 Hz. The blinking red LED pulls approximately 20 mA at 3 volts dc. Connect the positive lead of the LED to the positive battery terminal.

Construction

Drill all holes in the small black case. Center and drill a 1/16-inch hole where the piezo buzzer will be mounted. Do the same for the other end of the case, except use a 11/64-inch hole for the blinking LED. Drill a 1/8-inch hole below the LED hole for the wires to connect to Q1.

Cut a piece of 1/4-inch masonite 5 3/4 inches long and 2 inches wide. Round the ends and rough the edges with sandpaper or a file. Drill a 11/64-inch hole at the far end to mount Q1 (FIG. 5-10). Spray both sides of the board with black enamel to match the case. After all wiring and mounting the parts, epoxy the masonite to the back of the black case.

Glue on a piece of foil over the top of the thin masonite piece to reflect the laser beam. Place duct tape over this piece to cover Q1's lead wires (FIG. 5-11). Cement the piezo buzzer to the other end of the box. Mount the flashing LED in the hole and apply rubber cement on LED1 and Q1 to hold them in position. Now, connect all of the parts in series. Keep the polarity of each component marked on the schematic.

Testing

Test the infrared device by holding it directly in the sun's rays, under a lamp, or in fluorescent light. The buzzer should sound intermittently and

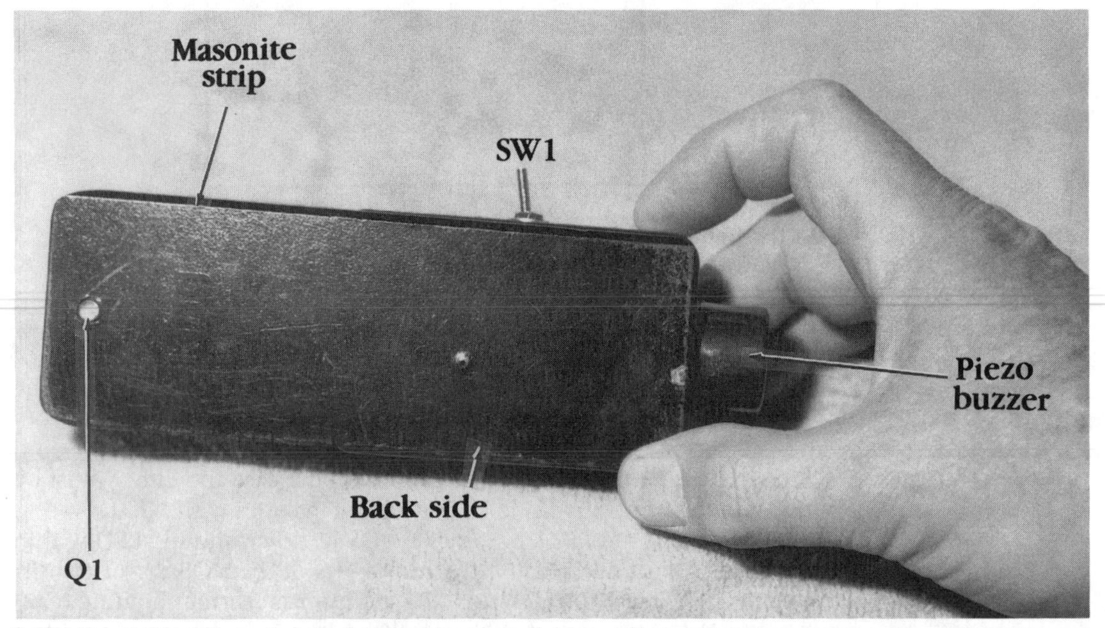

5-10 Mount Q1 at the end of the flat masonite or plastic piece with the piezo buzzer at the other. All parts except PB1 and Q1, are contained inside a 3 1/4- x -2-inch plastic box.

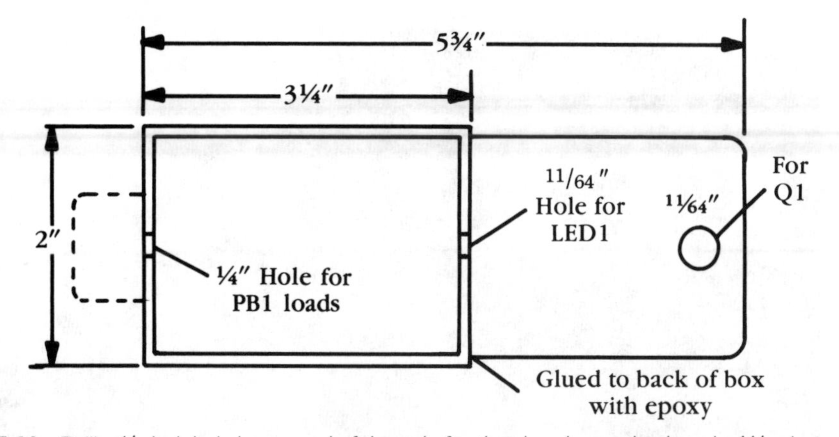

5-11 Drill a 1/4-inch hole in one end of the strip for the piezo buzzer leads and a 11/64-inch hole in the other end for the flashing LED. Cement the strip to the box with epoxy cement.

the LED should flash. The stronger the infrared beam, the stronger the sound and output of the checker. This small checker only indicates that the laser diode is operating from the optical lens assembly; it does not measure the current that the laser diode is pulling.

If the checker does not sound when held under a lamp, suspect that a component is wired in backwards. Doublecheck the polarity of each part and correct the wiring. Measure the voltage at the red terminal of the battery and at the emitter terminal of Q1. If the checker is still not operating, connect the battery wires directly to the buzzer. It should produce a loud sound.

Now, check the condition of Q1 and of the flashing LED with the DMM in the diode test position. Check Q1 with the red (positive) probe at pin 3 and the black (negative) probe at pin 1 and then pin 2. You should note a close reading on both the emitter and collector terminals. If one terminal shows a very low ohm measurement, the phototransistor is leaky. Check the LED across the positive and negative terminals in the same manner.

Checking the CD laser

The photo diode laser should be located under the flap or clamp assembly. The optical laser lens looks upward to the bottom side of the compact disc. On portable CD players, the laser optical assembly is out at the opening (FIG. 5-12).

Slip the photodetector (Q1) under the flap or clamp assembly for testing. Remember, the small laser indicator must be directly over the laser lens assembly. Slowly move the indicator into position with the CD player operating. Move the indicator until you hear the loudest pulsating sound from the piezo buzzer. In some players, you might have to shunt a test clip

5-12 Place Q1, the photo detector diode, over the laser beam of a portable CD player. Short the interlock with a clip or piece of solder. Do not look directly into the CD player laser beam.

Parts list

Q1	Infrared phototransistor detector, 276-142 or equiv.
SW1	SPST toggle switch, 275-645 or equiv.
PB1	1 kHz to 6.5 kHz piezo-element speaker, 273-065A or equiv.
LED1	Blinking red LED, 276-06C or 276-036 or equiv.
Batt.	A 9-V battery.
Cabinet	3¼″-X-2″ plastic box with metal lid.
Misc.	Battery clip harness, hookup wire, masonite or plastic, solder, etc.

lead across the interlock before the CD player will operate. A piece of solder or a paper clip might turn on the interlock in a portable battery-operated CD player.

If the signal is very weak, suspect a weak laser assembly that should be replaced. If there is no signal at all, the laser assembly is defective. Try the checker on several different CD players to acquire the required sound of a new or old laser assembly. This little infrared indicator can also be used to test infrared remote controls for the TV, VCR, or compact disc player.

INFRARED REMOTE-CONTROL CHECKER

You can check hand-held remote transmitters in minutes with this instrument. Just place the tester a few inches from the remote and press any of its buttons (FIG. 5-13). You should hear a chirping or audio tone depending on the type of remote. This checker will test all infrared remotes of TV, VCR, and compact disc players.

5-13 Check the remote control tester with a large, combination RCA VCR and TV remote transmitter.

Simply move the remote transmitter back and forth in front of the checker until the loudest sound is attained. Check the remote for weak conditions by moving it from the tester. Some remote transmitters might sound this checker from 2 feet away. Keep the remote pointed toward the checker.

The circuit

The infrared remote control checker circuit is built around an npn infrared photo transistor, which can be picked up in most electronic supply part stores (FIG. 5-14). Some are single-packed and others come as a set

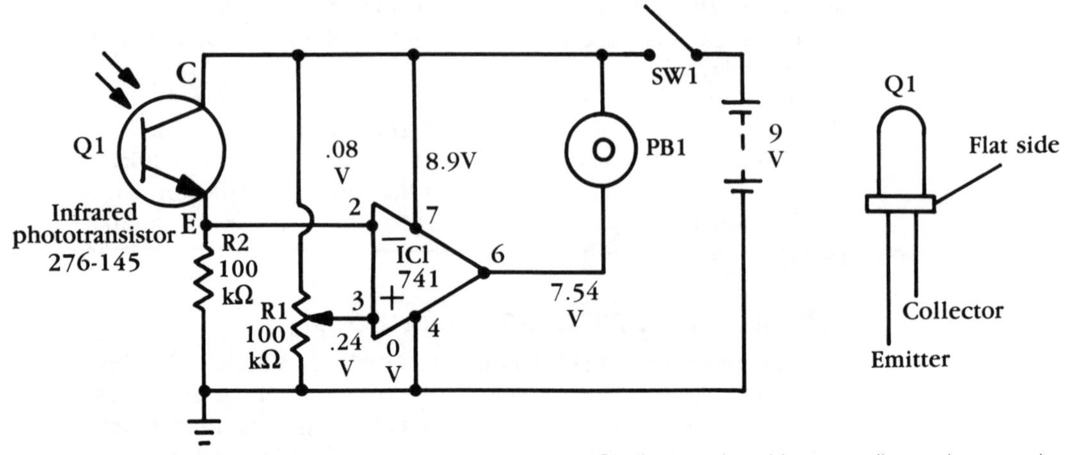

5-14 The circuit of the small infrared remote control indicator. IC1 allows testing with greater distance between the remote and the checker.

with the transmitting infrared diode. The infrared beam is picked up by Q1 with the input signal applied to IC1. This infrared npn silicon phototransistor provides high-speed photo sensitivity.

IC1, a general-purpose 741 operational amplifier with low power consumption, can be found anywhere. The 8-pin IC amplifies the audio signal to a piezo buzzer. R1 adjusts the input signal. IC1 increases the pickup distance.

PB1 is a loud 76-dB piezo buzzer for pc mounting (273-065). The buzzer draws only 7 mA of current at 12 volts ac. The 2.8 kHz sound is quite loud and can operate from 3 to 20 volts dc. A regular 9-volt battery provides power for the infrared tester. The infrared circuit is quite simple with only a few parts, and it can be mounted on a pc or perfboard chassis.

Preparing perfboard

Cut a 2¹/₄-×-1³/₄-inch piece of perfboard for the small chassis. Mount all parts except the piezo buzzer and the battery on the perfboard (FIG. 5-15) at the center toward the end of the chassis. Bend Q1's leads so the photo transistor will stick through the plastic case. Mount IC1 and R1 toward the middle to leave room for the 9-volt battery. Set the perfboard chassis close to the front hole in the plastic case. Except for Q1, no exact parts layout is necessary.

Drill a 11/64-inch hole in one end of the 3¼-×-2-inch plastic case. Center the hole for the infrared photo transistor. Drill a ¼-inch hole in the rear end of the case for the piezo buzzer leads. If you use a pc-mount buzzer, drill two 1/16-inch holes for the pc terminals to stick through the plastic case (FIG. 5-16).

Apply epoxy or rubber cement to the back side of PB1. Mark the pos-

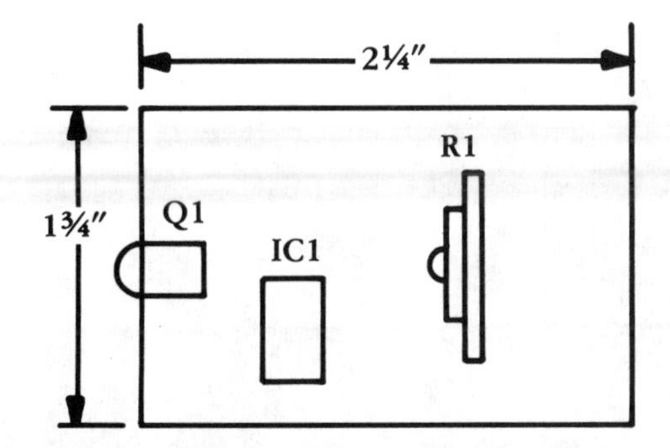

5-15 The chassis uses a 1¾-×-2¼-inch perfboard. Mount all parts through the perfboard holes and solder on back side.

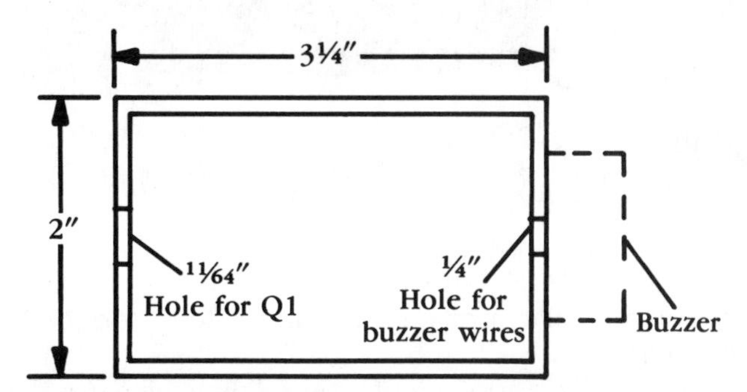

5-16 House the infrared remote checker in a 2-×-3¼-inch plastic case. Cement the piezo buzzer on the outside of the case.

itive (+) lead terminal on the case so correct polarity can be observed when you connect the circuit. Place a weighted object on the buzzer and let the cement dry while you wire the components on the perfboard.

Wiring

Start by connecting Q1 into the circuit. Do not place IC1 into its socket until the project is ready to be tested. Use a piece of bare solid hookup wire for all ground terminals. Connect terminal E of Q1 to pin 2 of IC1 and R2. Next, wire R1 and R2. Pin 7 of IC1 connects to the B+ and C terminals of Q1. Ground terminal 4 of IC1. Solder a lead to pin 6 and connect it to PB1. Make sure that the positive terminal of the piezo buzzer goes to the B+ side (FIG. 5-17). Wire the battery terminal leads—the red lead connected to one side of SW1 and the black lead to ground. Make all connections as short as possible.

Testing

Before you connect the 9-volt battery, recheck each wire connection. It only takes a few minutes to look over all of the wiring in this small circuit.

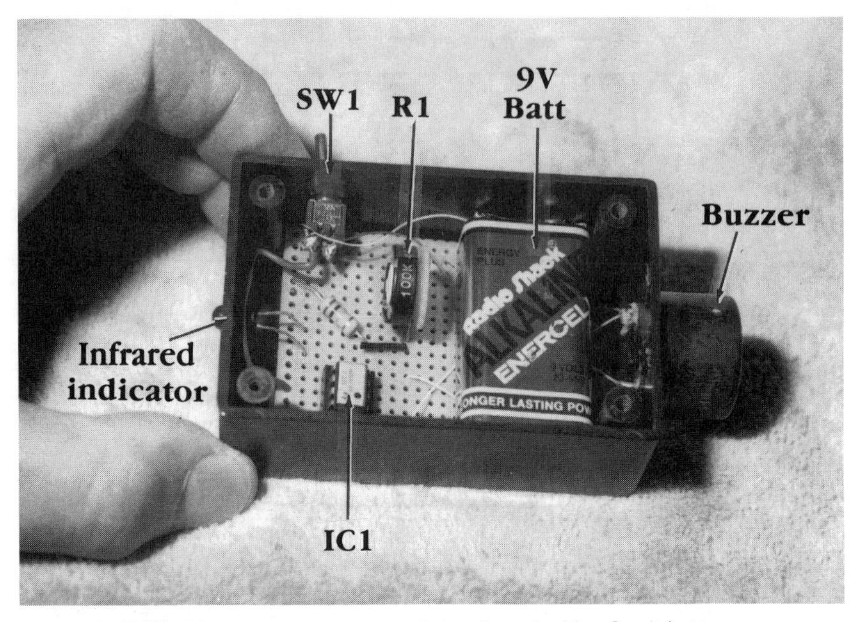

5-17 Mount all small parts on the periboard with a 9-volt battery.

Make sure that the polarity of PB1. Q1, and IC1 are correct. Now, insert the small battery.

Turn on SW1. Place a TV remote control directly in front of Q1. Press one or two buttons on the remote and you should hear a loud tone in the small buzzer. If not, take voltage and resistance measurements across the circuit. Measure the 9 volts at SW1. Low battery voltage at the switch might indicate a poor battery or overloaded circuits in checker.

Rotate the DMM to the 20-milliampere range and clip the leads across SW1 in the off position. If the meter pulls more than 20 mills, a short or defective component exists in the tester. Doublecheck all wiring to IC1. Is IC1 inserted properly? Make sure that pin 1 is at the dot area of IC1. Take accurate voltage measurements at each pin and compare them to those on the wiring diagram.

How to use

You should be able to hear a tone when SW1 is on and the infrared tester is in front of a suspected remote. Move the tester back and forth with your finger pressed on any button of the remote. Now, pull the tester one foot away from remote and adjust R1 for maximum buzzer signal. R1's adjustment does not vary the signal strength when the remote is close to the checker.

Some of the larger TV and VCR remotes produce a chirping or warbling sound when they are activated. See how far you can place the checker from the remote and still hear the sound. Usually, the battery is

5-18 A VCR remote is being checked with the infrared remote tester.

weak if the sound stops when the remote is only a few inches from checker (FIG. 5-18). Check each button with the tester. Most remote transmitters eventually become weak, dead, or intermittent. If the remote must be tapped or slapped, suspect intermittent or poor battery terminals. A dead remote might be caused by a defective IC or transistor. Do not overlook the possibility of a broken battery terminal or wire for dead conditions. Of course, a defective infrared receiver in the TV, VCR, or compact disc player might also cause weak, intermittent or dead operations. However, this little infrared checker can show you if the remote is functioning properly.

Parts list

Q1	Npn-type infrared photo-transistor, 276-142 or equiv.
IC1	741 general-purpose op amp. IC 8-pin dip, 276-007 or equiv.
PB1	Pc-mount piezo buzzer, 273-065 or equiv.
R1	100-kΩ pd board pot with thumb or screwdriver adjustment, 271-220 or equiv.
R2	100-kΩ ½-W resistor.
SW1	Submini SPST pc-mountable toggle switch, 275-645 or equiv.
Batt.	9-V battery.
Case	3¼"-X-2⅛"-X-1⅛" hobby box with metal lid, 270-230 or equiv.
Misc.	Perfboard, 9-V battery lead, hookup wire, 8-pin dip socket, solder, etc.

DELUXE SIGNALTRACING AMP

You can use this deluxe signaltracing amp to locate missing, weak, and dead signals in practically all audio or video units found in the commercial electronic field (FIG. 5-19). This tester is ideal when checking for distortion and weak signals in audio amplifiers. This project can check the radio and pre-amp circuits of cassette players for missing signals. Even the

5-19 Build the deluxe signal-tracing amp inside a commercial speaker cabinet. Some of these speakers can be picked up as close-outs for a few dollars at electronic outlets.

motor circuits in camcorders and CD players can be signaltraced with this amp tester. This circuit is quite similar to that of the hand-held audio signaltracer.

The circuit

The audio amp circuit consists of a high-impedance FET transistor (Q1) and several stages of audio amplification (IC1). Q1 was selected for its high impedance, low cost, and availability. By using two different test probes, rf and audio signals can be detected. Notice that C1 and C3 have higher working voltages than the other capacitors. Both the rf and the audio probes are used in input 1. Input 2 is a direct audio jack for high-wattage amplifier circuits. Use input 1 for tracing weak signals (FIG. 5-20).

Q1 amplifies low rf or audio signals and is coupled to SW2 with C3. SW2 switches in the two different audio signals. R5 controls the audio volume. IC1 (LM 386), an 8-pin IC, amplifies the audio circuit with C9 coupling the audio signal to the 8-Ω speaker. The maximum audio output is approximately 850 milliwatts with a 12-volt regulator source.

The low-wattage power supply consists of SW1, T1, D1, C10, and IC2. T1 steps down the power line voltage to 12.6 volts and applies it to a

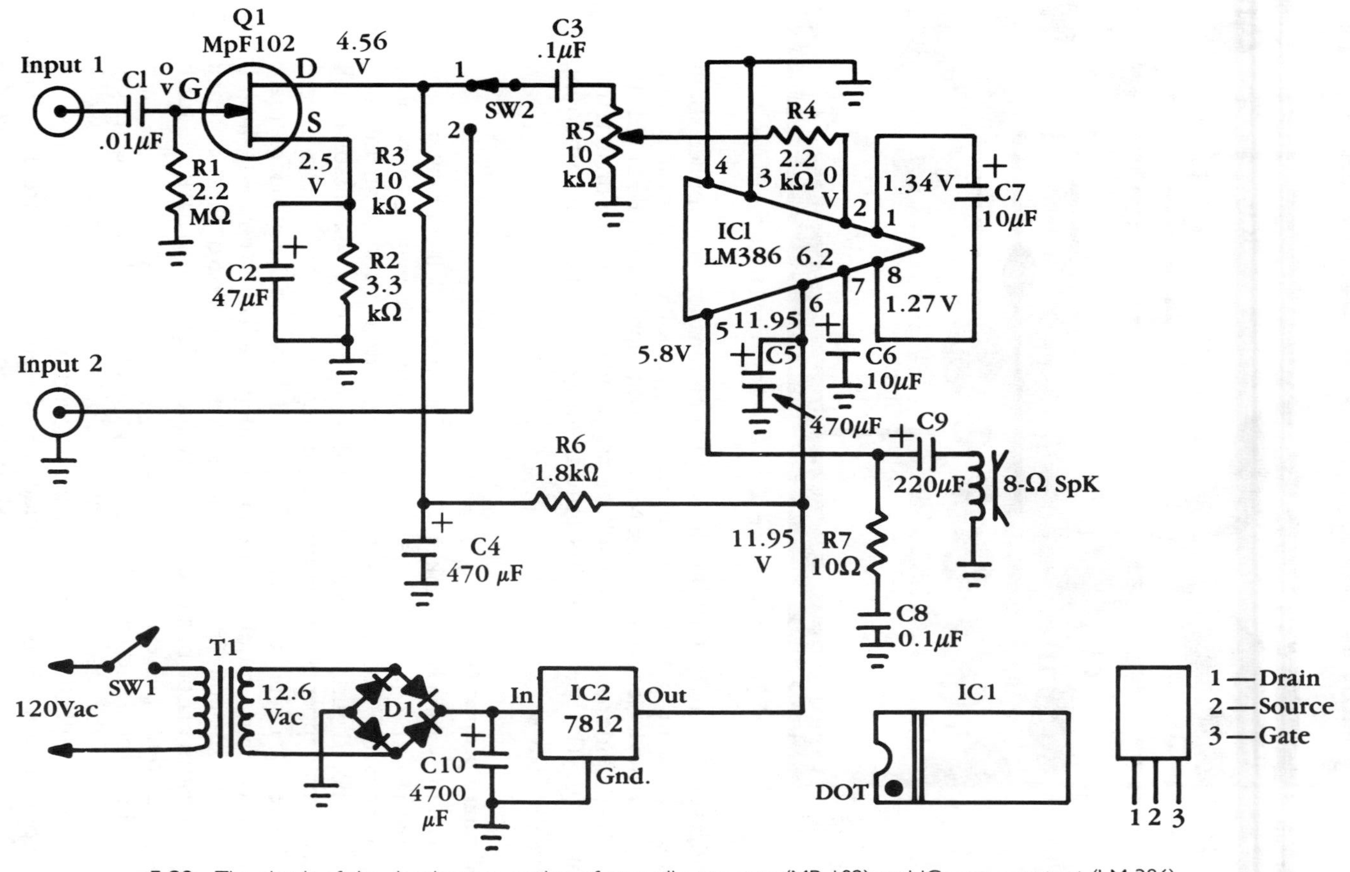

5-20 The circuit of the signaltracer consists of an audio pre-amp (MP 102) and IC power output (LM 386).

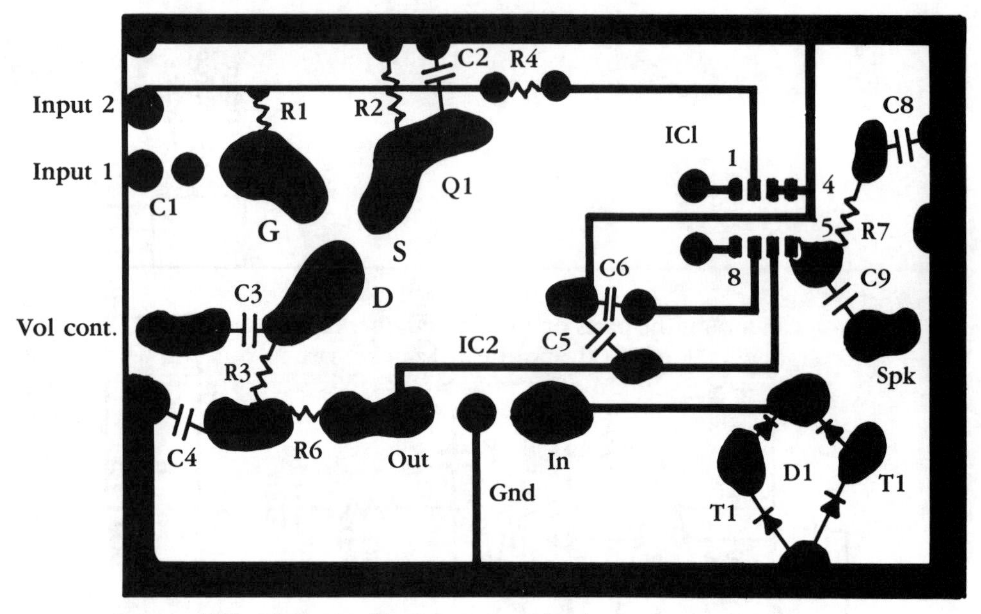

5-21 The actual size of the pc board wiring for the signaltracer.

bridge rectifier (D1). C10 provides ripple elimination and applies about 19.9 volts to the 12-volt regulator (IC2). The 12-volt regulator can be purchased at most electronic stores. Now, 12 volts applies to pin 6 of IC1. R6 and C4 provide a decoupling voltage-dropping circuit for the drain terminal of Q1. The schematic lists critical voltages for the troubleshooting procedure.

PC board preparation

Lay out the wiring on a blank single-sided pc board. Select or cut a piece of board that is 4½ × 3 inches. Clean the copper side with steel wool or soap and water. Some boards might have oxidation marks or an adhesive on the board surface. Clean the surface so etchant will dissolve any uncovered copper. Now, lay out the pc circuits.

After the circuit has been designed and drawn on the pc board, place it in the etching solution. Pour on just enough etchant to cover the board. It should take about 35 to 45 minutes to etch the board. Place a pencil underneath a plastic box or etching tray and rock the ends back and forth to speed up the process.

Check the board every 10 minutes with a toothpick or wooden pencil. Pour off the solution after the etching has been completed. Look closely to see if all copper has been removed between the small lines or component holes. Wash board under a water faucet. Make sure that all of the etching solution has been washed down the drain. Now, remove the black ink or tape with steel wool or a cleaning pad.

Drill all holes for component mounting using the smallest drill bit possible; some etching kits include a very small drill bit. Drill larger holes for the regulator IC and filter capacitor (C10). Be very careful when you drill the holes for the IC pin socket so as not to damage the pc foil. Doublecheck the 8 holes for the IC socket. Mark pin 1 of IC1 on both sides of the pc board. Drill 1/8-inch holes at the end of each board to mount it.

Mounting parts

You can mount the parts on the board rather rapidly. First mount all small components. Mount Q1 and the IC1 socket last so you don't break the small leads. Mount IC1 after you have completed all wiring on the 8-pin socket. After you have mounted all parts on the pc board, extend the leads for the switch and volume control (FIG. 5-22).

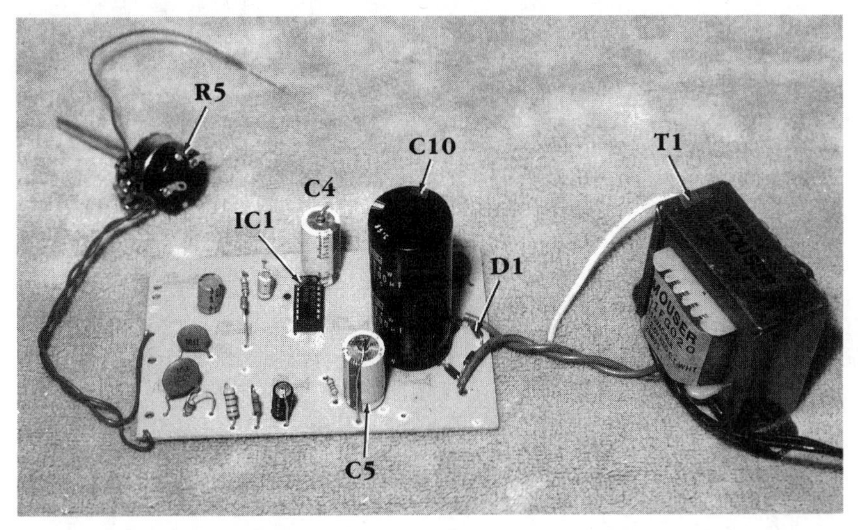

5-22 Mount all parts on the pc board after drilling the board holes.

Wiring

Solder a shielded cable, for the input of the pc board, to common ground or twist the leads together. Connect a shielded-type female plug to both inputs. Solder a piece of hookup wire from the center tap of the control to pin 2 of IC1. Solder input 2 and C3 to correct terminals of SW2 (DR). Connect the other side of SW2 (amp) with a piece of hookup wire to the output of C3 on the pc board. Twist all solder leads, except the shielded ones, to eliminate hum (FIG. 5-23).

Clean off any solder flux between the points of wiring on the pc board. Inspect the board with a magnifying glass to see if any terminals are soldered together on the IC1 terminals. Check each IC terminal—from

5-23 A rear view of the mounted components.

the top side of the socket to the wiring underneath—with the low-ohm range of the ohmmeter.

Speaker cabinet

Construct this signal tracer inside a commercial speaker case, 15 × 9 inches. The speaker was a 5-inch PM-type. These speaker cases can be found on the market as surplus or overstocked items for a few bucks. The cabinet is an ideal place to mount the pc chassis and transformer. Mount the input jacks, volume control, and SPDT switch on the front grille side. Besides boosting the volume, the larger speaker cabinet provides greater tone quality.

Testing

Doublecheck all cable, pc board, and input wires. Turn R5 on and rotate the volume control wide open. You should hear a little hum and noise. Place your finger on input 1 and 2. The hum on input 1 should be loud and much lower on input 2 when switched in. Troubleshoot the circuit if no hum can be heard on input 2 (FIG. 5-24).

First, check the voltages at IC1. Place a screwdriver blade on this terminal for greater pickup. Check pin 6 for 12 volts dc. Now, check all other voltages at IC1. Inspect IC1 for correct installation. If the IC is plugged in backwards, it might be damaged. Pin 1 should be at the "U" or white band end of the IC.

Doublecheck the speaker wiring. Sometimes these small wires will snap when the cabinet is turned over for soldering and inspection. If 12 volts are applied to pin 6 of the IC1 and there is still no sound, suspect a

5-24 Hook up and check the audio circuits of an AM/FM cassette player, with the signaltracer.

defective LM386 IC. Check Q1 voltages only after IC1 has sound. Inspect the terminals of Q1 for abnormal voltages and a lack of sound from input 1.

If the speaker motorboats or changes oscillations, make sure that C5 is close to terminal 6 of IC1. Always install silicon diodes and electrolytic capacitors with correct polarity. Make sure that pin 1 of IC1 is at the dot on the board. Keep all component leads short and make good soldered connections on the pc board.

Signaltracing a radio

The AM radio might consist of a converter, two i-f, a detector, and an audio output stage. The FM radio might have rf, mixer, oscillator, three i-f, MPX, and audio circuits. The AM/FM phono/cassette player might tie all of these circuits together (FIG. 5-25). Usually, the rf and i-f signals can be picked up after the converter stages in the AM/FM receivers.

Check each transistor from base to collector until you note the loss of signal. When the signal quits, you have located the defective circuit. Now, take voltage and transistor leakage tests at the suspected transistor. The signal should get louder as you trace it through the various circuits. Signaltrace the radio and audio amplifier by the numbers.

Cassette audio

Signaltrace the audio output signal (before the function switch) if the cassette audio is dead or weak. Go directly to the tape head wind and trace the signal through the pre-amp circuits. You should clean the tape head before taking signaltracing tests. A dirty cassette tape head can cause a

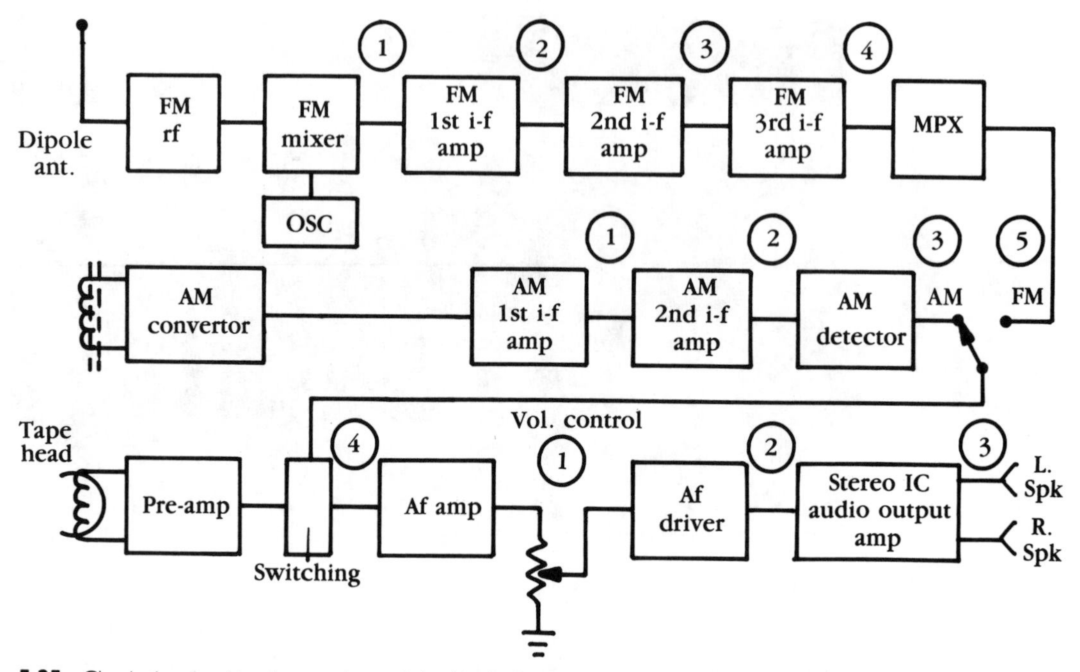

5-25 Check the signal by the numbers of the block diagram in the radio, cassette, and audio circuits of a typical AM/FM/MPX phono cassette player.

weak, distorted, or dead channel. You might find a dual IC in the pre-amp circuits of the latest cassette player.

Audio amp signaltracing

When the audio amp is dead, weak, or intermittent, clip the signal-tracer test leads to the center terminal of the volume control and common ground. If the signal is the same at this point, you know the trouble lies ahead of the volume control. Signaltrace the audio amp to the speaker. Gradually turn down the audio of the signal tracer as you work through the amp circuits. Each stage should have greater amplification. Switch to input 2 when volume becomes too great.

Test from the coupling capacitor to each base terminal. A loss after the capacitor indicates an open or dried-up capacitor. Next, proceed to the collector terminal of the same audio transistor. Keep going through each audio stage until you lose the signal. A radio signal, phono record, or cassette tape can be used as the signal source. Signaltrace the audio signals by the numbers. In today's audio circuits, you might find that the audio stages are inside one large IC. Always keep the signal on a signal tracer as low as possible.

Parts list

C1	0.01-μF ceramic 450-V capacitor.
C2	33-μF 25-V electrolytic capacitor.
C3	0.1-μF 500-Volt ceramic capacitor.
C7, C8	0.1-μF 50-V ceramic capacitor.
C4, C5	470-μF 25-V electrolytic capacitor.
C6	10-μF 25-V electrolytic capacitor.
C8	0.047-μF 50-V paper capacitor.
C9	220-μF 25-V electrolytic capacitor.
C10	4700-μF 50-V electrolytic capacitor.
R1	2.2-MΩ ½-W resistor.
R2	3.3-kΩ ½-W resistor.
R3	10-kΩ ½-W resistor.
R4	2.2-kΩ ½-W resistor.
R5	10-kΩ volume control and switch.
R6	1.8-kΩ volume control and switch.
R7	10-Ω ½-W resistor.
T1	120-Vac primary/12.6-V secondary power transformer.
D1	1 or 2 A bridge diode or four separate 1 A silicon diodes.
IC1	LM 386 IC power amplifier.
IC2	7812 12-V regulator.
Q1	MP102 FET transistor.
PCB	4½"-X-3" copper single-sided pc board.
SPK	5" or larger, in speaker cabinet (see text).
Input jack	2 male shielded jacks.
SW2	STST toggle switch.
Test probes	See chapter 6.
Misc.	Hookup wire, ac cord, standoffs, bolts, nuts, solder, etc.

Troubleshooting ICs

Locate the input terminal of the IC by checking the schematic. No signal at the input terminal might indicate trouble ahead. Now, locate the output terminal of the IC and check the audio signal. Suspect a defective IC if you can hear audio on the input and not the output terminals. Before removing the IC, take supply voltage measurements. A low supply voltage might reveal a defective power source or IC. Replace the IC if one of the channels is weak, dead, or distorted.

Signaltracing motors

A noisy motor in a camcorder or CD player can be checked with the signal tracer. Start at the motor terminals and work back toward the motor driver IC. If noise can be heard, proceed to the output terminals of the control microprocessor. No signal here might indicate a defective control component. If you trace the noisy signal to the motor terminals, but there is no rotation, suspect a defective motor. C1 and C3 prevent damage to the signal tracer by blocking dc voltages. The signal tracer can also locate warbling or garbled audio playback in the VCR circuits.

SINE/SQUARE-WAVE FUNCTION GENERATOR

The sine/square-wave function generator provides stable and accurate square and sine waveforms with a frequency range of 20 Hz to 20 kHz using a single control. This function generator is ac-powered and utilizes an 8038 IC. The sine/square-wave generator can be used to service most any commercial amplifiers (FIG. 5-26) and it can be used as a sweep generator.

5-26 Front view of the sine/square wave form function generator.

The circuit

This generator is built around the functional 8038 IC. Although the IC can also generate a triangle waveform, it is not used in this project. The output frequency range is controlled by R1 and C2. The square-wave output

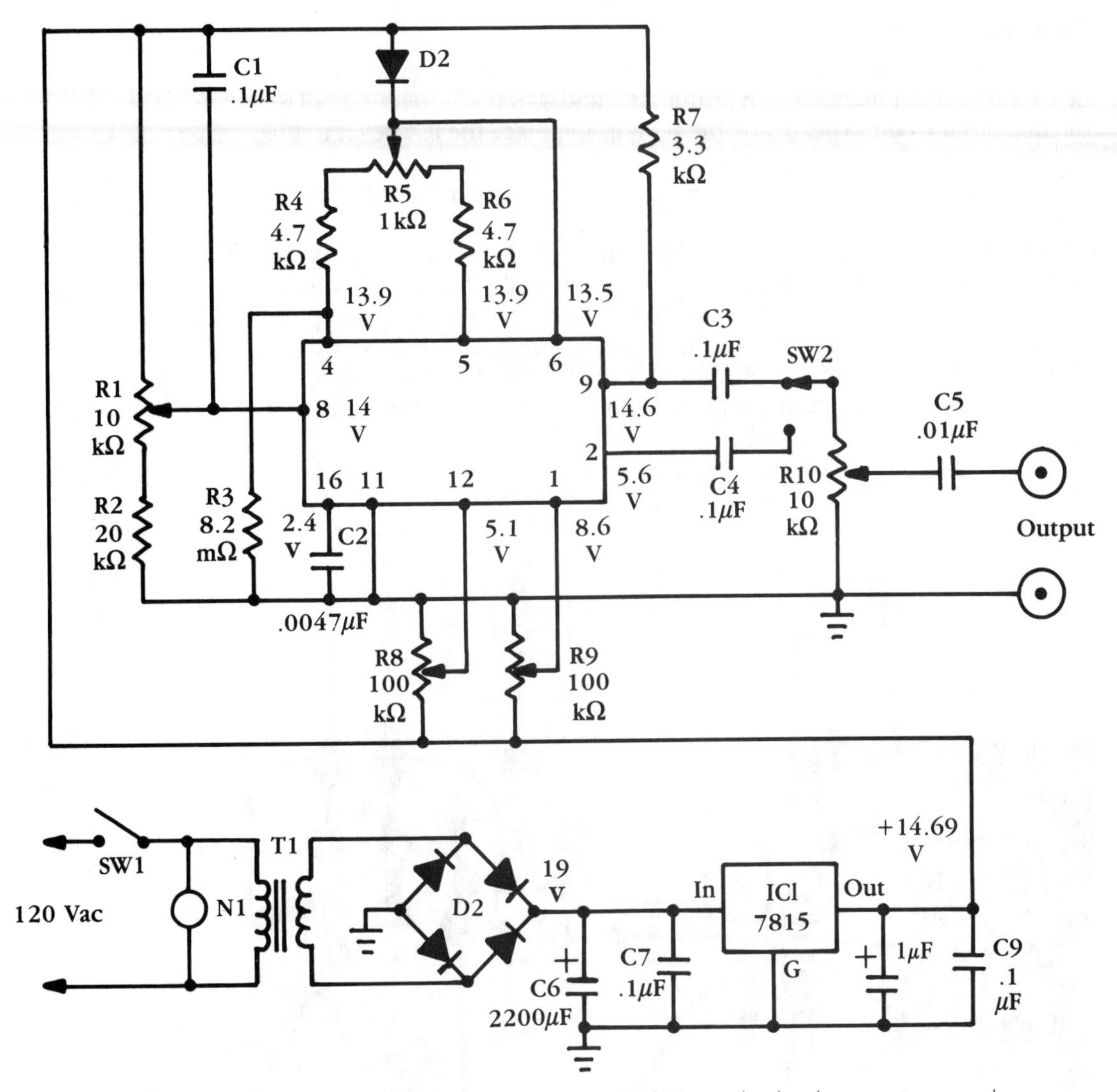

5-27 A simple function generator circuit with a controlled-output signal and an ac power supply.

terminal is pin 9 and the sine-wave output is pin 2 (FIG. 5-27). Each signal can be switched and controlled by R10 in the output circuit. C5 has a higher working voltage if it is applied to higher-voltage circuits.

Pin 6 of IC2 is supplied with a 15-volt source from a small power supply. T1 steps down the 12.6 volts to a bridge rectifier (D1). IC1 regulates the output voltage to 15 volts. N1 is a neon light that indicates when the unit is turned on. SW1 is on the back side of R1. The ac-operated signal generator is very stable.

Copper pc wiring

Cut a 4½-×-3-inch copper-clad pc board. Lay out the etched wiring and be careful not to touch lines on terminals 2 and 9 of the IC (FIG. 5-28). Use the pc-marking pen to get inside the IC symbol. Place the IC socket to the top left of the copper board. Mount all components, except R1 and R10, on the board. Of course, mount the power transformer directly on the cabinet chassis.

The board should take about 35 minutes to etch. Use the smallest bit possible to drill the mounting holes. A 1/16-inch drill bit is furnished with most pc-board kits. Be careful when drilling holes for the IC socket pins. They should be in the center of the symbol terminals and in a straight line. Drill a hole in each corner to mount the board. Before installing the parts, lay the pc board in the bottom of the case and mark the mounting holes.

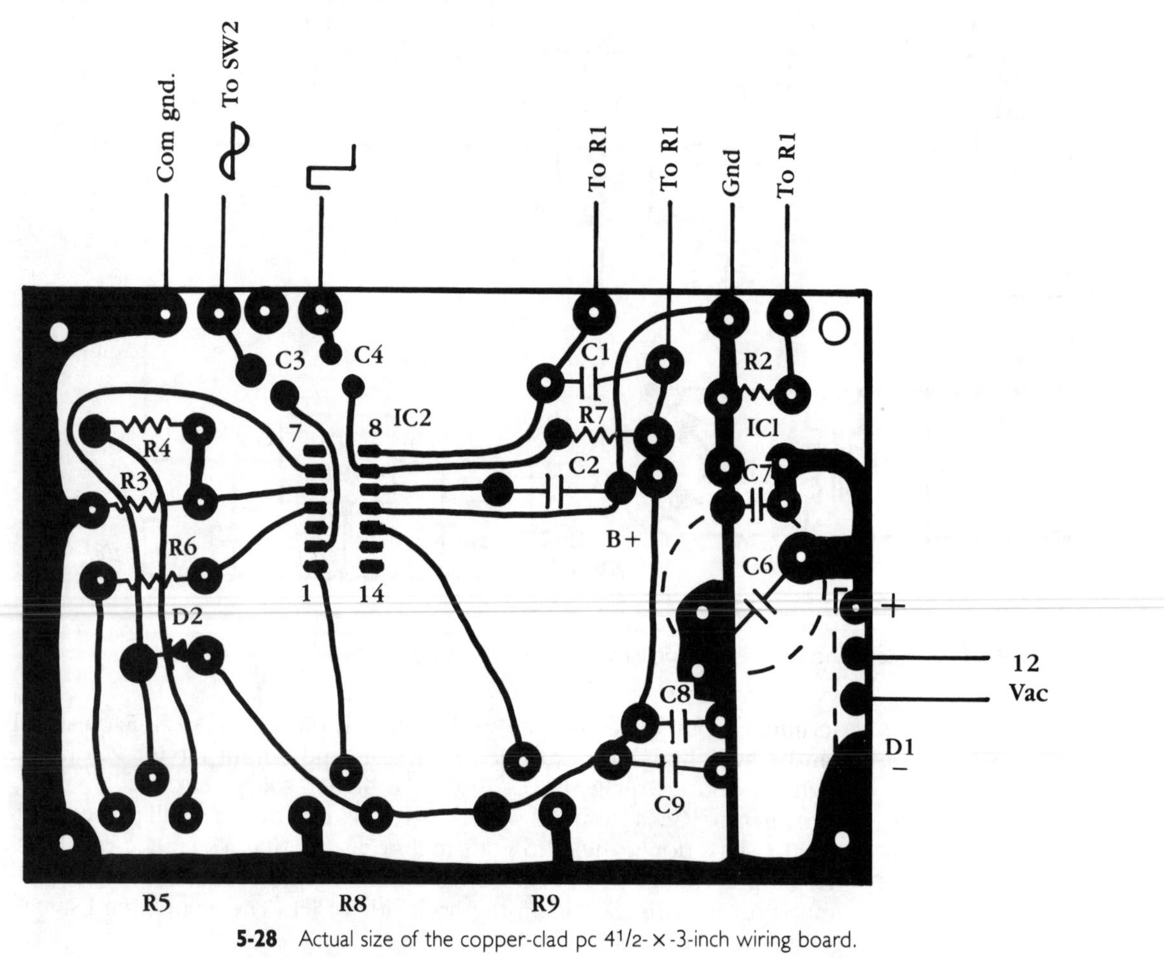

5-28 Actual size of the copper-clad pc 4½-×-3-inch wiring board.

Mounting parts

Solder in each small component and make sure that the polarity is correct for all diodes and electrolytic capacitors. Mount the small IC socket. Check so that each connection is good and no two terminals are soldered together (FIG. 5-29). A hand-held magnifying glass is ideal for checking small pc wiring and connections. Mark pin 1 on the top and bottom of the pc board with a felt-tipped pen. Next, mount R5, R8, and R9. Mount C6 last because it is taller than the rest of the components. Doublecheck all connections. Use a 35-watt or a small-pointed iron to make pc board connections. Make sure each part is in the right set of holes. Scrape the rosin off between the soldered pins of IC2.

5-29 A view of the pc board with all parts mounted, except the controls, jacks, and transformer.

Board construction

Solder a 4-inch piece of hookup wire from each waveform connection. Cut three 4-inch leads and connect them to the board for R1. Solder a black lead to the outside foil (common ground). The power transformer's 12.6-volt wires will be soldered directly to D1 when the board and transformer are mounted. Solder SW2, R10, C5, J1, and J2 together on the front panel.

Front panel preparation

With the ABS case, you can order a satin-finished aluminum front panel that is 1/8-inch thick. Not only does this front panel provide a professional appearance, but it is also quite sturdy (FIG. 5-30). Drill three 9/32-inch holes for the two controls and the switch. Center and space the holes 1 1/4 inch apart inside the aluminum panel. Drill a 7/16-inch hole in the center/top area for the 120-volt ac neon lamp. At the bottom, drill two 5/16-inch holes for the banana jacks. Mount N1, R1, R10, J1, and J2 on the front panel. Cut and tape the cover sheet over the front panel for the hole layout and to prevent scratches. Before mounting the parts, place the labels and numbers on the front cover. Spray on a couple of coats of clear finish.

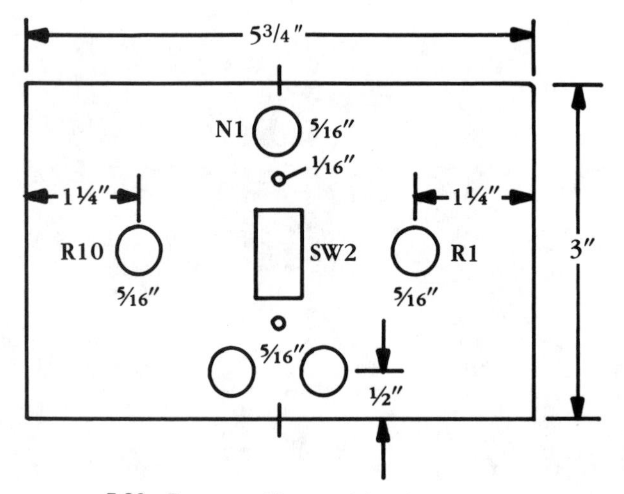

5-30 Front panel layout of function generator.

Testing

After the generator has been completely wired and all parts have been mounted, test it. Doublecheck all wiring connections. When R1 is turned on, N1 should light. If N1 comes on when the unit is merely plugged into the outlet, you have wired one terminal lead before the on/off switch. Disconnect and place it on the other switch terminal.

If the transformer makes a loud hum and begins to smoke, check the polarity of C1 and look for a possible leaky diode in the bridge circuit (D1). Usually, one or two diodes in the bridge rectifier will short out and become leaky if a heavy load is placed across the voltage output terminal. Make sure that the bridge rectifier is connected properly. The positive terminal should go to the input of IC1 and the negative terminal to common ground. If C6 is warm or hot, the working voltage might be too low or it might be inserted with the wrong polarity. Check the suspected bridge rectifier with the DMM in the diode test position.

Check for output voltage on the output terminal of IC1. If there is no voltage, IC1 might be backwards or no voltage might be at the input terminal. If 19 volts is at the input terminal and none is at the output (15 V), replace the defective voltage regulator. You might measure 14.95 volts at the output terminal with some regulators.

When the correct voltage is applied to IC2 and there is no output signal, check the voltages at each terminal and compare them with those on the schematic (FIG. 5-31). Supply pin 6 should measure 13.5 volts. Check the output signal at pins 9 and 2 with the oscilloscope. Remember, the signal wave forms seen will not look like a square or sine wave until the generator is adjusted correctly.

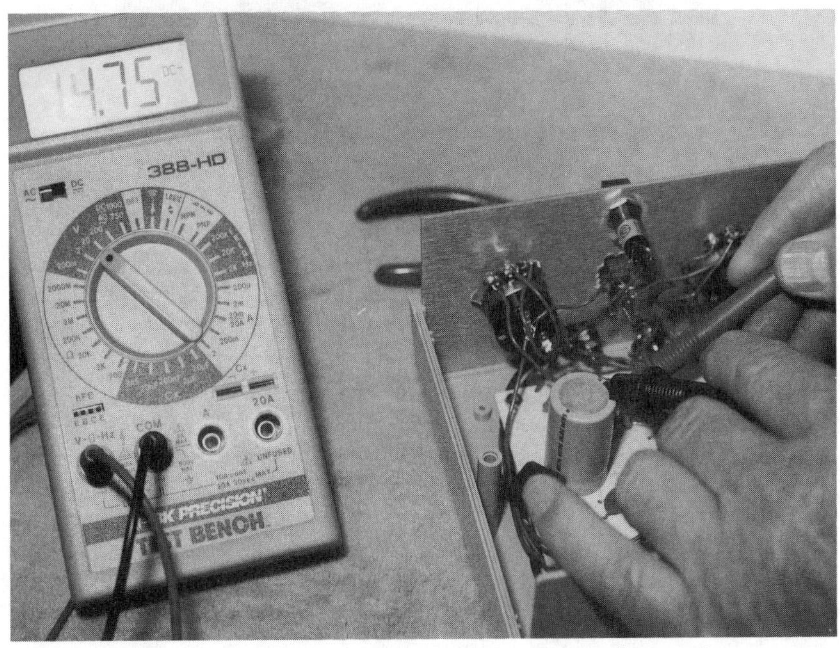

5-31 Take voltage measurements with the generator in operation.

Adjustments

Rotate the output control (R10) wide open and connect the scope to jacks 1 and 2. Connect the ground to the black jack and the scope probe to the red jack. Readjust the scope to obtain some kind of wave form. Adjust R1 to change the frequency of the waveform. Have at least three waveforms on the scope before you start to adjust the square and sine waves. Of course, the square wave is almost normal before any type of adjustment.

Flip SW2 to the sine waveform. Adjust R8 and R9 for minimum distortion. The sine wave should be fairly round at the apex of each waveform. Readjust both controls until you have a perfect sine wave (FIG. 5-32). Now, flip the switch to the square waveform and readjust R5 for a clean reading (FIG. 5-33). Recheck the sine waveform. You will find that R8 and

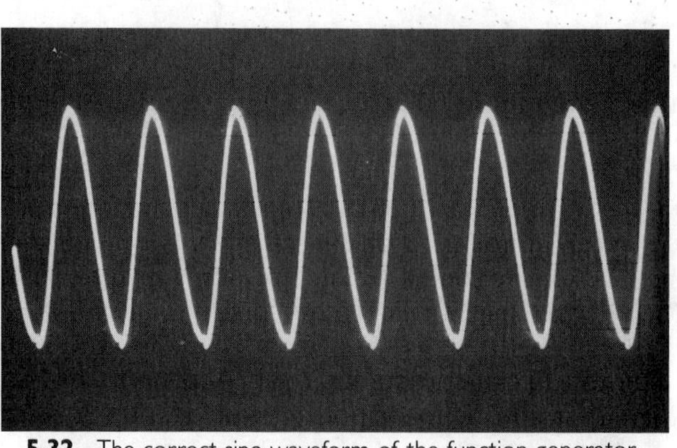

5-32 The correct sine waveform of the function generator.

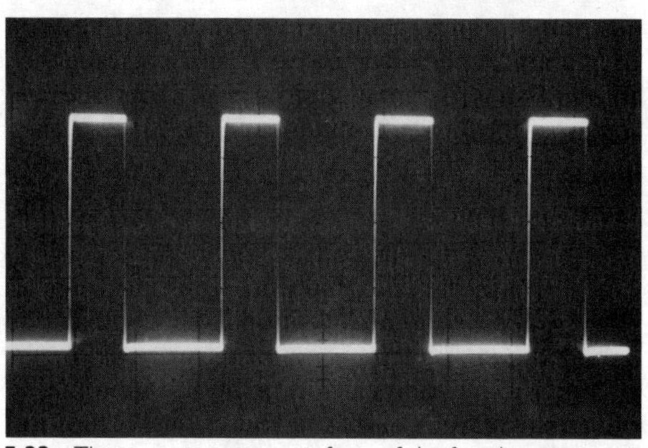

5-33 The correct square waveform of the function generator.

R9 effect the sine wave the most. By readjusting all three thumb-type controls, you should be able to reach perfect sine and square waveforms.

If you cannot find the correct resistance for R2, wire two 10 kΩ resistors in series. For R7, place a 2.2-kΩ and a 1-kΩ resistor in series to total 3.3 kΩ. Connect two 4.7-MΩ resistors in series for R3 if a 8.2-MΩ resistor is not available. Although this resistance is over 8.4 MΩ, the function generator is still stable with two clean waveforms.

DELUXE SINE/SQUARE-WAVE FUNCTION GENERATOR

This sine/square-wave generator is very stable and covers from 20 Hz to 20 kHz. Besides being a stable generator, the gain output can be monitored with an LED display (FIG. 5-34). The function generator circuit is built around a 14-pin monolithic IC. IC1, an 8038, delivers sine, triangle, and square waveforms with very little distortion. However, only the sine and square waveforms are used in this function generator.

Parts list

IC1	7815 15-V regulator.
IC2	8038 function generator, D C Electronics, P.O. Box 3203, Scottsdale, AZ., 85271-3203.
C1, C3, C4, C7, C9	0.1-μF 50-V ceramic capacitor.
C2	0.0047-μF 50-V monolithic or high-Q ceramic disc capacitor.
C5	0.01-μF 500-V ceramic or paper capacitor.
C6	2200-μF 35-V electrolytic capacitor.
C8	1-μF 35-V electrolytic capacitor.
R1, R10	10-kΩ linear control.
R2	20-kΩ ½-W resistor.
R3	8.2-MΩ ½-W resistor.
R4, R6	4.7-kΩ ½-W resistor.
R5	1-kΩ trimmer screwdriver- or thumb-variable resistor.
R7	3.3-kΩ ½-W resistor.
R8, R9	100-kΩ trimmer screwdriver- or thumb-variable resistor.
N1	120-Vac red neon indicator, 272-704 or equiv.
T1	12.6-Vac 450-mA stepdown power transformer, 273-1365 or equiv.
D1	1-A bridge rectifier.
D2	1N914 switching diode.
J1, J2	Banana jacks.
Case	MB-3C beige instrument enclosures, All Electronics.
Pc board	3″ × 4½″ copper-clad board.
SW1	On back of R1.
Misc.	Ac cord, grommet, hoodup wire, bolts, nuts, etc.

The circuit

The complete schematic diagram consists of a variable control sine- and square-wave generator, an LED display, and two low-voltage power sources. R1 changes the frequency and R6, R8, and R9 change the shape of the waveform (FIG. 5-35). SW2 switches either the sine- or square-wave signal to the emitter output control, R10. R10 sets the gain of the function generator.

The signal from R10 is fed to the output terminals and also to the LED display. From here, the signal receives full-wave rectification from D3 and D4. Q2 is an emitter/follower-type circuit connected to pin 5 of IC2. R14 adjusts the signal to compare with the setting of R10. The LED display

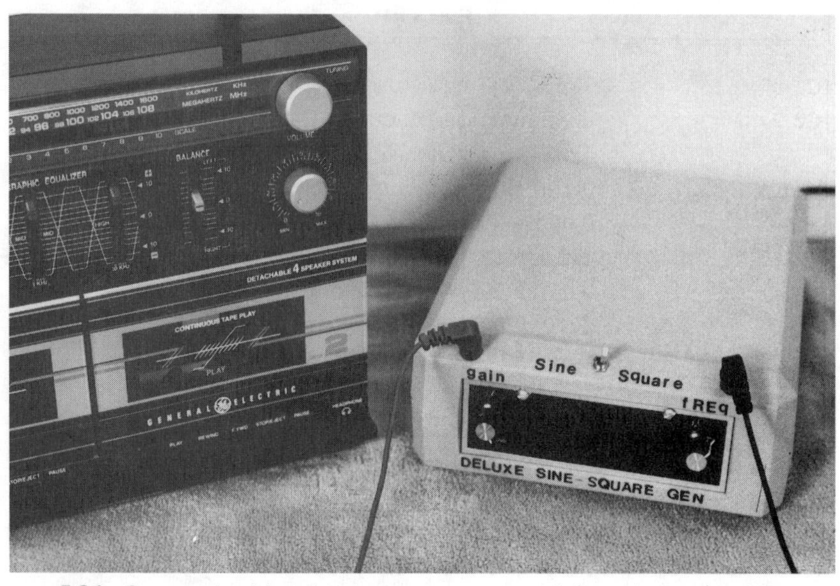

5-34 Connect the deluxe function generator to a radio/record player/amplifier.

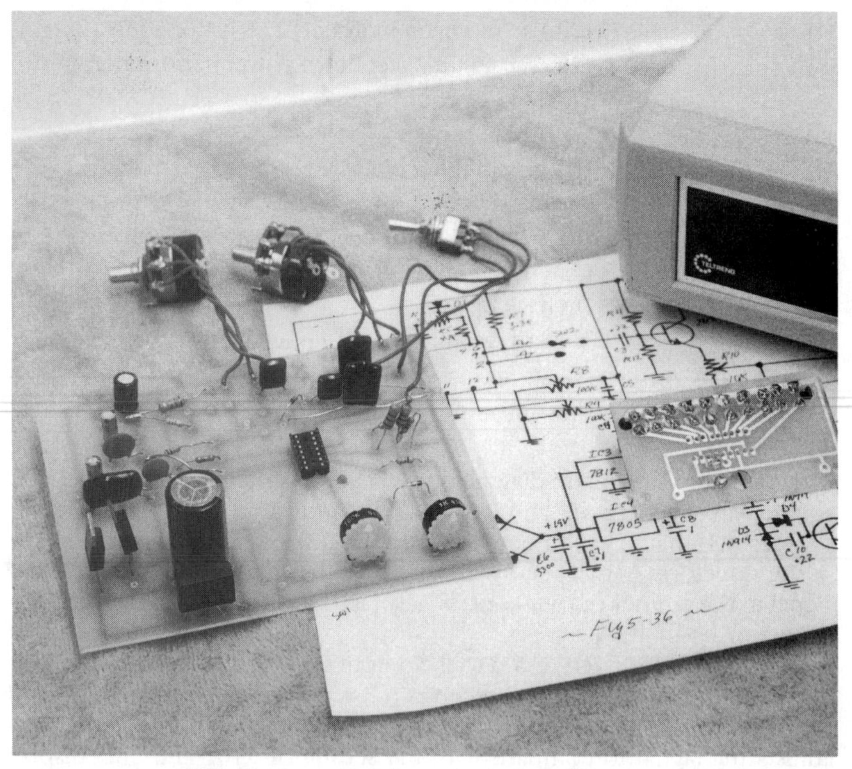

5-35 The front view of complete unit, ready to operate.

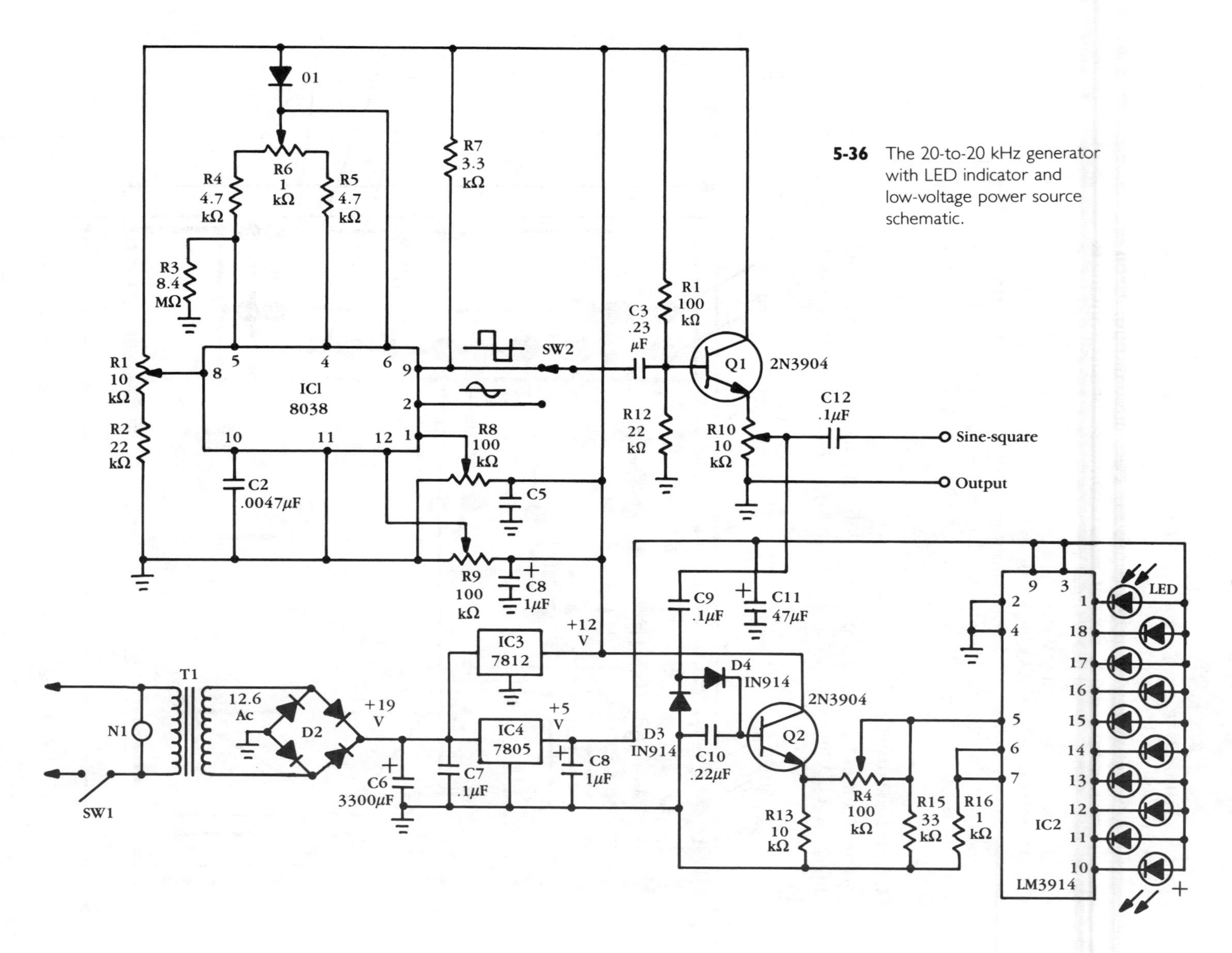

5-36 The 20-to-20 kHz generator with LED indicator and low-voltage power source schematic.

consists of 10 LEDs in one component. The positive side of the LEDs is connected to a 5-volt source (FIG. 5-36).

The low-voltage power supply consists of two different output voltage sources, rated at 5 and 12 volts. The 12.6-volt ac secondary winding is rectified by a full-wave bridge rectifier, D1. C6 provides adequate filtering with the two IC regulators. Voltage regulator, IC3, provides 12 volts for IC1, Q1, and Q2. IC4 provides a 5-volt source to operate the LED assembly, IC2. C11 mounts at the 5-volt output terminals on the main board to save room on the LED pc board. The deluxe sine/square-wave function generator is a great project to add to the service bench. Now, let's put it all together.

PC board construction

This project uses two separate pc boards, one of which is very small (containing the LED processor and LED assembly). The main pc board con-

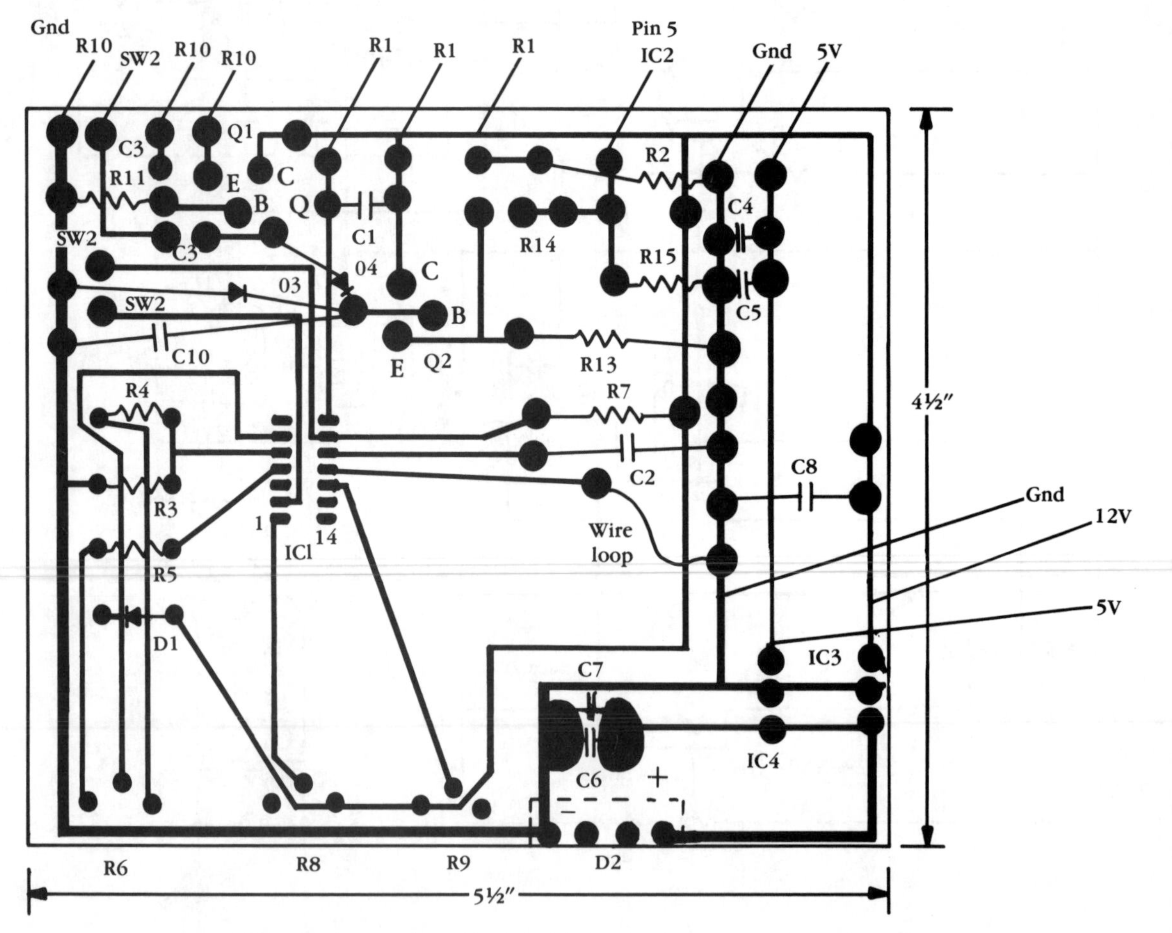

5-37 The large pc board layout with function generator and power supply.

tains most of the generator components (FIG. 5-37). Mount IC1 to the left and center, with the power supply and generator parts strung over a wide area. The power transformer is not mounted upon the cabinet chassis with secondary leads soldered to the pc board. Mount IC2 and the small components on a separate board so it can attach directly to the front panel. Only pin 11 has a loop bare ground wire.

Select the smallest bit possible to drill the holes for the component leads. Be careful when drilling the IC holes; keep the lines straight and not too close to the next terminal. The nine top holes that connect to the outside components can be drilled with the next larger size of bit. Make sure that the small bit goes completely through the board so the parts can be mounted. Check the fine-line pc wiring for possible breaks with the low scale of the DMM. Cut the large pc board to fit in the slotted area of the cabinet.

The small LED board consists of an 18-pin IC socket and 10-pin LED assembly. Be very careful when laying out the IC and LED assembly holes. The LED has terminals on each side. Draw a pencil line where the LED assembly is to be mounted. Then place a pin socket hole on each side of line, without touching the line (FIG. 5-38). Drill twenty small holes to mount the LED assembly and two larger holes on each end for the plastic clips to snap into the pc board. The only parts that you need to mount on this small pc board are the LED assembly, a 1-kΩ resistor and a 47-μF electrolytic capacitor. Lay the capacitor flat for mounting.

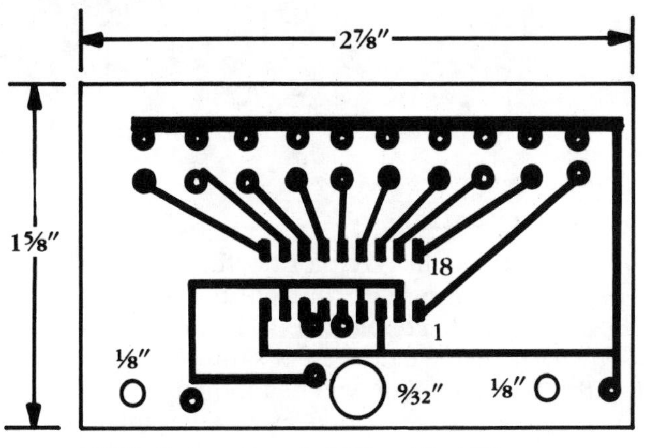

5-38 Pc board layout of IC2 and the LED array.

Mounting components

Inspect the board for possible breaks and poor connections after etching. Sometimes the fine-line pc wires will almost be etched away, when a complicated board is etched by hand. First, mount all small resistors, capacitors, and diodes. Leave C6 and D1 until last. You can solder better with the pc board flat. Solder the power transformer leads after you mount the board. Make sure that the polarity of diodes, capacitors, transistors, and regulators is correct.

Connect 3-inch hookup wire leads to common ground, and to the top and center terminals of R10. Solder extension wires to the top, center, and bottom of R1 (FIG. 5-39). Connect a flexible wire from the pc board to pin 5 of the small pc board. Solder a common ground wire to the small pc board to connect all grounds of IC2. Connect a 3-inch flexible lead from B+ 6.5 volts to pin 3 and pin 9 of IC2. Mount IC2 on the front cover bezel area so the 10 LEDs will be visible when controlling the gain control.

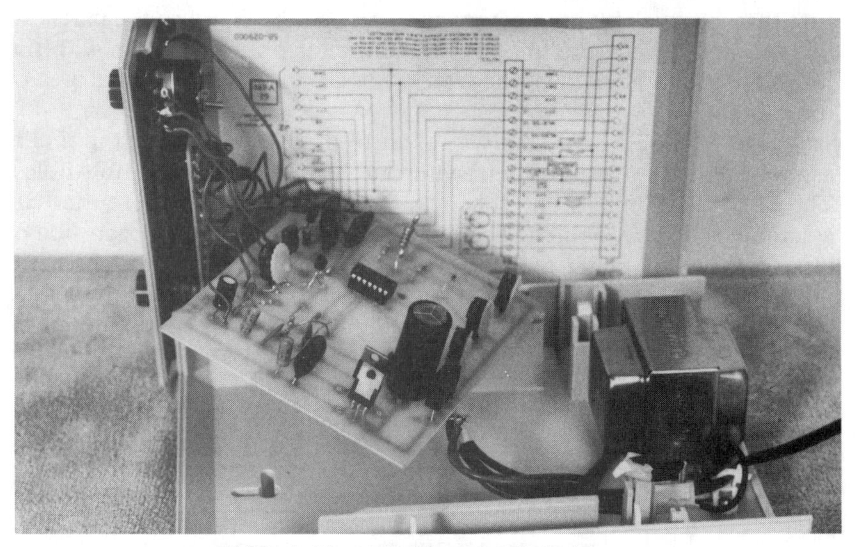

5-39 Inside view of the connected wiring.

Wiring the indicator board

After drilling all holes in the indicator board, mount the IC2 socket and the LED assembly. Make sure that the positive side of LEDs goes to the side where all are connected together. If the positive side is not marked, test it with the DMM in the diode-test position. When you get an ohmmeter reading, the black probe tip will be on the positive side of the LEDs. The LED will register like any diode, except that the resistance might be over 1.8 kΩ, and the regular diode would be closer to 0.55 ohms.

Spread and form the small LED leads to fit into each hole. Make sure that the negative LED sides are where all common diodes connect to the 5-volt source. If reversed, the LEDs will not light. The two plastic tabs at each end lock into the 1/8-inch holes. Drill a 9/32-inch hole for the pilot light, N1. N1 will mount on this board, but it is not connected to the LED circuits. Drill two 1/8-inch mounting holes at the bottom side of the pc board. Solder the LEDs and the IC2 socket. Drop R16 in its hole and solder it. Tie the input signal to pin 5 and R16 to pins 6 and 7. Run the 5-volt source to the B+ or common LEDs, and the common ground to pins 2, 4, and 8.

Front panel preparation

Mount the small LED indicator pc board in the center of the front panel. In this particular cabinet, the front bezel is made of smoked plastic and will let the LED light shine through. Of course, the 10 small LED holes can be drilled through the front cover if you use an opaque cabinet panel. Drill two small 6/32-inch bolt holes in the bottom side of the panel to hold the pc board in position. Mount N1 in the center hole of the pc board. Mount both the indicator light and the LEDs behind the smoked plastic panel. Two 1/2-inch plastic spacers hold the pc board away from the front panel.

Connect the extended wiring from the large pc board to the indicator board before mounting. Solder the 5-volt output to the proper pin 9 and pin 3 connections. Connect the output of Q2 to pin 5 of IC2. Solder the common ground to pins 2, 4, and 8.

Center and drill two 5/16-inch holes on each side of the indicator board. Mount both controls, R1 and R10, directly on the plastic front panel. Drill two 1/4-inch holes on the bottom side of the pc board in the plastic bottom cabinet for the generator output terminals. Drill a 1/4-inch hole in the top center for SW2. Letter controls R1, R10, and SW2 on the top panel and the output terminals on the bottom, before mounting the components (FIG. 5-40).

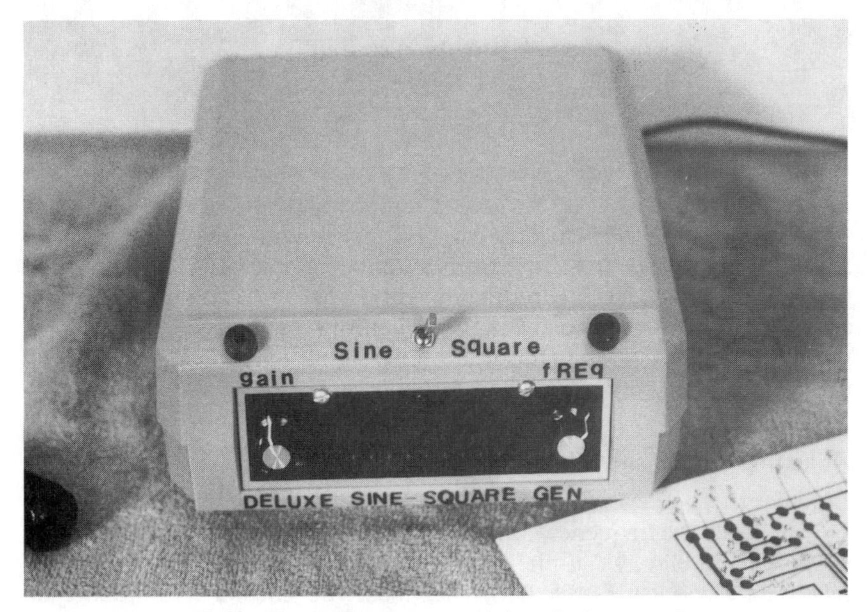

5-40 The components mounted on the front panel.

Testing

Doublecheck all connections and wiring before powering generator. N1 should light when R1/SW1 is turned on. The function generator can be checked by turning up the gain control, R10, and watching the LED display. If the display does not light, take critical voltage and resistance measurements.

Check the output voltage at the positive terminal of D1 and C6. If there is no voltage, suspect the connections at D1 and T1 (FIG. 5-41). If you find voltage at the input and no voltage on the output of IC3, suspect that it is installed backwards, is defective, or that there is a poor connection to the common ground wire. The voltages at IC3 and IC4 should be 12 and 5 volts, respectively. Check the IC regulator if one voltage is normal and the other has no voltage.

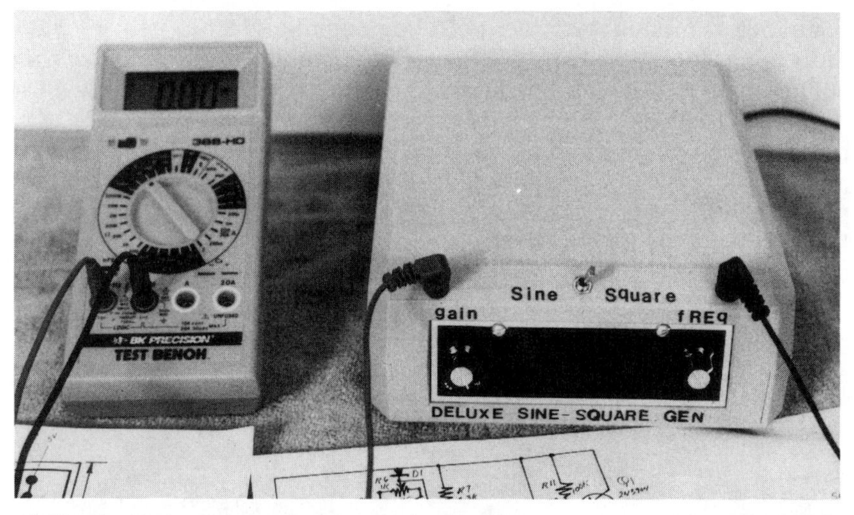

5-41 Troubleshooting the generator circuit with voltage measurements from the DMM.

The output of the generator can be checked with a scope or an audio amplifier. Signaltrace the audio from SW2 to the base of Q2 with a signal tracer. If you rotate R1, the frequency will change and can be heard in the amplifier. Use the scope to check the waveforms and make adjustments. If the generator functions without any LED lighting, readjust R14. Check the voltages applied to pins 9 and 3.

Adjusting R6, R8 and R9

Connect the scope input terminals to the function generator and apply power. Rotate the frequency control until about three waveforms are on the scope screen. Flip SW2 to the sine-wave position. Adjust R8 and R9 alternately for a good sine wave (FIG. 5-42). Make the sine wave round at each apex. You might need to touch up R6. Now, switch to the square-wave sig-

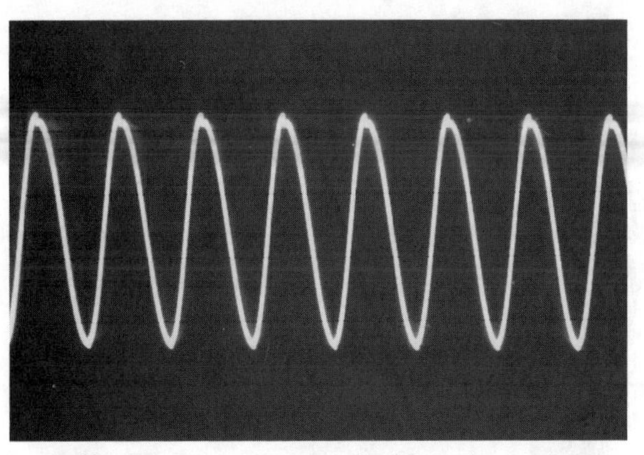

5-42 The output sine-wave signal from the generator.

nal. Readjust R6 for a good square-wave signal (FIG. 5-43). Go back and check the sine waveform. You might need to finely adjust R8 and R9, making sure that the sine wave is not distorted. Once R8 and R9 are properly adjusted, the waveforms will not change when the frequency is varied.

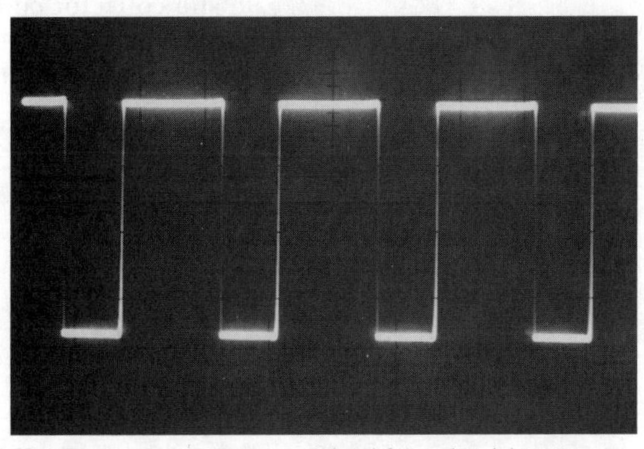

5-43 The output square-wave signal from the deluxe generator.

Readjust preset control R14 for maximum generator gain. Keep the gain control low enough to apply adequate signal to the amplifier to be checked. If you apply too much volume, the waveforms might begin to clip. You might wish to insert distortion rather than locate distortion in the amplifier.

Sine-wave distortion

If you inject a sine wave from the generator into a stereo amp and the wave is flattened on the top and bottom, the amp stage is overloaded (FIG. 5-44). Negative clipping at the bottom of the sine wave might be caused by excessive transistor gain and collector current. When the positive sine

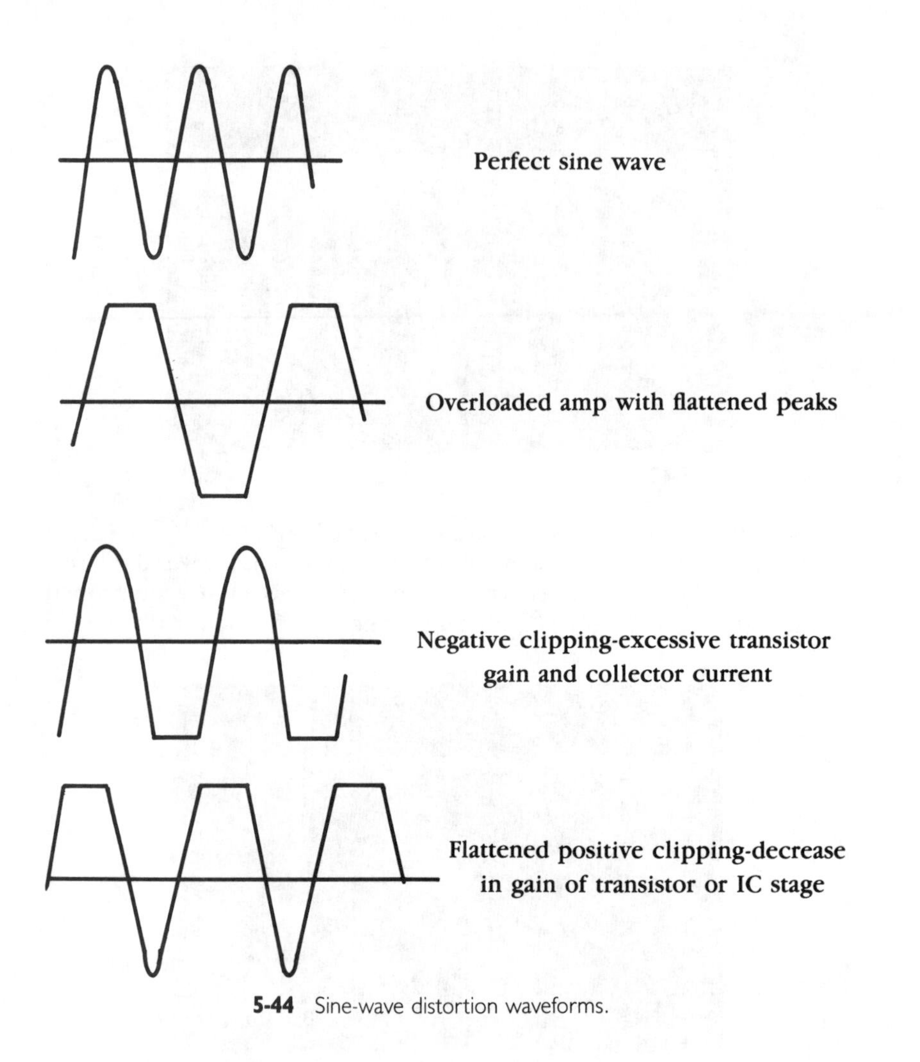

5-44 Sine-wave distortion waveforms.

wave is flattened, suspect a decrease in transistor or IC gain. Distortion in the amplifier caused by low-level stage overdrive produces waves or noisy peaks at the top of each waveform.

Square-wave distortion

With a square-wave signal injected into the amplifier, a poor low- and high-frequency response might resemble a sine wave with a sloping waveform (FIG. 5-45). A very poor low-frequency response might result in a sway-back square waveform. Poor high-frequency response almost looks like an inverted fish hook. A poor phase-shift waveform is shaped like a tilted square with unstable or parasitic oscillations in the middle top and body.

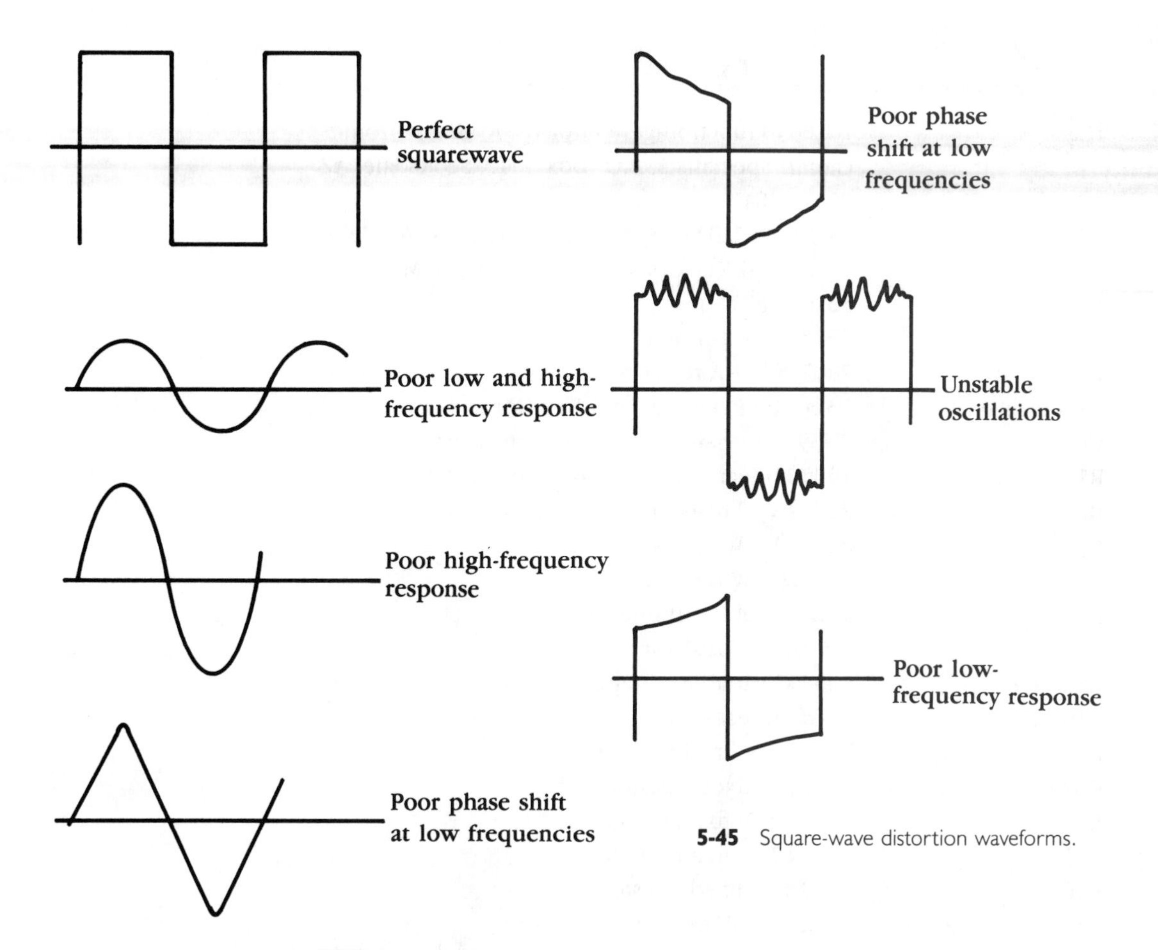

5-45 Square-wave distortion waveforms.

DC MOTOR TESTER

The external motor power source can quickly determine if a small dc motor is defective, has open windings, or will not rotate. Sometimes these small motors will overheat, become intermittent, or will just stop rotating. By connecting the correct external voltage source to the suspected motor, you can see if either the motor or the motor drive source (FIG. 5-46) are defective.

Within the camcorder, the small-loading, capstan, zoom, and auto focus motors might operate at a fraction of a volt or close to 5 volts. These simple camcorder motors can be checked with this 0-to-5 volt source. Start at zero and slowly raise R1 to the correct operating voltage of the motor.

Like the camcorder, CD motors can operate from a fraction of a volt up to 12 volts. The loading motor can operate from 0 to +6.5 volts. Most CD player motors operate from an IC driver or a transistor.

Parts list

IC1	ICL8038 function generator IC. Available from: Circuit Specialists, P.O. Box 3047, Scottsdale, AZ 85271-3047.
IC2	LM3914N LED driver bar graph display, Available from: Digi-Key Corp. Thief River Falls, MN 56701-0677.
IC3	7812, 12-V 1-A regulator.
IC4	7805, 5-V 1-A regulator.
LED display	P321 10-LED array, Digi-Key Corp.
Q1, Q2	2N3904 low-signal npn-type transistor.
R1	10-kΩ linear control with on/off switch.
R2	22-kΩ 1-W resistor.
R3	8.2-MΩ 1-W resistor.
R4, R5	4.7-kΩ 1-W resistor.
R6	1-kΩ variable trimmer resistor.
R7	3.3-kΩ 1-W resistor.
R8, R9, R14	100-kΩ variable trimmer resistor.
R10	10-kΩ linear control.
R11	100-k 1-W fixed resistor.
R12	22-k 1-W fixed resistor.
R13	10-kΩ 1-W fixed resistor.
R15	33-kΩ 1-W fixed resistor.
R16	1-kΩ 1-W fixed resistor.
C1, C5, C7, C9, C12	0.1-μF 100-V ceramic capacitor.
C2	0.0047-μF 100-V ceramic capacitor.
C3, C10	0.22-μF 100-V capacitor.
C4, C8	1-μF 50-V electrolytic capacitor.
C6	3300-μF 50-V electrolytic capacitor.
C11	47-μF 50-V electrolytic capacitor.
T1	12.6-V secondary 2- or 3-amp power transformer.
D1, D3, D4	1N914 diode.
D2	2-amp bridge rectifier.
N1	120-Vac neon light indicator.
J1, J2	2 red and black banana jacks.
Cabinet	MB-2C 3 × 6 × 6.25 plastic cabinet or equiv. Available from: All Electronics Corp., P.O. Box 567, Van Nuys, CA 90408
SW1	On back of R1 SPST.
Misc.	14-pin IC socket, 18-pin IC socket, pc board, bolts, nuts, solder, ac cord, etc.

5-46 The dc motor tester connected to a larger cassette motor.

Cassette motors operate anywhere from 3 to 15 volts. Small cassette motors can operate from the 0-to-5-volt source, larger motors can operate from 10-, 12-, and 15-volt sources, and dc phono motors operate from 9 to 12 volts. Check the motor schematic for the correct voltage operation. This motor tester voltage source will supply voltage up to 1 amp for camcorders, CD, phono, and cassette dc motors. You can even test small solar-powered or toy motors with this voltage source (FIG. 5-47).

Motor problems

Check the defective motor with ohmmeter continuity tests to see if the motor winding is open. Most small motors have a low-ohm measurement, under 10 ohms. Some schematics list the motor resistance—erratic ohmmeter readings might indicate an intermittent or dirty commutator.

Measure the voltage applied to the motor. If voltage is present and there is no movement, the motor is defective. Replace the motor if the voltage is correct and the motor rotates intermittently. If no voltage registers, there is either an open isolation resistor or a poor voltage source. Often small motors in camcorders and CD players have a driver IC or transistor.

By applying external voltages to the intermittent motor, you can isolate the motor or voltage source. Rotate the motor shaft by hand when the motor stops. If the motor starts again, replace it. Sometimes intermittent motors malfunction when a different voltage is applied to the motor ter-

5-47 The front panel of the motor tester.

minals. Try to isolate the defective motor by applying the correct voltage from the motor tester.

Be careful of the polarity when applying voltage to the suspected motor. The dc motor will reverse direction when the voltage polarity changes. It's best to remove small motors from CDs or camcorders before applying an external voltage. For instance, if a video camera zoom motor is focused completely out or in, the wrong polarity could make the motor bind or strip its fine plastic gears. Also, disconnect the motor plug or one wire from the motor so that the IC or transistor driver is not damaged. Do not apply a long external voltage if the motor does not rotate. Reverse the connections and see if the motor will turn over.

External voltage source

This little motor tester contains five different voltage sources. The voltage source is variable and can be used to operate small motors in camcorders and CD players. The 5-volt source operates 5-volt motors in camcorders, CDs, and small cassette players. The 10-volt source provides voltage to dc phono motors and auto cassette motors. The 12-volt source operates motors found in the auto entertainment field and the 15-volt source provides power for standard cassette motors.

The circuit

A stepdown power-line transformer, T1, provides the different power sources. SW1 turns the tester on as N1 lights, indicating that the tester is operating (FIG. 5-48). The 18-volts ac is applied to a 2-amp bridge rectifier, D1. Here, the dc voltage is filtered with C1. Each IC regulator provides a regulated dc voltage to each voltage jack.

IC3 has a small zener diode, Z1, in the ground circuit of the 8-volt regulator to increase the voltage to 10 volts. The 5-volt output voltage is

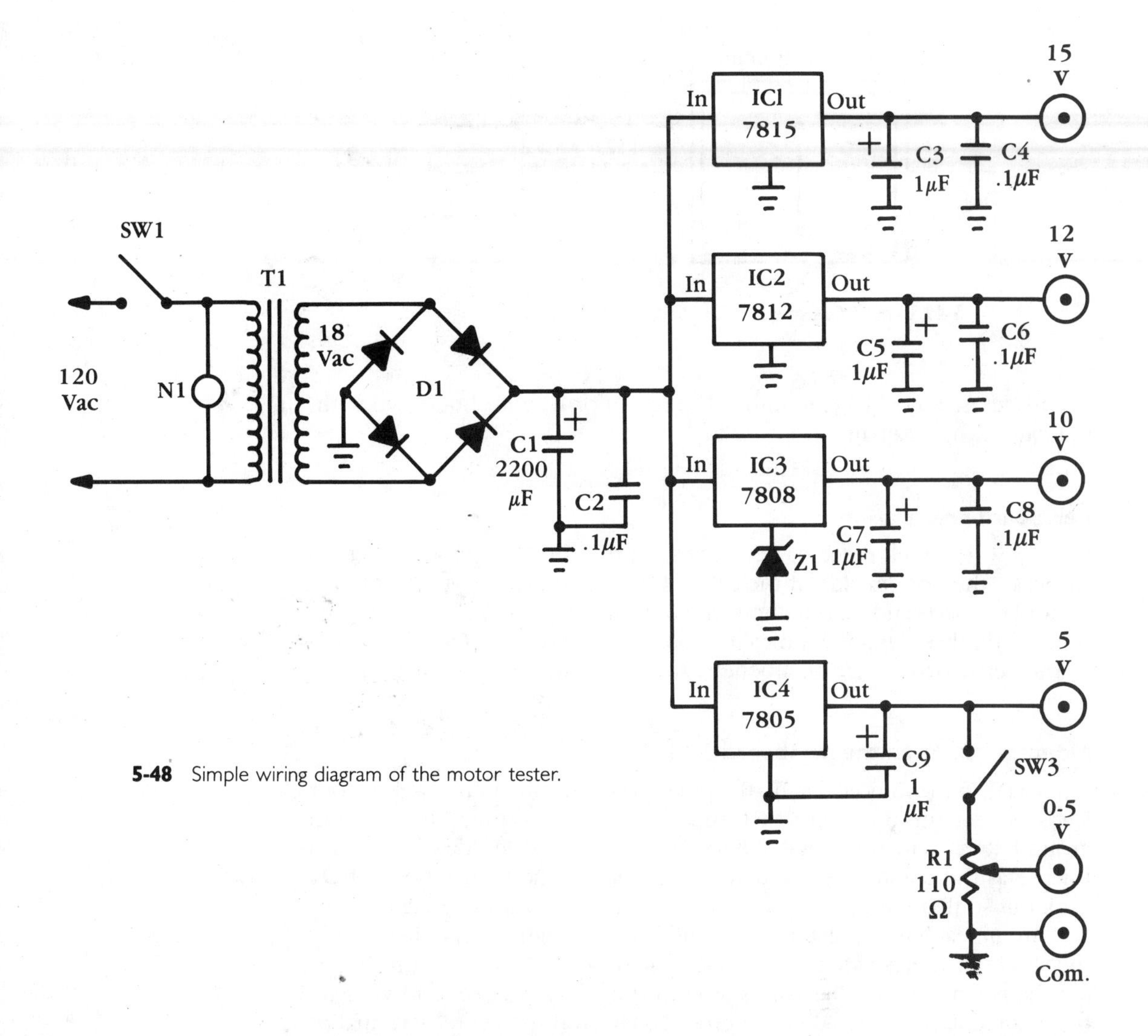

5-48 Simple wiring diagram of the motor tester.

fed to a high-power 110-ohm 25-watt rheostat and a jack. R1 can vary the voltage from 0 to 5 volts dc. SW3 disconnects R1 because it pulls extra current when it isn't in use.

Meter circuit

The meter circuit consists of a 0-to-15-volt dc panel meter, a 15-kΩ resistor, and a 5-position switch (FIG. 5-49). The 15-kΩ resistor is included with a Radio Shack meter to adjust the voltage at 15 volts full-scale. Connect the positive terminal of the meter to R2 and SW2. Connect each voltage source to its respective match on the terminal of SW2. Now, you can mount each voltage, including the 0-to-5-volt source. When a dead short

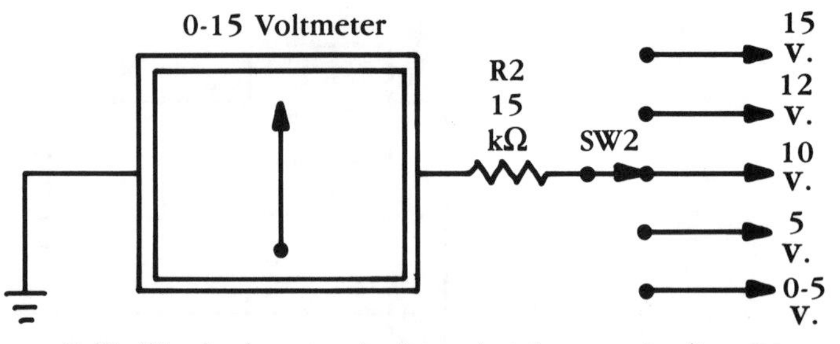

5-49 The simple meter circuit to select the correct voltage.

occurs across any voltage terminal, the regulator IC will shut down without causing any damage.

Perfboard preparation

Cut a 3-×-4½-inch piece of perfboard. Mark 2 or 3 mounting holes. In this case, the bottom side of the cabinet has small raised plastic supports to hold the parts and boards above the bottom. Line and drill the required holes. Drill three 9/64-inch holes to mount the perfboard. Grind all edges of the perfboard square on a bench sander or with a course file.

Mounting parts on the perfboard

Mount D1 in the middle, facing the power transformer. Bend over all four leads. Solder the B+ lead to C1 and the negative terminal to common ground. Run a bare piece of #22 hookup wire down one side of the board for use as the common ground wire. Let the two ac connections of D1 stick out so that the yellow transformer leads can be soldered later.

Mount each IC regulator in line and leave enough room to later apply the slip to the heatsinks. Do not attach the heatsinks until you finish wiring the perfboard. The heatsinks tend to snag or bend over and you can break a regulator terminal in the process. Ground the center terminal of each regulator. Make sure that the input of the regulators are tied to the positive terminal of C1. Now, all regulators should face the same direction. Solder each small bypass capacitor on the output side of regulators. Run a wire tie to the edge of the perfboard for each voltage source.

Preparing the front panel

Lay out all holes on a sheet of paper before drilling them. Draw all jack holes and SW1's hole at the bottom of the front panel. Drill 9/32-inch holes for the banana jacks and a 7/16-inch hole for SW1 (FIG. 5-50). Keep these holes 5/8 inch up from the bottom edge of the metal panel. Drill a 3/16-inch hole for N1 and place it in the top left-hand corner. Tape the

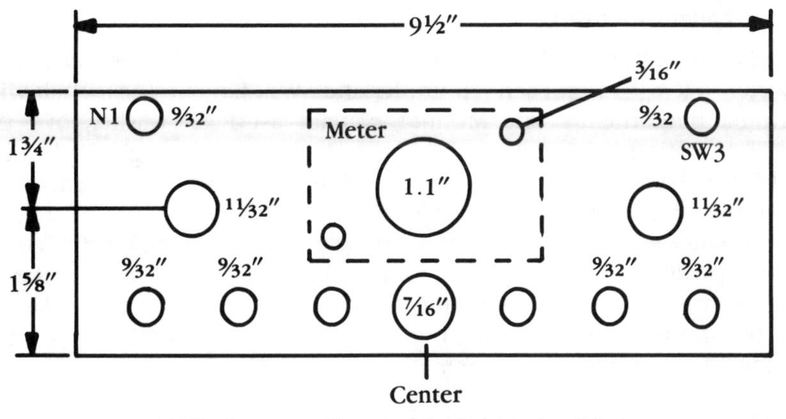

5-50 Front panel layout of the holes to be drilled.

paper-hole layout to the front panel and leave it there so you won't mar the front panel.

Place the voltmeter in the center; it will require a large 1 1/8-inch hole, 1 3/4 inches from the top panel. Use the meter template found on the back side of most meters to drill the large holes and the two 3/16-inch holes to hold the meter on the front panel. Drill a 11/32-inch hole for SW1 and R1. Place the control in the holes and rotate it back. The small stop will show where a 1/8-inch hole should be drilled to keep the controls from turning.

Drill small holes with a 1/8-inch bit and proceed with a large bit until the large hole is finished. This method keeps the larger bit from skating and marring the front panel. Cut the large meter hole with a circle cutter or by drilling several small holes in a circle. If you take the small-hole route, make sure that the 1/8-inch hole is on the outside of the drawn circle. Continually dust and bounce the small drill burrs from the front metal piece.

Lettering front panel

Letter and number the voltage jacks on the front panel. Draw a fine pencil line on the metal front piece so that the lettering is straight. Select larger transfer lettering for the name of test instrument. You can buy these rub-on transfer letters and numbers at most stationery and typewriter stores.

Outline R1 and SW2 with direct-etching dry transfer lines. These lines help dress up the project and make lines for switches and control settings. After you finish the lettering, rub-off the pencil lines with an eraser. Be careful not to destroy or throw the lines out of order.

Now, spray on three coats of clear lacquer. Leave each coat dry for twenty minutes before you apply the next coat. These coats will protect the front cover finish and lettering. Finish the front panel lettering before you mount any parts. Be careful not to get numbers or letters too close to the component holes or edges or they might slip into a groove.

Front panel mounting

First, mount all the jacks and SW1 in the bottom row. Tighten each jack washer so it will not loosen. Mount R1 and SW2. Line up the switch with terminal 1 and the first line. Mount the meter last. Place washers over the small mounting bolts so that the nuts will not pull through. Drill these holes larger so you level the meter with the front panel. Dab enameled paint over each nut so it will not loosen.

Mount the power transformer and the perfboard to the bottom cover. In this particular cabinet, place molded plastic stand-offs to raise the pc boards raised 3/8-inches above the bottom of the plastic chassis. Mount the transformer directly on the bottom of the front cover. Now, you can connect the various components.

Connecting

Connect one side of the ac cord to the terminal of the on/off switch. Solder a wire from other switch terminal to one side of T1 and N1. Connect the other side of N1 to the ac cord and the black wire of the power transformer. Tape the three wire connections.

Solder the two yellow leads of T1 to the extended bare leads of D1. Solder a longer piece of hookup wire from the pc board to each voltage jack. Make sure the voltage mark on the perfboard corresponds with the correct voltage jack. Solder a black wire from the perfboard to common ground.

Connect a short piece of hookup wire between the center terminal of R1 and the 0-to-5-volt jack. Solder the top terminal of R1 to the 5-volt jack. Ground the other terminal to common ground.

After you connect the perfboard and the transformer, solder in the meter circuit. Select the first five switch terminals of SW2. Solder each voltage jack to its respective terminal on SW2; start with the 0-to-5 volt jack and finish with the 15-volt jack. Connect the 15-kΩ resistor, furnished with the meter, in series with the positive terminal of the meter. Ground the negative terminal of M1. Doublecheck all connections and wires.

Voltage and resistance tests with the DMM

Measure the resistance at each voltage jack to common ground. The 5-volt jack should have a 110-ohms reading. 24-kΩ should be measured from the 10-volt jack to common ground, 10-kΩ from the 12-volt jack, and 6.7-kΩ from the 15-volt jack to the common banana-jack terminal. If the voltages measure many ohms lower than the values listed here, suspect a leaky regulator.

Now, push on the ac switch and N1 should light. Rotate the DMM to the 20-volt position and measure at each voltage jack. Often, the voltage is a fraction of a volt lower than what is numbered. For instance, the voltage at the 15-volt jack might only measure 14.97 (FIG. 5-51). Check the con-

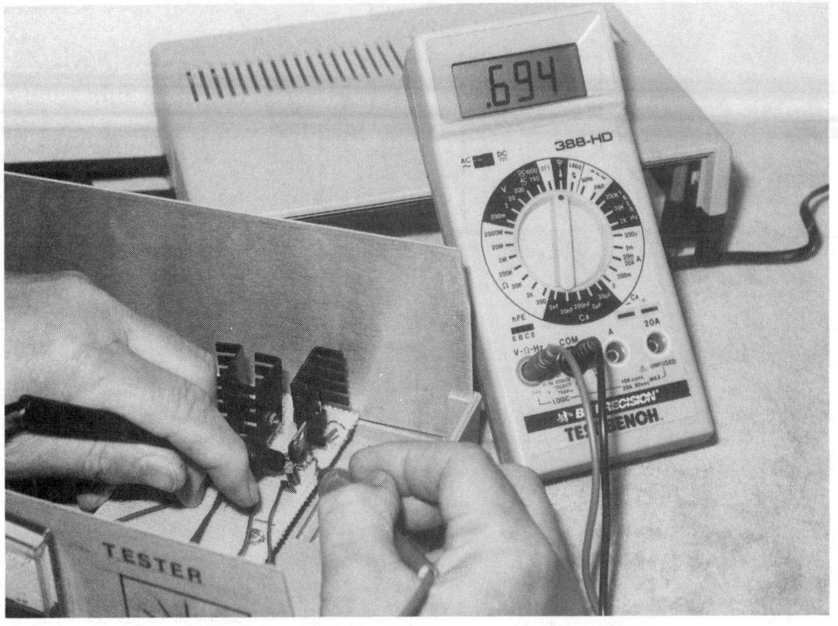

5-51 Check each voltage source with the DMM.

nected regulator IC when one or more of the voltages are not present. Notice if the IC regulator is connected backwards.

If you measure no voltage at any jack, check the voltage at the positive terminal of C1 (+26 volts). If no voltage is here, D1 might be connected wrong. Doublecheck the polarity at terminals C1 and D1. If N1 does not light, suspect that SW1 or wiring connections are defective. Sometimes new molded ac plugs have internal broken plug connections. If the transformer groans and the bulb dims after being switched on, suspect that the bridge rectifier is shorted or that either the bridge rectifier or capacitor C1 have reversed polarity. If polarity is reversed, C1 will quickly become hot and damage could result.

Checking cassette motors

The motor in a cassette tape deck with AM/FM/MPX can be operated from a regular voltage source or from a regulator circuit (FIG. 5-52). In this case, the regulator transistor, TR211, receives the 12.7-volt source and provides 9.6 volts, regulated. To quickly determine if the motor or the regulator circuit is defective, place the 10-volt external source across the motor terminals.

If the motor operates at full speed, suspect a defective regulator circuit. Replace the motor if you find voltage at the motor terminals or if an external voltage is applied and it will not rotate. Most tape deck motors operate from 9 to 15 volts. If the motor runs fast with the regulator circuit

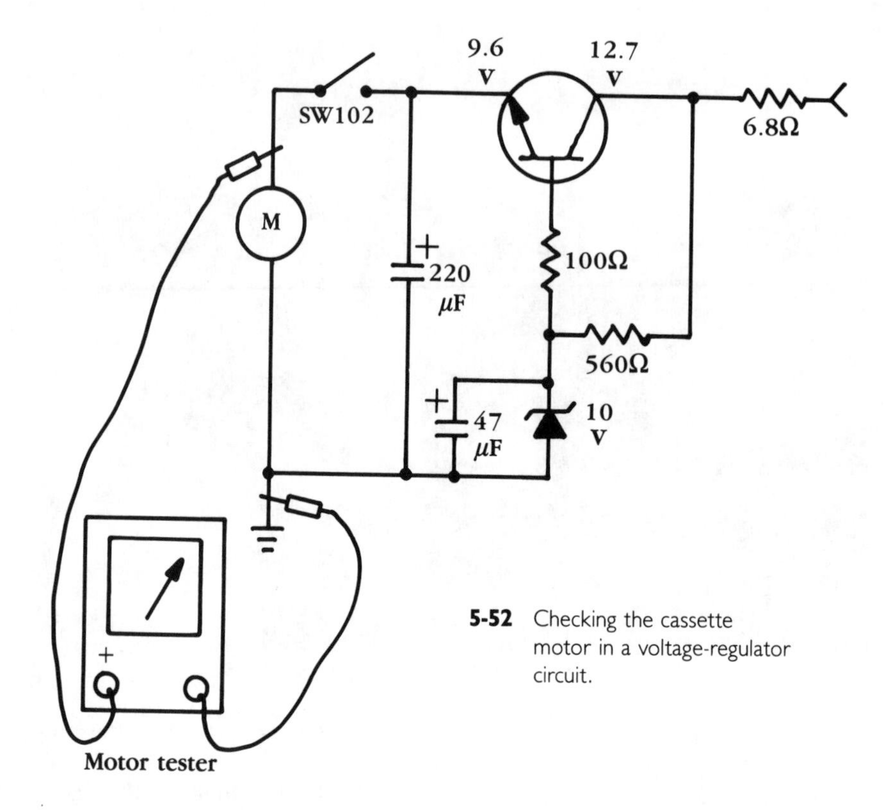

5-52 Checking the cassette motor in a voltage-regulator circuit.

and normal with the external voltage, check for a leaky transistor or zener diode.

Portable or battery-operated cassette players usually operate from 3 to 12 volts. Connect 3-to-5-volt motors to the 0-to-5-volt source for motor tests. Clip 10 or 12 volts to the portable cassette motor if the voltage is higher. Check the battery source to determine the voltage applied to the motor. This voltage is usually less than the total battery voltage. Sometimes these voltages are not listed on the motor shell or in the schematic.

Auto cassette motor

The auto cassette motor can operate from a 14.8 or 13.8 dc power battery source. Some auto cassette motors have regulator circuits, but many do not. Place the motor switch and isolation resistor in series with the positive motor lead (FIG. 5-53). Check the voltage that is applied across the suspected motor. If the voltage is normal and the motor will not rotate, suspect that the motor is defective. Apply 12 or 15 volts dc, from the external source, to determine if the motor will not rotate. Often, if higher voltages are applied to the motor, it might go into the intermittent mode.

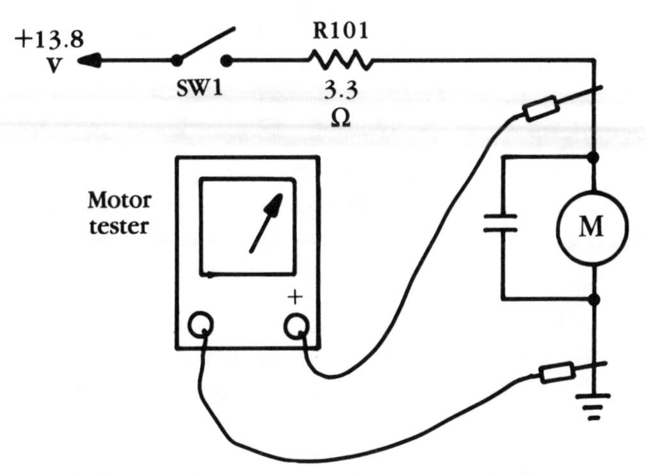

5-53 Checking the motor in the auto cassette player.

CD player loading motor

Most loading and capstan dc motors have an IC or transistor driver circuit. Connect the motor to the output terminals of the IC driver (FIG. 5-54). Often, the voltage supplied to the IC or transistor driver will indicate what voltage is applied to the motor. Check the CD loading motor voltage when you load or unload the disc.

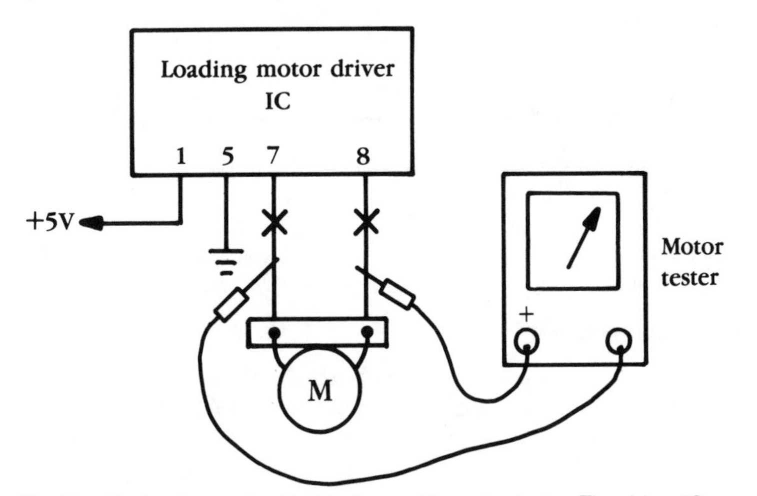

5-54 Checking the loading motor in CD player with motor tester. The driver IC was disconnected from the motor terminals.

Often, the motor voltage reverses when the disc pulls into the CD player. Check the disc tray gear assembly to see if motor is locked. Apply the correct 0-to-5 voltage source across the loading motor terminals. Remove the lead going to the IC driver. Now, reverse the polarity. If the motor reverses and runs normally with the external source, check the motor drive circuits.

Camcorder motors

Many camcorder motors operate at a fraction of a volt. Always, use the 0-to-5-volt source when trying to isolate the motor in the camcorder. It's best to remove the motor terminal wires. Some of these cables unplug from the motor. Practically all camcorder motors operate from a motor drive circuit (FIG. 5-55). Capstan camcorder motors operate with a forward rotation of +2.7 volts and in reverse with −2.4 volts. Check the continuity of the motor winding with the low scale on the ohmmeter. If there are no signs of continuity, clip the 0-to-5-volt source to the motor and slowly raise the voltage.

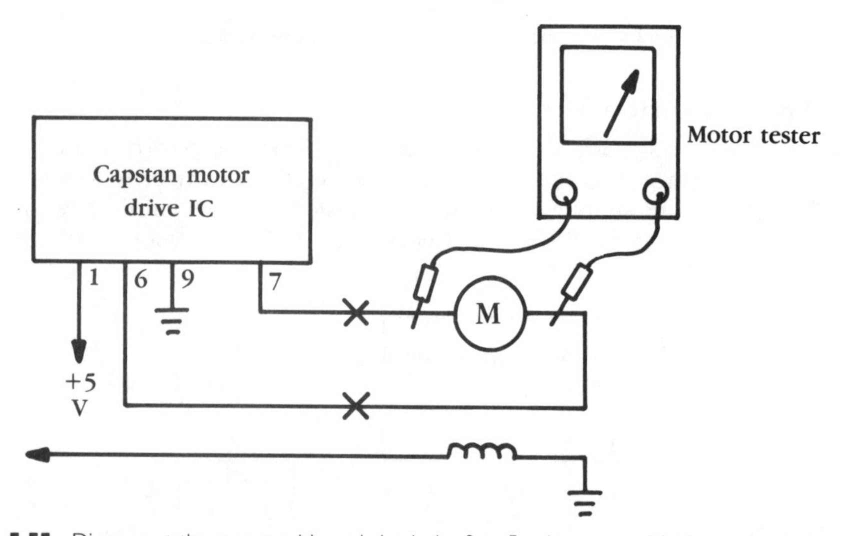

5-55 Disconnect the motor cable and check the 0-to-5-volt source with the motor tester.

Small dc motor tests

You can check toy and solar motors with the 0-to-5-volt dc source (FIG. 5-56). Slowly raise the voltage to see if motor will run. If the motor operates erratically with a lot of sparking, discard it. Do not apply more voltage than normal. Discard the motor if it has noisy bearings or intermittent operation.

Check the applied voltage at the motor circuit. Sometimes the voltage specifications are stamped on the motor. Check the continuity of the motor windings with the ohmmeter. If you measure a high resistance between the motor terminal and ground, the motor might be grounded. You can locate a defective motor with resistance, voltage, and external voltage tests. Apply an external voltage if the continuity is good and if the

5-56 Checking a small dc motor found in toy automobiles.

5-57 Checking the motor in the cassette player.

motor doesn't operate when voltage is applied to the motor terminals from the motor tester (FIG. 5-57). Always reverse the motor leads when you want to reverse the motor direction of small dc motors. The motor tester can also be used as a power supply for other projects.

R1 has metal-type clip ends. Obtain clips to go over all three square ends or solder the center brass terminal and use miniature alligator clips on the outside terminals. The two outside spade ends are made of a hard material that cannot be soldered.

Parts list

N1	120-Vac neon bulb.
T1	18-V 2-A power transformer, 273-1515 or equiv.
D1	2-A bridge rectifier.
C1	2200-μF 35-V electrolytic capacitor.
C2, C4, C6, C8	0.1-μF 100-V ceramic capacitor.
C3, C5, C7, C9	1-μF 35-V electrolytic capacitor.
R1	110-Ω 25-W rheostat, cat. #RHE-110, Available from: All Electronic Corp., P.O. Box 567, Van Nuys, CA 90408.
R2	15-kΩ, furnished with meter.
M1	15-V panel meter, 270-1754 or equiv.
SW1	Push on/off SPST switch, 275-1565 or equiv.
SW2	5-position single-pole rotary switch, 275-1385 or equiv.
SW3	SPST toggle switch.
Cabinet	7″-X-10″-X-4″ cabinet, CTB-2. Available from: Global Specialties, P.O. Box 1405 New Haven, CT 06505.
Jacks	6 banana jacks, 274-725 or equiv.
IC1	7815, 15-V regulator.
IC2	7812, 12-V regulator.
IC3	7808, 8-V regulator.
IC4	7805, 5-V regulator.
Z1	3.3-V 1-W zener diode.
Misc.	Line cord, perfboard, nuts, bolts, etc.

13.8-VOLT 5-AMP REGULATED POWER SUPPLY

This 13.8-volt 5-amp power source was designed to power today's auto radios. The power supply provides an adequate-to-normal dc voltage for servicing auto radios, mobile CB rigs, ham receivers, and other consumer electronic equipment (FIG. 5-58).

The small power source is enclosed in a two-piece metal cabinet with insulated output voltage terminals. The 13.8-voltage source is cabled to the auto radio with a 4-foot rubber cord and large insulated alligator clips to connect all of the input wires and connections. A 10-amp dc meter can be used to measure the operating current of the auto or CB radio. The push button ac switch is easy to operate with an ac neon indicator.

The circuit

N1 indicates that the power supply is operating when SW1 is switched on. The 12.6-volt power transformer supplies ac voltage to the full-wave bridge rectifier, D1. C1 smoothes the ac ripple in the dc voltage using IC1

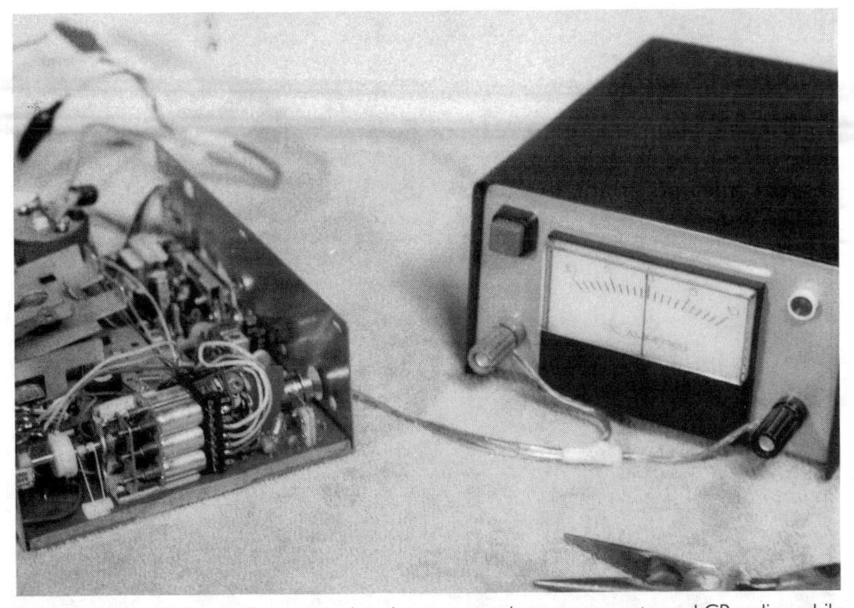

5-58 Use the 13.8-volt, 5-amp regulated power supply to power auto and CB radios while you service them on the bench.

as the 13.8-volt regulator (FIG. 5-59). IC1 resembles the usual horizontal output transistor. The case of the regulator is at ground potential.

The 5-amp 50-watt positive regulator provides 13.8 volts to the amp meter. The 13.8-volt regulator can be obtained from electronic suppliers of dealers as RCA SK 9342 or Sylvania ECG 934. A low-priced 0-to-10-amp meter was chosen from Hosefelt Electronics, Inc. to measure the current drawn from the power supply.

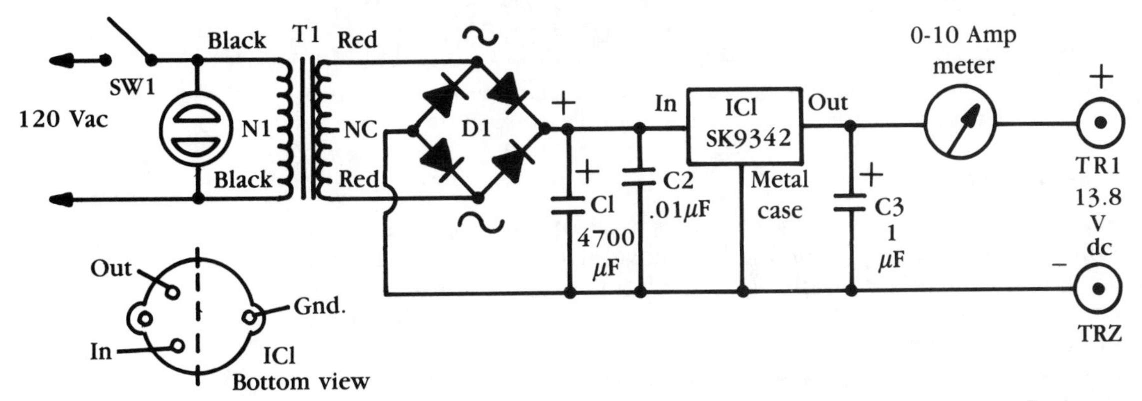

5-59 Build the full-wave bridge rectifier circuit around an SK 9342 or ECG 934 13.8-volt 50-watt regulator. The power source is powered by a 12.6-volt 3-amp power transformer.

Cabinet layout

A two-piece metal enclosure was chosen, with a vented steel top and an easy-to-work aluminum front, bottom, and rear panel, to house the regulated power source (FIG. 5-60). The top steel panel has a black crackle finish. The neon indicator, switch, meter, and terminal post will be mounted on the front panel and the rest of the parts will be mounted on the bottom panel, inside the cabinet. This project not only looks great, but you can't tell that it isn't a manufactured product.

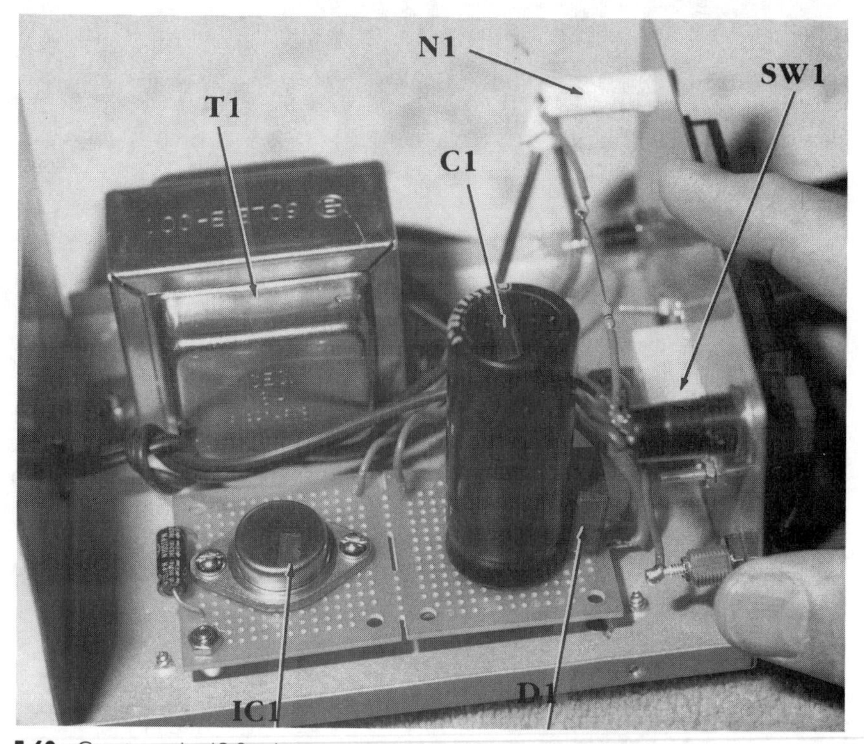

5-60 Construct the 13.8-volt power source inside a two-piece professional-looking cabinet. The bottom side and panels are made from aluminum for easy drilling.

Front panel layout

Since the bottom section of the cabinet is made of aluminum, it is easy to drill. Draw the hole layout on the front cover (FIG. 5-61). If you are using an electric hand drill, place the lip of the front cover over the edge of the bench for easy drilling. If you are using a drill press, lay a cloth down to protect the cabinet finish. A small paint brush is ideal to keep small chips and turnings off the drill press or bench and protect the cabinet finish.

Place the dc amp meter on the front panel. Draw the outline of the meter and the holes to be drilled with a pencil. Pencil marks are easy to

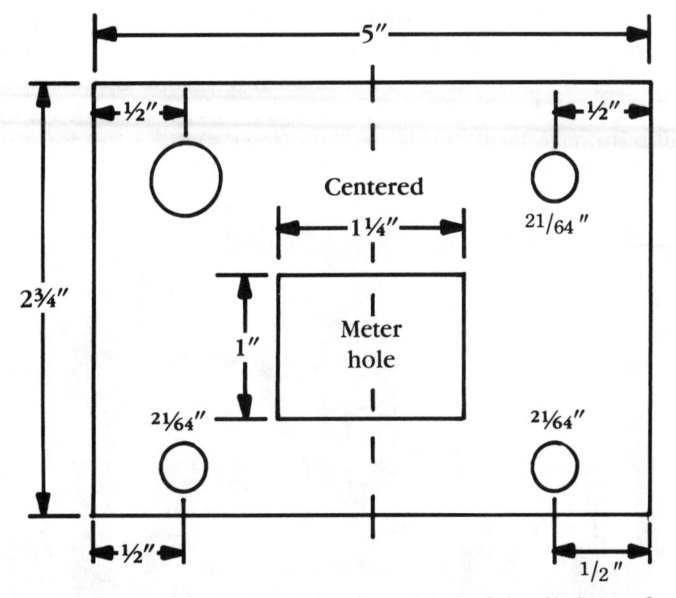

5-61 Layout the front panel with the meter dimensions and drawing layout. Start with a small drill and finish with an exact-size drill bit.

erase from this type of surface. The large meter hole can be cut out with a circle cutter or drill several small holes around the section to be removed. Enlarge each hole with a larger bit and break out the remaining piece of metal. File the opening to remove the rough burrs on the front and back panels.

Centerpunch each hole in the metal front panel before drilling. Start with a small 1/4-inch bit and enlarge the remaining hole with the bit shown on the front panel drawing. Be careful not to let the bit skate across the finished panel. Clean the aluminum ridge from around the holes with a pocket knife or rat-tailed file.

Bottom cabinet layout

Mount the power transformer to the back right side of the bottom chassis. Lay the transformer close to the outside edge and mark each mounting hole. Next, lay the dual general-purpose board beside, with a 1/2-inch clearance on each side. In this case, a regular perfboard can be used instead of a predrilled solder-ringed board. Drill all small holes with a 1/4-inch drill bit. Mount all electronic parts on the dual board, except those on the front panel and the power transformer.

General-purpose board layout

The dual general-purpose board was chosen because of its size and mounting area. The board has 2 sets of indexed solder-ringed holes (FIG.

5-62) to help keep the parts in place. The $1^3/_4$-x-$3^5/_8$-inch pc board can be purchased from Radio Shack. Mount the 3-amp 13.8-volt power regulator at one end. Mark the in and out terminals and the end-mounting holes on the pc board. Drill the indexed holes slightly larger than those for the in and out terminals. Drill two $^1/_4$-inch holes at each end to hold the heatsink and the regulator.

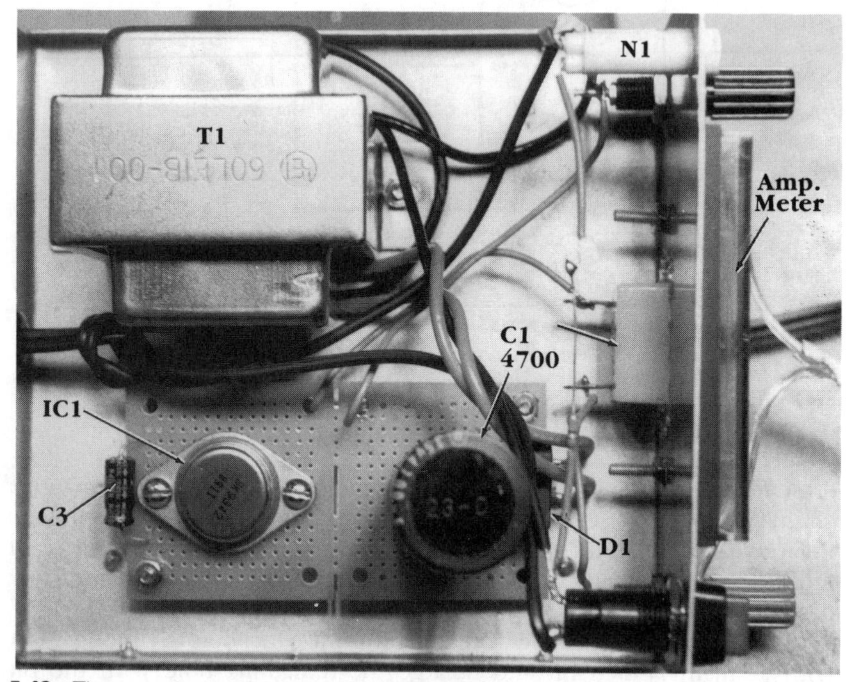

5-62 The general-purpose board has indexed solder-ringed holes for easy mounting. Holes for C1 and D1 might need to be enlarged with a small drill bit.

Since the power regulator will be mounted directly on the heatsink, no transistor socket is needed. Apply heatsink grease between the bottom of the regulator and the heatsink. Now, bolt the regulator to the general-purpose board with $^4/_{40}$ bolts and nuts.

You might need to enlarge the small indexed board holes for C1 and D1. Mount the 6-amp bridge rectifier on the opposite end from the regulator. Mount and solder C2 and C3 after you mount IC1 and C1. Mount four $^1/_4$-inch metal spacers between the board and the metal chassis to prevent shorting the connections in the power supply. Dab rubber cement on all parts of the front panel so they will not loosen.

Doublecheck

After soldering the connections, doublecheck the wiring with the schematic. Check the pc board at least twice for the correct polarity of D1,

C1, and C2. Make sure the in and out terminals of the IC1 regulator are correct; if reversed, you might damage it. The positive (+) terminal of the amp meter will be connected to the output of IC1. Recheck D1 to see that the ac terminals of the power transformer are connected correctly. This is where the fun begins.

Testing

It's always exciting when you finish an electronic project and test time rolls around. Connect a DMM or a VOM to the connection clips of the power supply. Rotate the meter to the 20-volt scale. Notice that the red lead is positive and the black lead is negative. Plug in the ac cord and turn on SW1. D1 should light with a voltage reading slightly above +14 volts, without a load attached to the power supply (FIG. 5-63).

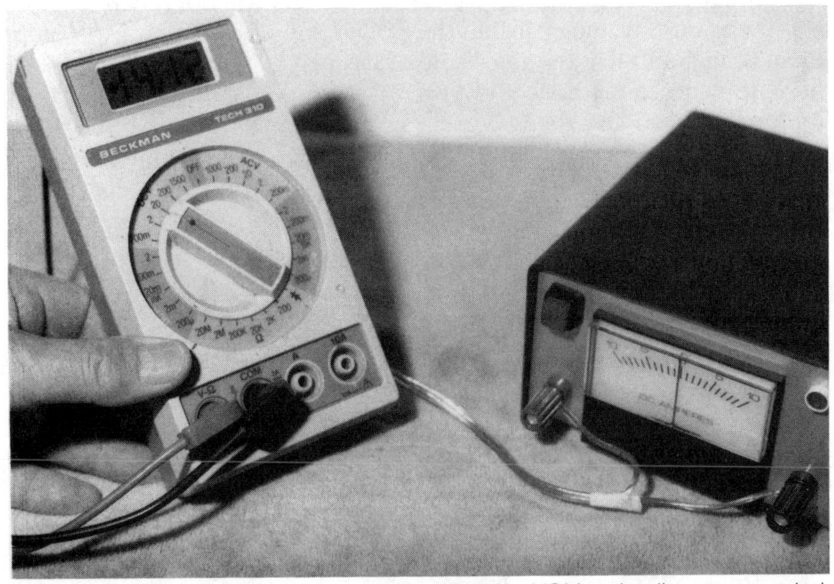

5-63 Check the power supply by connecting a DMM or VOM to the clip output terminals. The voltage should be slightly over +14 volts without any load.

If the power supply transformer groans with low or no dc voltage, the power supply is wired wrong or a component is defective. Recheck the wiring. Measure the ac voltage at the secondary terminals of D1. If it is low, suspect that D1 is leaky or is wired incorrectly. Remove one red lead and measure the ac voltage across the two red leads (12.6 volts ac).

If you find 19.9 volts dc at C1 and the in terminal of IC1, suspect a leaky IC regulator with low output voltage. Make sure that the dc output voltage from C1 is applied to the input terminal of IC1. Measure the dc voltage at the output of IC1. If normal voltage is found here, expect that a meter winding is open or that a terminal connection is poorly soldered. Check the ammeter by connecting a 10-ohm 10-watt resistor momentarily across the alligator clips The meter should read near 3 amps.

Auto connections

Most American-made auto radios have a negative ground and European radios have a positive ground. Some larger trucks have radios with positive grounds. Check the car radio before you connect it to the power supply terminals. Some of these radios have the polarity stamped on the metal chassis.

Connect the power supply to the car radio with the positive lead attached to ignition or to the battery lug. Clip the negative or ground lead to the chassis or ground lug. With the latest digital/quartz clock auto receivers, both the ignition and clock lugs must be connected to the positive terminal of the power supply or the receivers will not function (FIG. 5-64). If the current meter inside the power supply reads above 5 amps, there is no doubt that the auto radio or CD player is shorted. Recheck the hookup wires to the radio and power source. Measure the dc voltage at the positive lead and the radio chassis ground. Do not leave the power

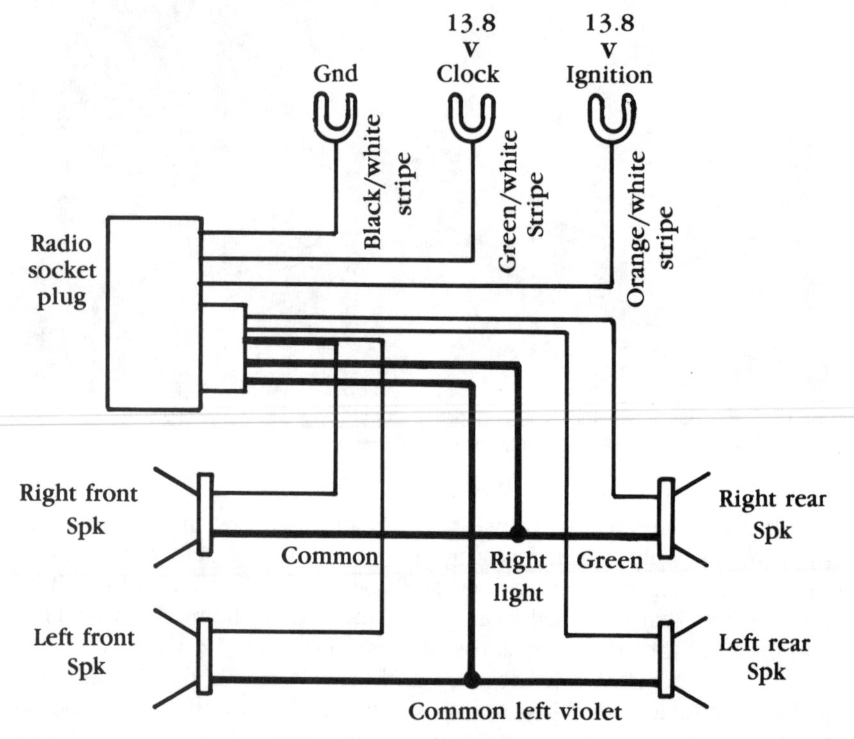

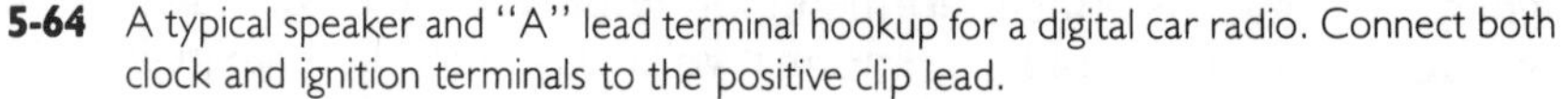
5-64 A typical speaker and "A" lead terminal hookup for a digital car radio. Connect both clock and ignition terminals to the positive clip lead.

Parts list

SW1	SPST push-button switch, 275-1565 or equiv.
N1	120-V neon pilot light, 272-712 or equiv.
T1	12.6-V CT 3-A power transformer, 273-1511A or equiv.
D1	6-A silicon bridge rectifier.
IC1	13.8-V 5-A regulator, RCA SK9342 or Sylvania ECG934 or equiv.
C1	4700-μF 13-V electrolytic capacitor.
C2	0.01-μF 400-V ceramic capacitor.
C3	1-μF 35-V electrolytic capacitor.
Meter	0-to-10-A dc meter.
TR1, TR2	Red and black terminal binding posts, 276-149 or equiv.
Cabinet	Two-piece cabinet, 3″ × 5¼″ × 5⁷⁄₁₆″, 270-253 or equiv.
Heatsink	TO-3 case heatsink, 276-1371 or equiv.
Misc.	4/40 mounting bolts, nuts, two ½-inch insulator spacers, ac cord, rubber grommet, solder, etc.

supply connected if pulling excessive current. Check the radio for a leaky output transistor or IC.

UNIVERSAL 3-AMP ADJUSTABLE POWER SUPPLY

This 3-amp adjustable power supply has a voltage source between +1.2 and 33 volts. The universal power supply output is monitored with a dc voltage and a separate current meter (FIG. 5-65). The circuit uses a 2-amp fuse and a 3-amp output circuit breaker.

The power source is housed in a 2¾-×-6¼-×-9½-inch Global Specialities plastic cabinet. This commercial cabinet has top and bottom plastic half shells that are fitted with polished aluminum front and rear panels, mounting hardware, and pc board bosses (FIG. 5-66). All controls are mounted on the front panel, except the 3-amp fuseholder.

The circuit

SW1 and the 2-amp fuseholder are found in each leg on the primary winding of T1. The secondary, 25.2 volts, is fed to a 4-amp full-wave bridge rectifier (FIG. 5-67). Single 3-amp silicon diodes can be used here, if desired, and C1 filters the dc input voltage to the voltage regulator, Q1.

Input terminal 2 of Q1 should measure 39 volts dc. The regulated output voltage is found at the metal case of Q1. Terminal 1 is tied to the adjustable voltage, R2. A 0-to-3 amp dc ammeter operates in series with the voltage output and a 3-amp circuit breaker. The adjustable output voltage is measured with a 0-to-50-volt dc voltmeter. R3 provides meter calibration for converting a 0-to-1 milliampere meter to a voltmeter. R3

5-65 The adjustable 3-amp power source in service. The output dc voltage varies from 1.2 to 37 volts.

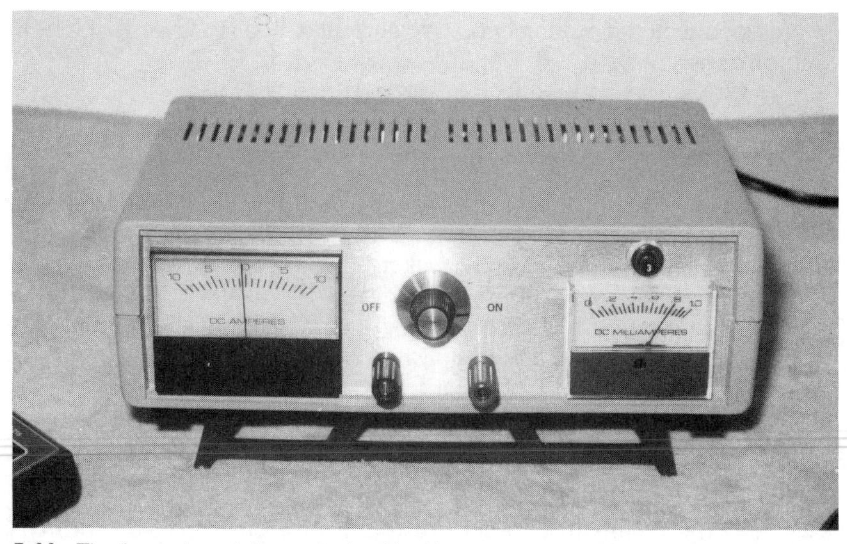

5-66 The front view of the professional-looking tester constructed in a Globe Specialists plastic cabinet.

should be eliminated from the circuit if a regular 0-to-50-volt meter is used.

In this circuit, the power supply transformer and bridge rectifier are protected by the 2-amp fuse, if D1 or C1 becomes leaky. The output of the dc power supply is protected from an overloading component outside the power source with the 3-amp circuitbreaker. The dc voltmeter should

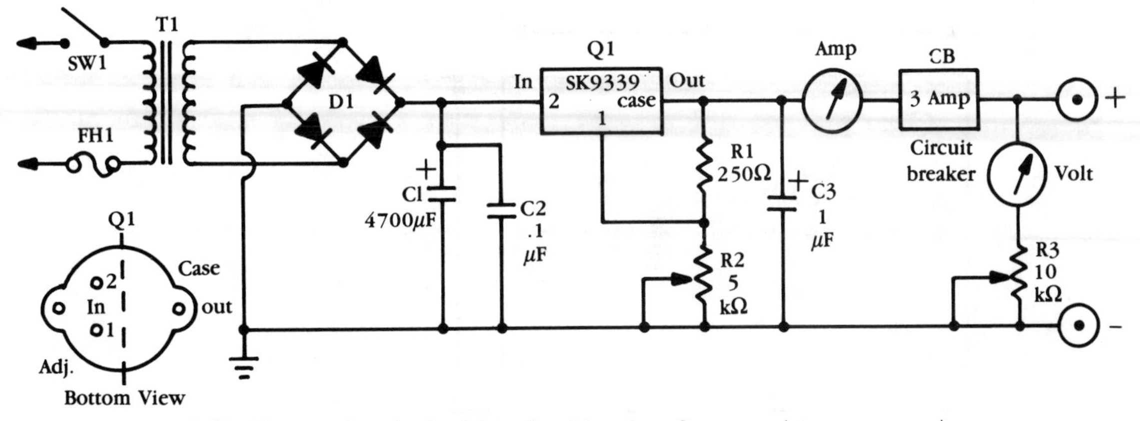

5-67 The complete circuit of the adjustable-voltage 3-amp regulator power supply.

be connected after the circuitbreaker so that when the breaker is open, no voltage is indicated. The dc voltmeter also serves as an indicator when the power supply is operating.

Regulation

Good voltage regulation is built around the adjustable voltage regulator, Q1. An RCA positive SK 9339 voltage regulator was used, although it can be replaced with a 3-amp Sylvania ECG970 or a LM350K regulator. Q1 has a TO-3 mount and it should be placed on a TO-3 heatsink. These regulator transistors can be obtained through electronic wholesalers, retail distributors, or your local radio and TV suppliers.

Dc meters

It's best to obtain a 0-to-3-amp or a 5-amp dc meter for measuring the current and a 0-to-50 or a 0-to-30 volt dc meter. The 0-to-3-amp dc and 0-to-50-volt dc panel meters can be purchased by mail order from Circuit Specialists. Check your local distributor and Radio Shack for meter requirements. The meters and the commercial cabinet are the most expensive components of this project.

Of course, a 0-to-1 milliampere meter can be picked up a lot cheaper from local electronic distributors or surplus stores. If you use a 0-to-1 milliampere meter instead of a 0-to-50-volt meter, it must have voltage calibration (R3) and voltage markings added to the meter dial.

Chassis construction

Cut a 3-×-4-inch piece of perfboard for the chassis (FIG. 5-68). Place the perfboard over the center pc board plastic bosses to drill the mounting holes. Drill three 5/32-inch mounting holes in the perfboard to secure the chassis to the bottom plastic cabinet supports. Drill two 8/32-inch holes to

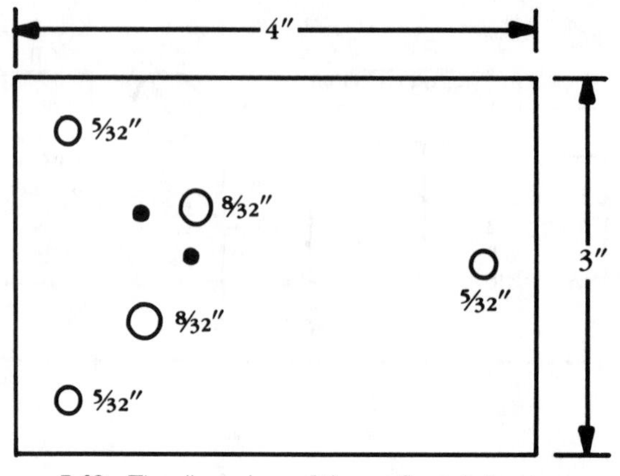

5-68 The dimensions of the perfboard chassis.

mount the TO-3 heatsink and power regulator Q1. Fit all other components through the small perfboard holes.

Now, mount the power regulator and the heatsink. Place washers at the bottom side of the perfboard. Mount the large filter capacitor, C1, next to the heatsink. Bend over the large capacitor terminal leads. Place the bridge rectifier opposite from C1. The rest of the small components can be mounted as you wire them into the circuit.

Preparing the front panel

Use extreme caution when drilling and mounting parts on the front panel to prevent scratching or marring the aluminum surface. Keep the meters inside the top and bottom lip of the plastic cabinet. Lay out the front panel, as shown in FIG. 5-69. Drill or knibble out the back side for mount-

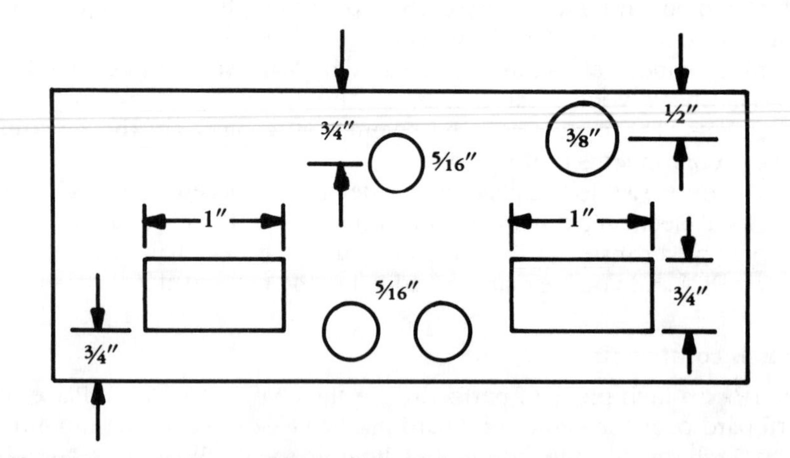

5-69 The front panel layout with hole sizes and dimensions. Be careful not to mar the aluminum finish.

ing the two dc meters. You might want to use a small 1/8-inch bit to drill holes around the outside of what will be the meter hole. After you drill the holes, break out the side and top holes. Then bend the cut piece out of the front metal panel. Drill two 5/32-inch holes on each side of the meter for securing them to the front panel.

Center the voltage adjust hole (R2) and the voltage output terminal posts. Drill a 5/16-inch hole for each test probe jack and for the variable resistor, R2. Next, drill a 3/8-inch hole above the voltmeter for the 3-amp breaker. From the back side, drill a small lock hole to keep the R2 control base from turning. Mount all front-panel components after you drill all of the holes. To keep the drill bit from skating across the aluminum finish, centerpunch all of the holes to be drilled, except for the meter holes.

Back panel layout

The ac fuse holder and power cord can be mounted on the back metal panel. If there isn't enough room to mount the 3-amp circuit breaker on the front panel, mount it beside the fuse holder. Center both holes over each other (FIG. 5-70). Drill a 17/32-inch hole for the 3-amp fuse holder. Drill a 11/32-inch hole for the rubber grommet that protects the ac cord. Tie a knot in the ac cord so that it cannot be pulled out of the back panel.

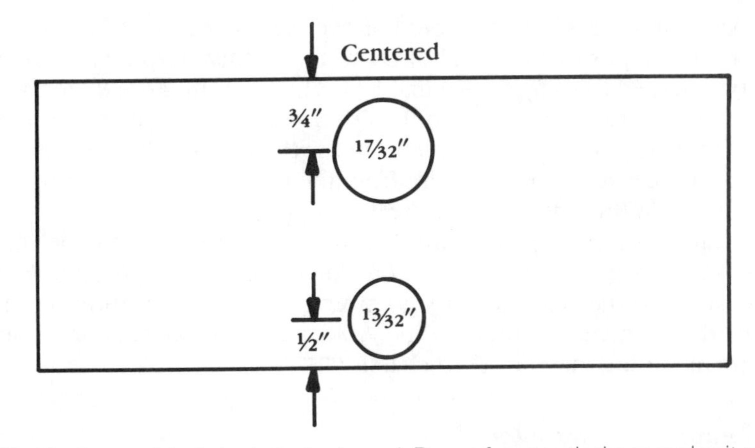

5-70 The layout of the holes in the back panel. Do not forget to tie the ac cord so it cannot be easily pulled out.

Wiring

Make all component leads as short and direct as possible. Wire one side of the ac cord to the switch and other side to the fuse holder. Solder the primary winding of T1 to one side of SW1 and the fuse holder. Notice that the ac on/off switch and the 2-amp fuse are in series with the power line and T1 (FIG. 5-71).

Twist the two ac terminals of the bridge rectifier out at the edge of

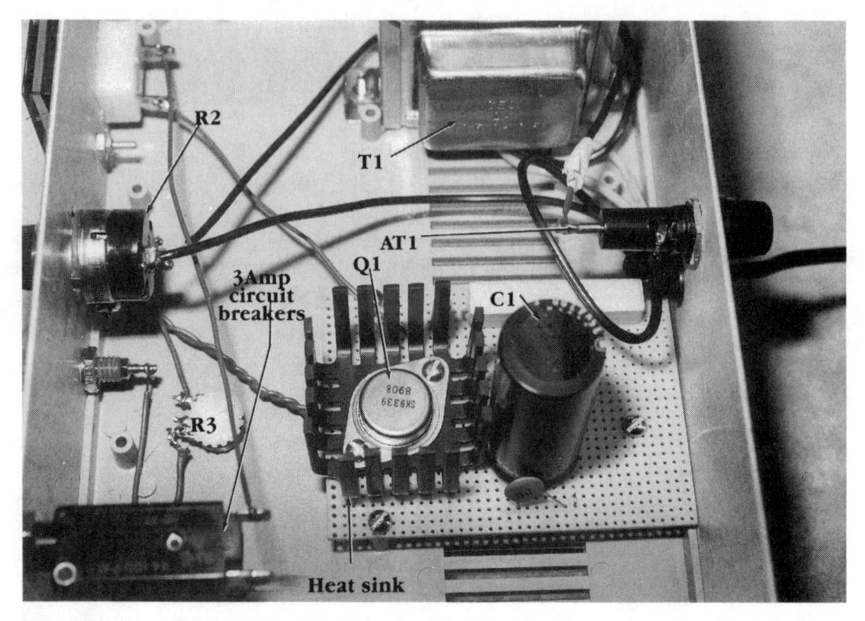

5-71 Close-up of the parts mounted on the perfboard.

the perfboard to easily connect to the two yellow secondary transformer wires. Tie the positive terminal of D1 to the positive terminal of C1. Solder two twisted leads from terminal 1 of Q1 and ground the R2 terminals. Run a red hookup wire between the Q1 case and the positive terminal of the amp meter. Wire the amp meter in series between the regulator and the circuit breakers. Solder a wire from the negative amp meter terminal to one side of the 3-amp circuit breaker.

Connect all front panel wires before you mount it on the perfboard chassis. Use a bare #22 hookup wire for the common ground. Always make sure that the meter polarity is correct. Connect the positive terminal posts of both meters to the B+ line. Solder the yellow transformer leads before mounting the perfboard chassis (FIG. 5-72).

Calibrating the voltmeter

If you use a 0-to-1-milliampere meter instead of a 0-to-50-volt dc meter, place R3 in series with the ground lead. Calibrate the meter at various voltage measurements (FIG. 5-73). You must connect the positive meter terminal to the positive adjustable output voltage. R3 can be a small ceramic or wafer-type preset control.

Rotate R3 open so that the entire resistance is in series with the meter. Connect a DMM or VOM across the power supply output terminals. Adjust voltage control R2 to total 36 volts on the DMM. Readjust R3 so that the meter hand reaches the end of the scale, then mark the end at 36 volts. Likewise, mark 25, 20, 15, 12, 9, and 3 volts on the meter when you

5-72 Complete layout and view the components inside the cabinet. Notice that T1 is bolted directly to the bottom cabinet.

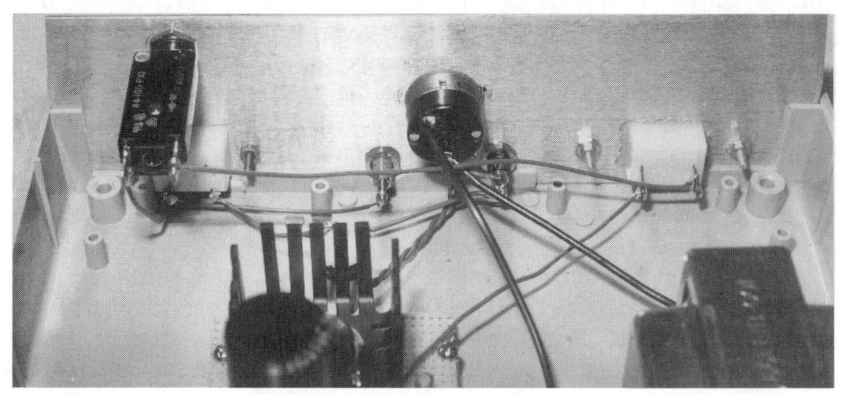

5-73 A back view of the front panel that shows the meter slots and part locations.

rotate R2 to these voltages on the DMM. Simply mark the voltages on the meter dial. Now, redraw the meter dial, place the numbers, and process the new meter dial, as described in chapter 7. Cement the new voltage dial to the voltmeter and replace the plastic meter lid.

Testing

After applying the foot mounts and the flip-up leg assembly, the adjustable power supply is ready to be tested under load (FIG. 5-74). If the power

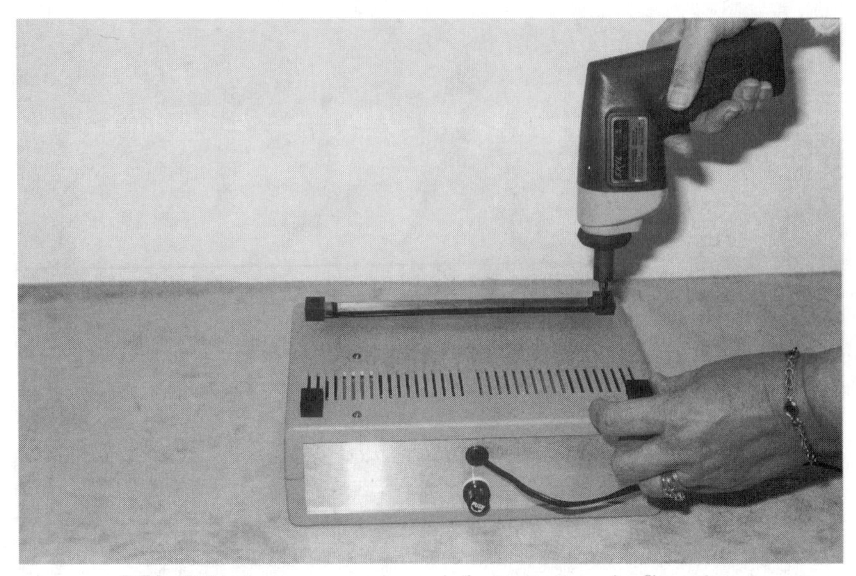

5-74 Finish the power supply: apply foot mounts and a flip-up leg.

supply groans when turned on, suspect a defective component or improper wiring. Doublecheck all wiring with the power unplugged. A leaky bridge diode or filter capacitor can overload the power transformer. Check the 2-amp fuse for open conditions.

Take resistance and voltage measurements within the power supply. Quickly measure the resistance across C1. The DMM ohmmeter numbers will charge up and slowly discharge down with a normal capacitor. If you note a measurement under 1 kΩ, check each diode within the bridge rectifier and check C1 and C2 for leakage. The voltage at pin 2 of Q1 should be near 38.8 volts.

If the voltage is normal at pin 2 and it is very low at the case output terminal, suspect an overload in the output circuit or a defective Q1. Inspect the wiring. Check the total resistance across the case of Q1 and the ground, with R2 wide open. The resistance should be around 5 kΩ. If you find an overload connected to the output, the amp meter will show a high reading or the circuitbreaker will kick out. Monitor the output voltage with an external DMM or voltmeter connected to the output terminals (FIG. 5-75).

Various circuit tests

This adjustable power source can provide dc power to any radio, receiver, experiments, or any device that does not require voltages over 36 volts and 3 amps of current. The various sections of a TV can be serviced with this supply such as horizontal oscillator circuits, with external dc voltage applied in a high-voltage shutdown chassis. The vertical and sound cir-

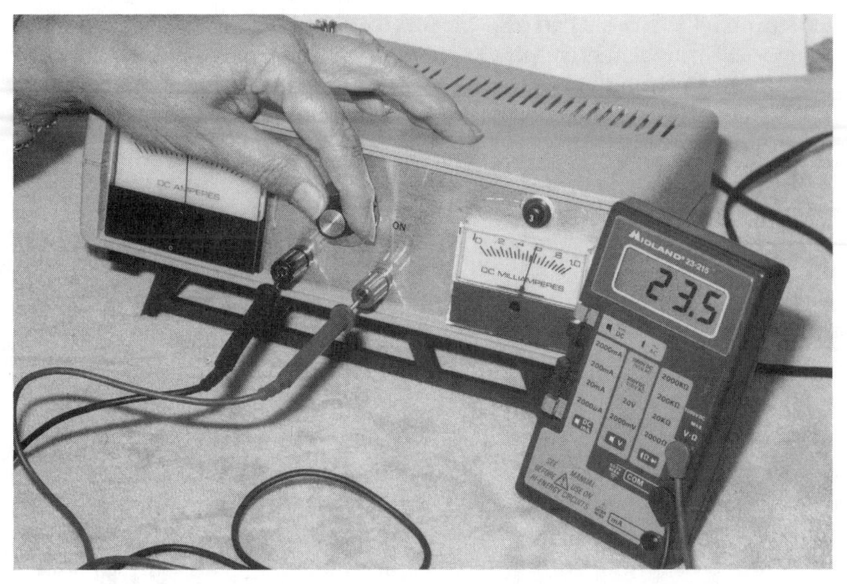

5-75 Check and compare the output voltage with an external DMM.

Parts list

T1	25.2-V 2-A power transformer, 273-1512 or equiv.
D1	4-A silicon bridge rectifier.
C-1	4700-μF 35-V electrolytic capacitor.
C2	0.1-μF 50-V ceramic capacitor.
C3	1-μF 50-V electrolytic capacitor.
Q1	RCA SK 9339 3-A voltage regulator, or Sylvania ECG 970 or equiv.
Post	Red and black insulated banana post terminals.
FH	3-A fuseholder and 2-A fuse.
CB1	3-A circuit breaker, 3 CB Hosefelt or equiv.
Amp meter	0-to-3-A dc 20-1118 panel meter, Circuit Specialists or equiv.
Voltmeter	0-to-50-Vdc panel meter, 20-1124 or equiv.
R1	250-Ω 5-W resistor.
R2	5-kΩ linear control with switch.
R3	50-kΩ ceramic trimmer variable resistor (use only with 0-to-1-mA meter).
SW1	SPST on rear of R2.
Cabinet	2.75″-×-6.25″-×-9.5″ cabinet from Global Specialities, part number P/N 110-0006 or equiv.
Misc.	ac cord, hookup wire, solder, 10-3 heatsink, etc.

cuits in the TV chassis can also be serviced with the external dc voltage when you find a defective power supply. You can easily troubleshoot the various stereo sound amp circuits by providing an external voltage from the 3-amp power source. This heavy-duty adjustable power supply is ideal for the technicians service bench.

TV TROUBLESHOOTER

Although this group of power supplies is intended for the electronic technician working on a TV chassis, it can be used to service high-powered amps or PA systems. The TV power supplies are identical with power supply 1 and 3, a positive variable source, while power supply 2 is a negative variable regulated power source. All of these supplies can operate many dc circuits in the electronic field.

When high-voltage shutdown or chassis shutdown occurs in the TV chassis, the low-voltage power supply and horizontal circuits must function before any other circuits can operate. The TV power supply can be subbed for the low-voltage power supply and derive secondary voltage from the flyback transformer (FIG. 5-76). The supply voltages separately can power the horizontal and vertical deflection circuits with high- or low-voltage shutdown problems.

5-76 Front view of the completed tri-polarity variable-voltage power supplies.

Isolation transformer

Always use a variable isolation transformer before attaching test instruments to the TV ac/dc chassis. If the TV set is power-transformer operated and test instruments operate from a power-transformer source, you will not encounter trouble attaching the test instrument to the TV chassis. Since most new TVs are of the ac/dc variety, the isolation transformer must be used.

Simply plug the ac cord of the TV chassis into the isolation transformer. Now, when you apply testers or meters to the metal or common-board chassis, the furs will not blow and other low-voltage regulation circuits will not be damaged. Besides the trouble already in the TV chassis, you can cause additional problems by simply attaching the ac-operated test instrument to the chassis. Play it safe and plug the TV chassis into a variable isolation transformer before attaching any outside test instruments.

Power supply 1 circuits

The low-voltage power supply employs a positive variable voltage regulator, IC1, to acquire a +1.2-to-37-volt source (FIG. 5-77). Connect SW1 to the primary winding of T1, along with N1. The bridge rectifier, D1, has 25.2 volts ac applied to the ac terminals. The dc output voltage is around +37 volts, as applied to C1 and IC1.

The variable positive regulator is a TO-220-type with a separate heat-sink. R2 varies the output voltage. Two banana jacks in the positive output source monitor the current with the external DMM or VOM. Simply pull out the male plugs and insert the meter terminals. Connect the meter switch, SW3, to terminals A and B of power supply 1.

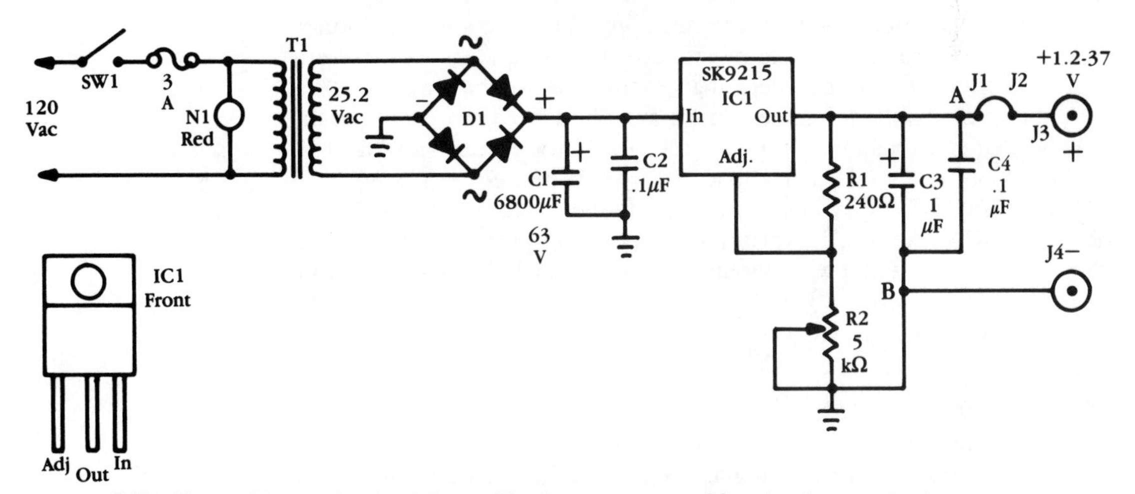

5-77 The positive-regulated +1.2-to-+37-volt power source. R2 varies the output voltage source.

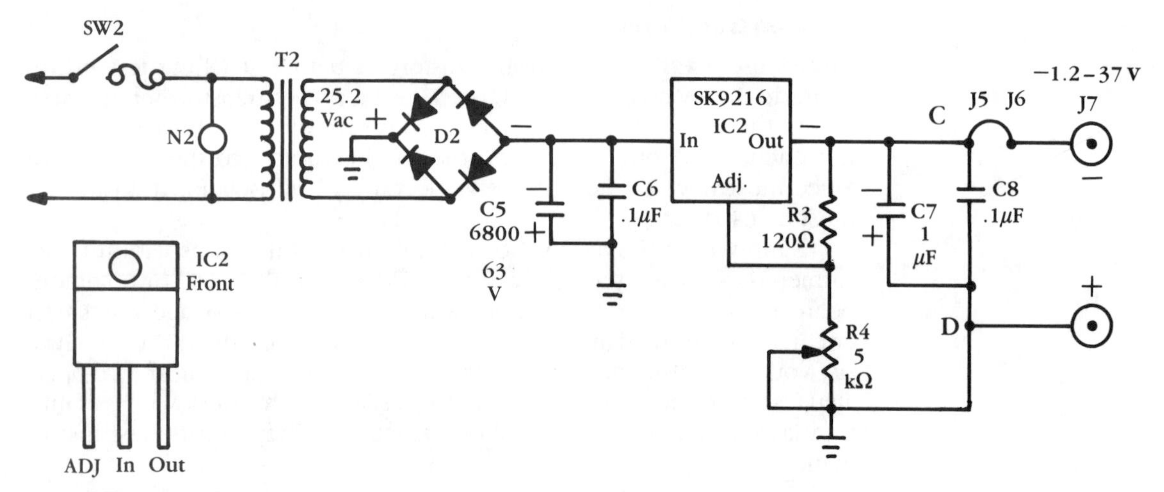

5-78 The negative-regulated power supply 2 (− 1.2 to − 37 V). R4 varies the negative output voltage.

Power supply 2 circuits

Power supply 2 is almost identical to number 1, except it has a negative 1.2-to-37-volt source. SW2 turns on the ac voltage applied to T2 (FIG. 5-78). N2 is a green indicator that shows if the power supply is turned on or off. T2 is identical to T1 and it connects to the bridge rectifier, D2. D2 connects backwards compared with D1. Connect the positive output terminal of D2 to ground and negative applies to IC2.

Tie C5's negative terminal to the negative dc source and to IC2. IC2 is a negative variable voltage regulator with a negative 1.2-to-37-volt output source. R4 varies the output voltage. Connect all electrolytic capacitors and negative-polarity diodes to the negative adjustable source. Connect two sets of banana jacks to the negative voltage source for external current measurements. Connect terminals C and D to the meter circuits.

The 0-to-50-volt dc meter can be switched between each variable voltage source with a DPDT-type switch, SW3. You can use a separate 0-to-50-volt dc meter for each voltage source, if desired (FIG. 5-79). Each voltage source will be switched into the dc meter circuit to monitor the dc voltage when serving the various circuits. Either positive power supply can be monitored with the meter and SW4. The small dc meter can monitor both voltage sources and it fills less front-panel space than if two meters were used.

Power supply 3

This power supply is identical to power supply 1, except that it uses a different power transformer and voltage regulator. T3 has a 19-volt secondary, running about 29 volts to the regulator (FIG. 5-80). IC1 is an LM 317T-type that can be purchased most anywhere. R1 (240 ohms) and R2

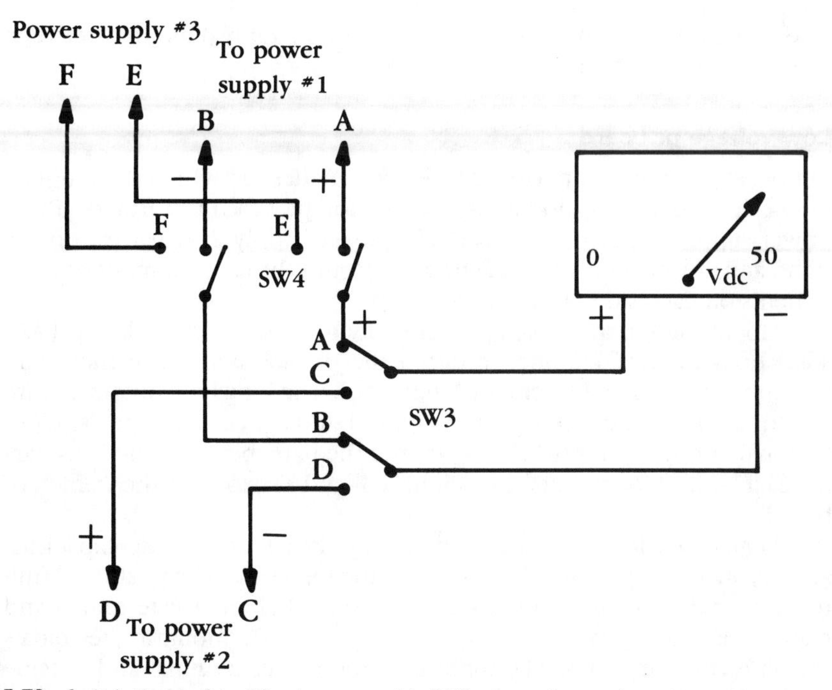

5-79 Switch the dc 0-to-50-volt meter with SW3 to monitor each positive and negative voltage source separately. Power supply 1 connects to A/B, power supply 2 to terminals C/D, and power supply 3 to E/F.

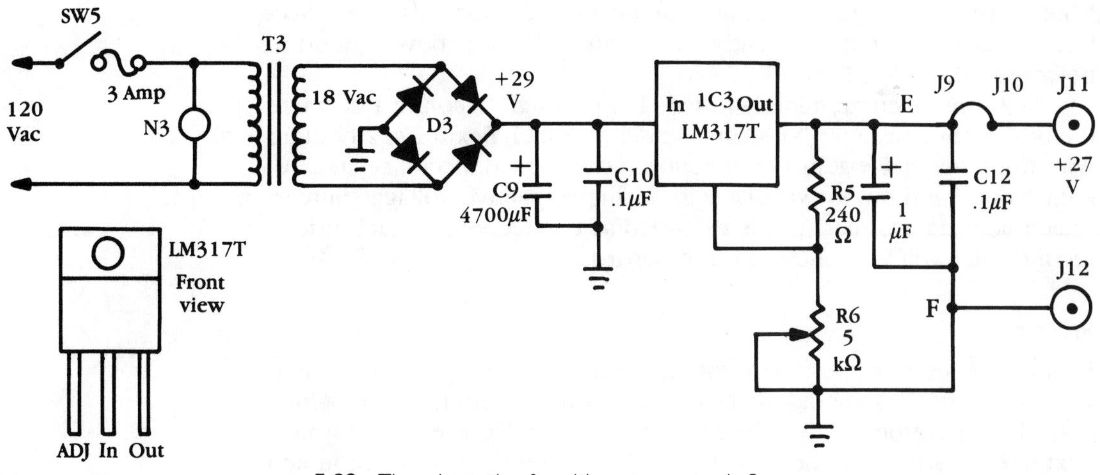

5-80 The schematic of positive power supply 3.

(5 kΩ) are the same as in power supply 1. Power supply 3 mounts between 1 and 2 on the perfboard.

Low-voltage pc board

Cut or select a piece of perfboard, 2¾ × 6 inches, on which to mount all of the low-voltage components, except for power transformers. Place power supply 1 (positive) on the left, power supply 2 (negative) on the right, and power supply 3 in the middle. Mount the exact number of components on each side of the board.

Mount the 2-amp bridge rectifier ½ inch in from each end. Run a #22 solid hookup wire 1½ inches down the outside and a 2½-inch bare hookup wire toward the center of the board. The long bare buss wires are the common ground (−) of power supply 1 and the common positive (+) terminal of power supply 2. Now, mount the parts between the buss bars on each side. Mount filter capacitors C1 and C5 toward the center of board.

Remember, in the negative power supply, all electrolytic capacitors and negative diode terminals connect to the high side. The positive terminals are on the common (+) buss wire. Look at IC2 and the terminals and make sure that you mount them correctly. The IC2 mounting terminals are different than IC1's. The input is at the center terminal and outside terminals are output and adjust. So you will need to mount IC2 differently than IC1.

Within the IC1 power supply, the negative terminal of D1 is the negative ground. The outside terminal of IC1 goes to the "in" connection of the positive terminal on D1 and C1. The center terminal of IC1 is the "out" terminal that goes to R3 and the positive terminal of C3. All electrolytic capacitors and the positive terminals of D1 are above ground in power supply 1 (FIG. 5-81).

The component terminal wires are long enough to connect between the positive and negative buss bars. Solder an 8-inch flexible #22 hookup wire to each variable-adjust connection. Likewise, connect a separate 6-inch ground and output voltage wire to the respective voltage sources. Attach both transformers to the extended ac connections of each bridge rectifier after you wire and mount the board.

Wiring

Drill four holes in the perfboard for mounting on the floor of the cabinet. Use ½-inch spacers to hold the chassis up from the bottom. Drill ⅛-inch holes in the bottom of the cabinet to mount the 3 power transformers next to each bridge rectifier, D1, D2, and D3. After all parts are mounted on the front cover, start wiring the tester.

Bring the ac power line through a grommet hole in the rear chassis. Tie the two ac wires into four-lug terminal strips. Run one of the ac wires from the strip to one side of the on/off switch. Solder a piece of hookup

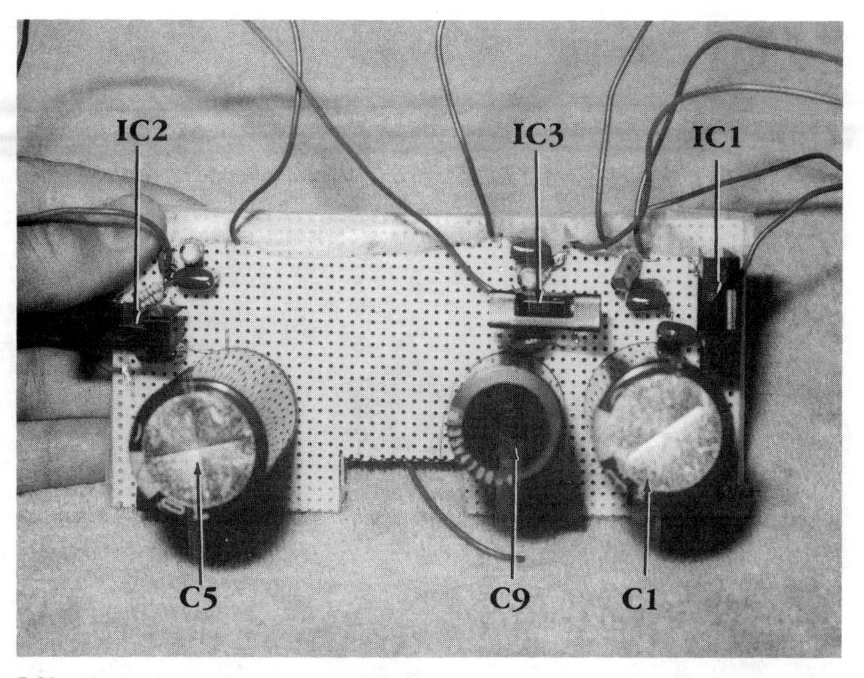

5-81 Mount all parts for power supply 1 on the left and those for negative power supply 2 on the right. Use solid #22 bare hookup wire for common buss connections.

wire from the other side of the switch to one primary terminal of the transformer. Connect the other primary lead (black) to the remaining side of the power line. Wire and solder both switches in the same manner (FIG. 5-82). Solder the secondary winding (25.2 volts) to both bridge circuit terminals and solder both bridge rectifiers in the same manner.

Run the adjustable terminal of IC1 and IC2 to the respective 5 kΩ linear controls, F2 and R4. Ground the center and other side of the control to the common ground of respective circuits. Connect the positive output lead (at A) to J1. Solder J2 to the positive output terminal. Likewise, solder the output negative wire of power supply 2 (terminal C) to J5. Connect J6 to voltage output terminal J7. Now, tie the common ground jack of power supply 1 to the negative banana jack, J4. Do the same with J8 of the negative supply to the common (+) ground and power supply 3. Doublecheck all wiring connections on both power supplies for the correct polarity and common ground.

Cabinet layout

Mount the 0-to-50-volt panel meter in the center of the front panel. Keep the meter as high as possible. Center the 3/8-inch holes for each variable voltage control between the meter and the front edge of the cabinet. Drill two 9/32-inch holes for the dial indicators (N1 and N2) above the variable

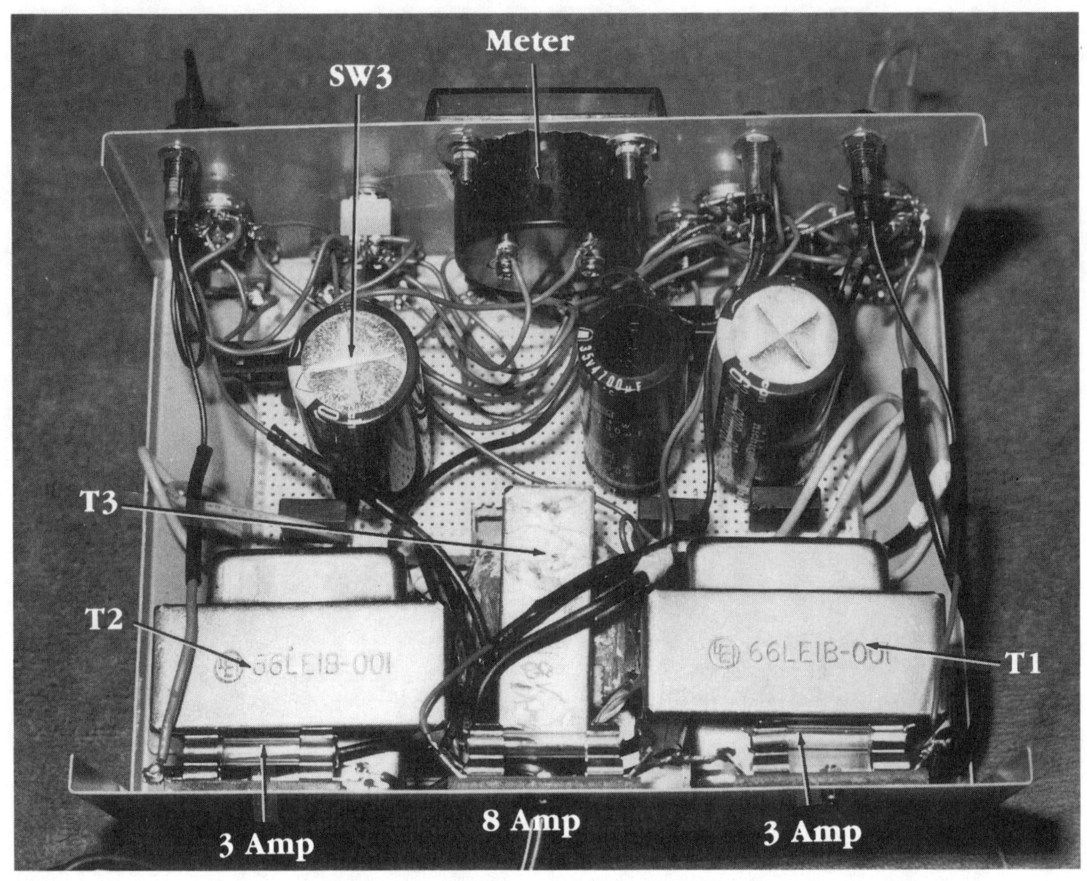

5-82 The top view of the wired components.

controls. Drill two 3/8-inch holes, for SW1 and SW2, between the variable controls and the meter mounting flange (FIG. 5-83).

Line up the bottom banana holes, 1/2 inch from the bottom of the front panel. Drill all eight banana jack holes 1/4 inch in diameter. Keep the jacks close under each power supply section. Drill 3/8-inch holes for SW3, SW4, and SW5, then mount the components.

Testing

Go over each connection once more before flipping on the switch. Connect a DMM or VOM to the positive output terminals. First, check power supply 1. Flip SW3 to the positive power supply. When SW1 is turned on, notice if the red indicator light is on. Now, rotate R2 from +1.2 to +37 volts. The panel meter and the DMM should be quite close with voltage readings. Rotate R2 down and notice how the voltage meter tracks.

Now, go to power supply 2. Switch on SW2 and N2 (green) should light. Flip SW3 to the negative power supply. Rotate R4 from 0 to −37

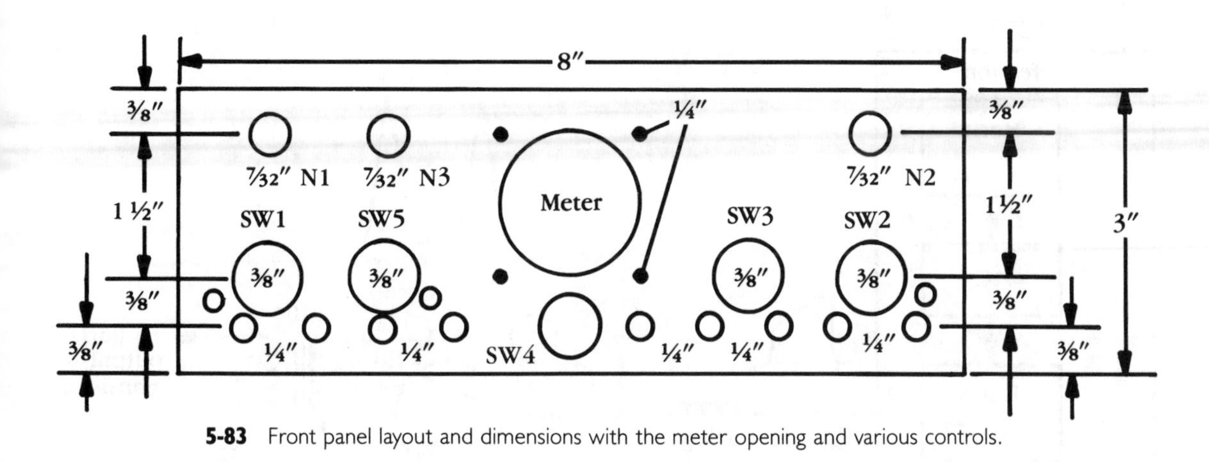

5-83 Front panel layout and dimensions with the meter opening and various controls.

volts. Doublecheck the voltage at the output terminals with the DMM or VOM.

If you find no voltage at the panel meter or at the DMM, suspect that IC2 is wired backwards. Measure the negative voltage at the input terminal (−38 volts). When voltage goes into IC2 and not out, suspect poor wiring or a defective negative regulator IC. Doublecheck the negative polarity of each electrolytic capacitor. If these filter capacitors are wired backwards, the bridge rectifier might short and the capacitors would become very warm.

Check the output dc voltage and the ac voltage at the input of the bridge rectifiers. Likewise, check the input and output voltages of IC1, IC2, and IC3. Remember, the current-meter jack-shorting wires must be plugged in from J1 to J2, J5 to J6, and J9 to J3 from each power source. Make sure that the positive terminal of D2 is grounded. Check power supply 1 in the same manner.

Injecting voltage into horizontal circuits

Locate the positive supply voltage source to the horizontal oscillator transistors or to the IC deflection circuit. The power source can be marked (VCC) on the voltage-supply IC pin. Likewise, the positive supply, connected to the collector circuits of the horizontal transistor, can be checked using the same procedure. Notice the correct B+ voltage applied to the horizontal output circuit (FIG. 5-84).

Slowly raise the dc voltage of the power supply to the required operating voltage. Notice if the panel meter pulls down or barely moves when voltage is applied. The horizontal circuits or the deflection IC might have a leaky component tied to the IC, or a leaky IC. Now scope the horizontal oscillator circuit for correct waveform. Usually, the horizontal driver transistor has a higher collector voltage, but the drive pulse can be traced to

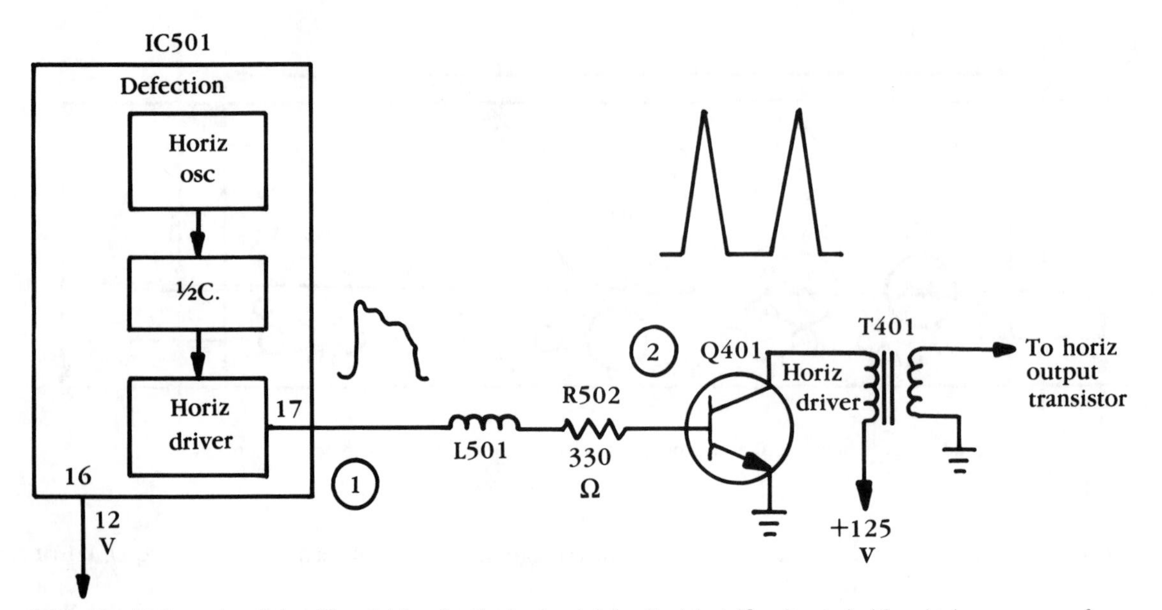

5-84 The horizontal oscillator IC and driver for the horizontal circuits. Inject 12 volts at pin 16 and take scope waveforms at the numbers.

the base terminal. If this occurs, you at least know that the horizontal oscillator circuit is operating.

If the chassis shuts down without any high voltage or if it keeps blowing fuses, it's difficult to locate the defective horizontal component. Check the horizontal drive and output circuits when the oscillator is operating. Check the horizontal oscillator circuits with the TV power cord pulled for this method of voltage injection.

Injecting voltage to the vertical circuits

To determine if the vertical circuits are loading down the flyback circuits, inject voltage from the dual power supply. Inject 11.3 volts at the vertical countdown circuit (36) to determine if these circuits are working (FIG. 5-85). Scope pin 23 at the vertical driver output for correct waveform. If the signal is normal, trace it to pin 4 of the vertical output (IC301).

To check the vertical output stage, monitor vertical output terminal 2 of IC301 with 26.2 volts applied to pin 8. If the pulse waveform is normal to the vertical flyback circuit, the vertical circuits are normal. If either stage does not produce the required vertical signal, repair the vertical circuits with injected voltages from the dual power supply. You can use a sine or square wave from the external sine/square-wave generator to provide vertical sweep. Although the waveforms might not be perfect, you can at least determine if the vertical circuits are functioning.

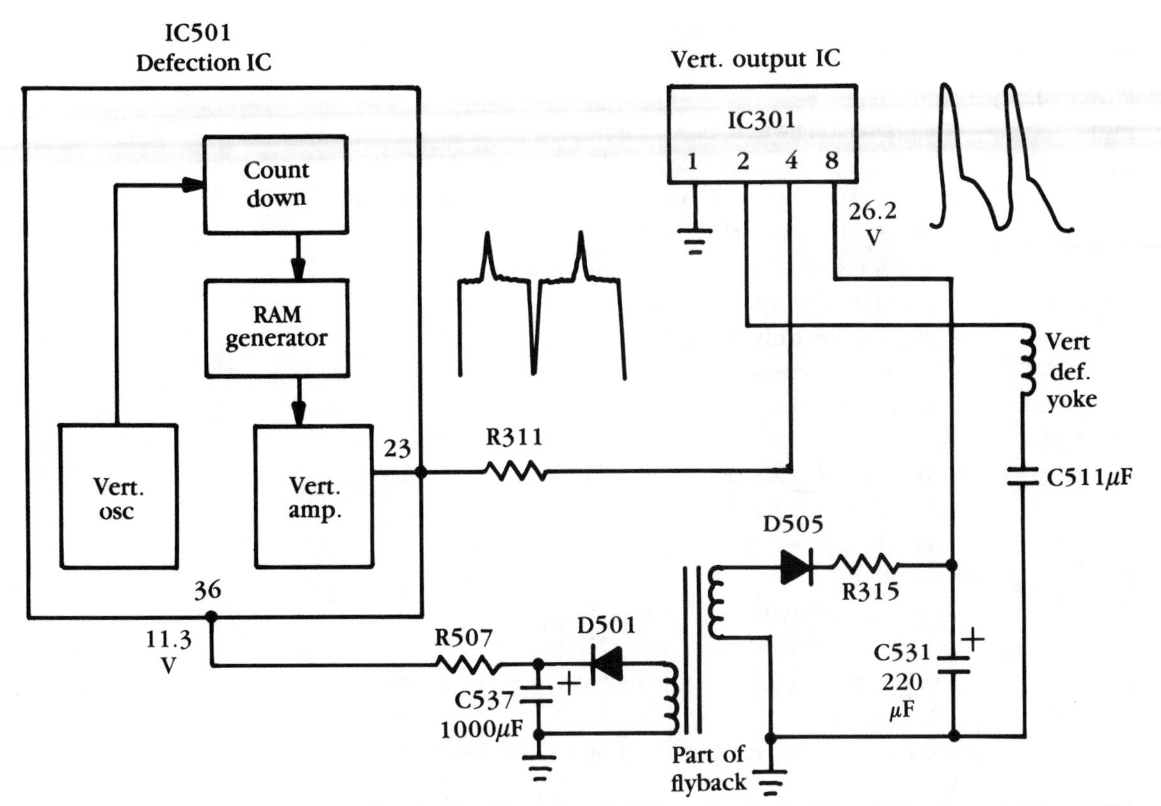

5-85 Block diagram of the vertical IC deflection, IC, and vertical IC output circuits. Inject 11.3 volts at pin 36 or IC501 from power supply 3 and 26.3 volts at pin 8 of power supply 1.

Attaching the tester to the chassis

Plug the ac/dc TV into an isolation transformer when attaching test instruments to the chassis. When injecting voltages from the dual power supply, pull the TV plug out. Do not plug the TV cord into the ac outlet. It's wise to use very fine needle-pointed alligator clips to attach the power supply source to the various circuits.

Clip the ground lead to common ground. Note if the circuits you are checking have a hot or common ground. Often, it's difficult to clip the voltage terminal to an IC pin (FIG. 5-86). Solder a bare wire to the voltage-supply pin and snap on the injected-voltage alligator clip. Usually, the voltage is left in place when problems are encountered inside the circuit. Likewise, solder a wire pin to the various circuits to be checked. Now, scope the different pins for the required waveforms.

Conclusion

This power supply unit is handy to inject voltage into the various circuits of the TV chassis, camcorder, CD player, cassette player, and amplifier.

Parts list

Power supply #1 (positive)	
SW1	Toggle off/on switch.
T1	25.2-V 2-A power transformer, 273-1512 or quiv.
IC1	RCA SK9215 or ECG956 variable-power positive regulator.
C1	6800-μF 63-V electrolytic capacitor, NC802 Hosefelt Electronics, Inc.
C2, C4	0.1-μF 100-V ceramic capacitor.
C3	1-μF 50-V electrolytic capacitor.
R1	240-ohm 1-W resistor.
R2	5-kΩ variable linear control.
J1, J2, J4	Black banana female jacks.
J3	Red banana female jack.
N1	120-Vac red neon indicator.
D1	2-amp bridge rectifier.
Power supply #2 (negative)	
SW2	Toggle on/off switch.
T1	25.2-V 2-A power transformer, 273-1512.
IC2	SK9216 or ECG 957 variable negative power regulator.
D2	2-A bridge rectifier.
C5	6800μF 63-V electrolytic capacitor, NC802 Hosefelt Electronics, Inc.
C6, C8	0.1-μF 50-V ceramic capacitor.
C7	1-μF 50-V electrolytic capacitor.
R3	120-ohm 1-W resistor.
R4	5-kΩ linear control.
J5, J6, J8	Black female banana jacks.
J7	Red female banana jack.
N2	120-Vac green neon indicator.
Cabinet	3¹¹⁄₁₆" × 8¼" × 6⅛", 270-274.
Misc.	2¾-×-3.6-inch perfboard, power cord, hookup wire, solder, nuts, bolts, etc.
Meter circuit	
M1	0-to-50-Vdc panel meter, #20-1124 Electronics Specialists.
SW1, SW2	DPDT paddle switch, #SW54 Hosefelt Electronics, Inc.
SW5	SPST switch.
N3	120-Vac green neon indicator.
T3	18-V 2-A secondary power transformer, 273-1515 or equiv.
D3	2-A bridge rectifier.
IC3	LM317T +1.2-to-+27-V voltage regulator, 276-1778 or equiv.

Parts list cont.

C9	4700-μF 35-V electrolytic capacitor.
C10, C12	0.1-μF 50-V ceramic capacitor.
C11	1-μF 50-V electrolytic capacitor.
R5	240-ohm 1-W resistor.
R6	5-kΩ variable linear control.
J9, J10, J12	Black banana jacks.
J11	Red banana jack.
Misc.	3-A fuses, 3 fuse blocks, etc.

5-86 Connect the finished power supply to a TV chassis to inject voltage to the various circuits.

Both negative and positive voltage sources can be used in some older vertical TV circuits. The positive and negative voltages can be injected into high-powered amplifier and camcorder circuits. Remember, when the supply voltage drops, suspect an overloaded circuit. The dual power supply can be used in project circuits calling for a positive and negative voltage source. Place a 3-amp fuse in the power line to protect all three power sources.

DUAL-CHANNEL 50-TO-100-WATT DUMMY LOAD

High-powered dummy loads are required for servicing present-day PA amplifiers and auto systems. Since many of the latest audio amps have high-wattage output systems, a speaker dummy load must be able to withstand up to 100 watts or more. The dual-channel audio load tester provides an adequate speaker load when the amplifier cuts out and quits after several hours of operation. Besides checking high-powered amps for intermittent breakdowns, the dummy load can be used instead of speakers with dc-coupled output amplifiers (FIG. 5-87).

Although the audio amp can be turned down when checking and making tests, a full load is required for the amp to become hot after several hours of operations. The dummy load can be used to protect the speaker when the audio output circuits become unbalanced in a dc amplifier. The speaker voice coil can be quickly ruined if dc voltage is applied across the coil in unbalanced dc output circuits (FIG. 5-88). If you do a lot of high- or low-powered amp repair, this test instrument is handy.

5-87 Connect the dual-load tester to the audio amp for servicing.

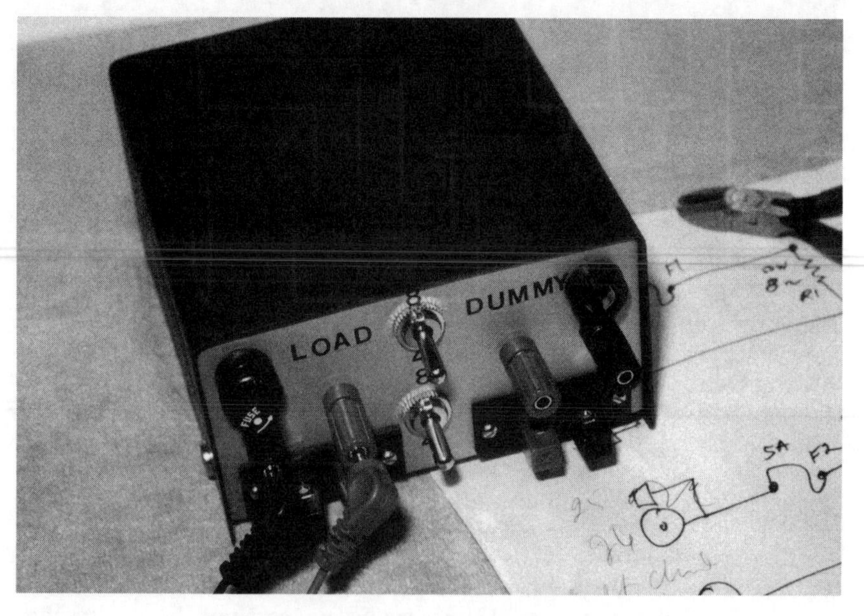

5-88 Front view of the dual-channel dummy load.

The circuit

Several 10- or 20-watt resistors can be placed in a series/parallel circuit to acquire the correct wattage for a speaker dummy load. In this circuit, four 50-watt 8-ohm noninductive power resistors are wired for 50- and 100-watt loads. Although the single high-wattage resistors might cost more than many individual series/paralleled resistors, the two-switch circuit is quite simple.

The dual-channel dummy loads are protected with two 5-amp fuses. The 50-watt 8-ohm fuse is wired into the circuit as a speaker load (FIG. 5-89). When the audio amp is connected to the stereo channel load, the load impedance is 8 ohms. Two different types of connections are provided for easy audio amp hookup.

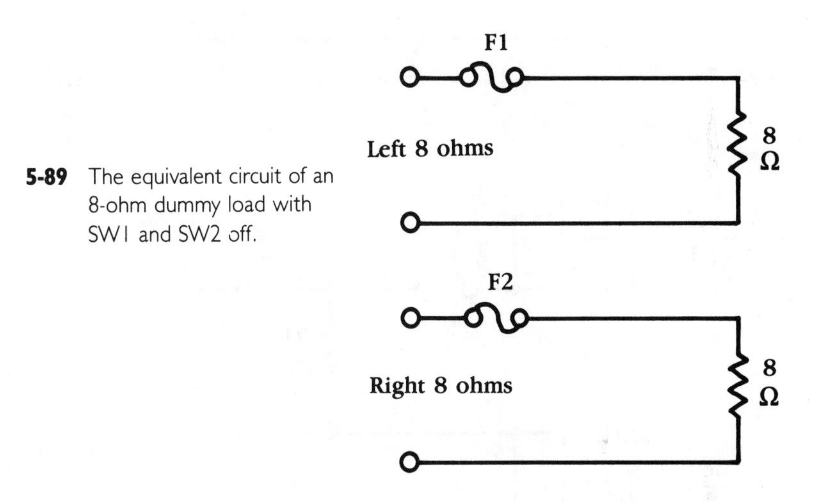

5-89 The equivalent circuit of an 8-ohm dummy load with SW1 and SW2 off.

When the dual-channel amp is connected to the 4-ohm load, SW1 and SW2 are switched on. This action provides two 50-watt 8-ohm resistors in parallel, making the load 4 ohms. Up to 100 watts can be applied (FIG. 5-90). Most of the large PA amplifiers and auto high-powered amps operate at 4-ohms impedance. If greater wattage and 2-ohm loads are needed, another 8-ohm 50-watt resistor can be switched into the circuit. Both load circuits are identical in this tester (FIG. 5-91). When one set of jacks are used for loading, the other set can be used to check the ac voltage.

Choose a panel-mount screw-type 5- or 10-amp fuse holder that can be mounted on the front panel for easy fuse replacement. Insert a 5-amp fuse in each holder. The four high-wattage resistors can be held into position with correctly placed terminal strips. The 8-ohm resistors should be held away from the cabinet when a load is applied for several hours (creating very warm resistors). These 8-ohm 50-watt resistors can be acquired from electronic parts stores or through mail-order firms.

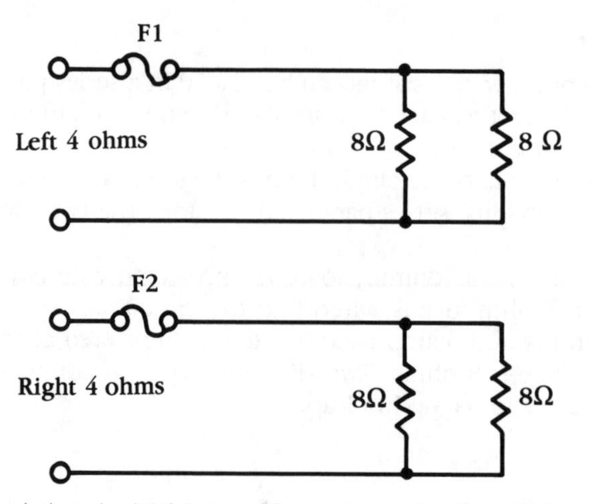

5-90 The equivalent circuit of 4-ohm 100-watt dummy load with SW1 and SW2 on.

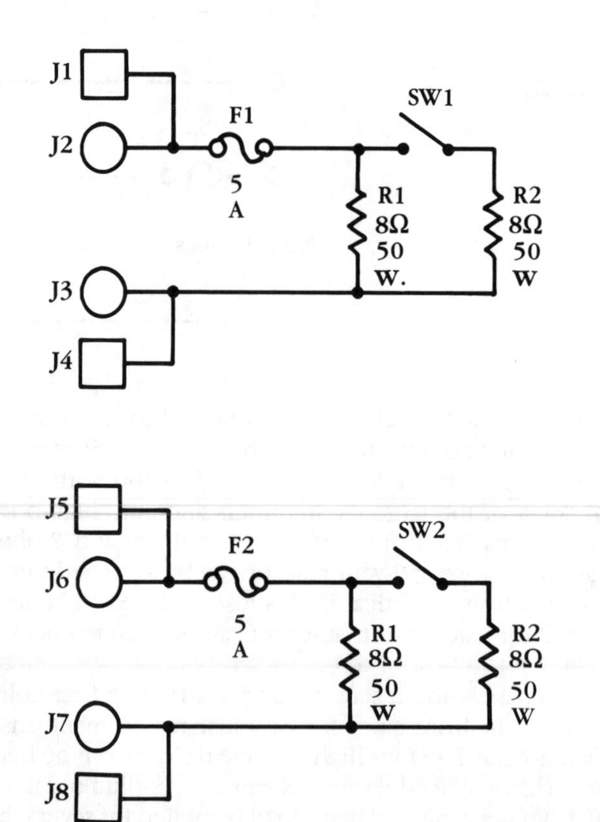

5-91 The complete wiring schematic of the dual-channel 50- and 100-watt load.

Front panel layout

Select a metal cabinet with a steel top and ventilation holes. The front, back, and bottom aluminum bottom panel is easily drilled. The 3-×-5¼-×-5⅞-inch metal cabinet is large enough to handle the 50-watt resistors and extra parts. Choose two-position push-button-type terminals and nylon binding posts for each audio channel.

Lay out the holes to be drilled on a white piece of paper, cut it to fit, and tape it to the metal front panel, to prevent scratching and damaging the finish (FIG. 5-92). In each top corner, drill a ½-inch hole to contain the fuse holders. Drill two 7/16-inch holes to center for SW1 and SW2. Mount and space the dual-channel input jacks on each side of the switches.

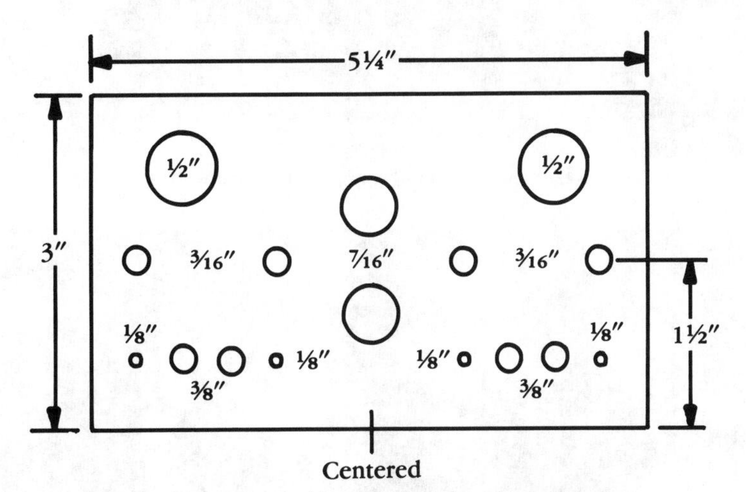

5-92 The front panel layout with hole sizes and locations.

After drilling all holes, remove the paper layout and mark the various components. Place all lettering and numbers on the front panel before mounting any parts. Now spray two or three coats of clear finish over the front panel. This finish not only protects the lettering, but it prevents scratches and marks on the front panel.

Wiring

Wire all components together with #22 hookup wire. Start with the left channel input jack and solder the ground (black) jacks together. Likewise, connect the above-ground (red) jacks (FIG. 5-93). Run the hookup wire to F1 from the red input jacks. Solder a length of wire from F1 to SW1. Connect a wire from the 50-watt resistor to the same side of the switch. Tie the ground side of both resistors together and run a lead to the ground (black) jacks on the front panel. Solder the other end of R2 to the other side of SW1.

Wire the right channel in the same manner. It's best to wrap several

5-93 The completed wiring of the dummy load.

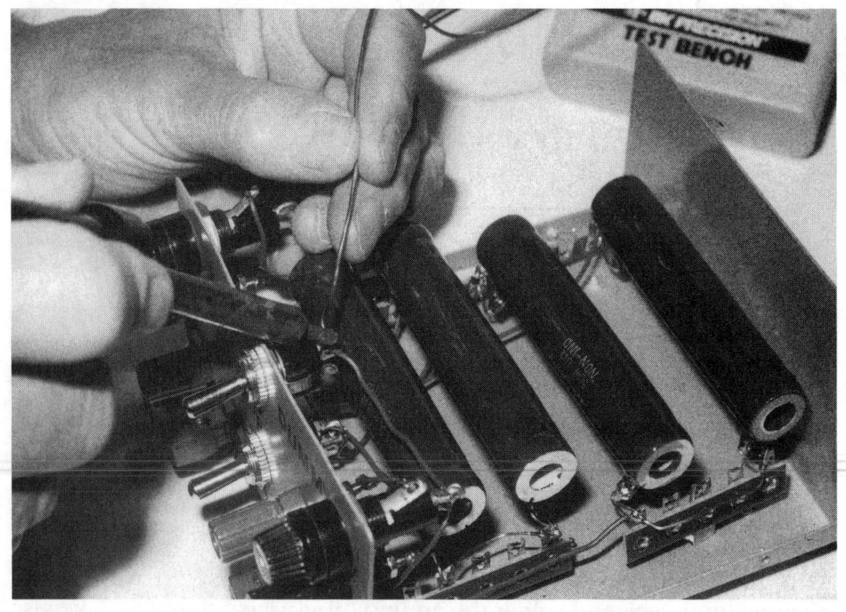

5-94 Solder the switches inside the metal front panel.

turns of bare wire around high-wattage resistors before soldering them to the terminal strips. Keep each channel away from the other (FIG. 5-94). Make no connections between a component in one channel and one in the other. Don't even connect the ground (black) jacks to the metal cabinet—in case the cabinet grounds against the amplifier to be tested.

Testing

Checking the dummy load is simple with no applied voltage. Connect a low-ohm meter to the left terminals and the resistance should be near 8 ohms. Flip SW1 to the 4-ohm load, which should be near 4 ohms. Check the right channel in the same manner. Check for poor switch contacts or connections if the resistance is not cut in half when you switch either SW1 or SW2. Inspect for properly soldered connections.

Dc output circuits

In many of the early and present high-powered amplifiers, the audio output circuits operate in a direct-coupled dc circuit. High-powered output transistors work directly with a balanced circuit. If the circuit is unbalanced with a defective transistor, resistor, or capacitor, dc voltage is found at the speaker output terminals. Of course, the moment this occurs, the speaker voice coil burns.

The same condition can exist in high-powered amplifiers that use IC components. Balance the input voltage to the power IC and if the voltage is higher or if the IC becomes leaky, then dc voltage is at the speaker terminals. With the dual-channel audio load tester, the dc voltage runs to the high-powered resistors, instead of the voice coil. Check the other channel jack to see if any voltage appears across the load.

In the high-powered amplifier, the power output transistor or ICs have higher applied negative and positive voltages (FIG. 5-95). The dc source has 40 volts, positive and negative, applied to terminals 2 and 9 of IC2. The speaker output is taken from balanced emitter terminals within the power IC. A positive and negative 1.4 volts are applied to pins 1 and 0 of the IC. An unbalanced input voltage or a leaky IC can place excessive

Parts list

SW1, SW2	SPST 5-A toggle switches, 275-603 or equiv.
F1, F2	10-amp panel/mount fuseholder, 270-364 or equiv. with 5-A fuses.
J1, J4, J5, J8	Speaker two-position push-button terminals, 274-315 or equiv.
J2, J3, J6, J7	Nylon binding post to accept banana plugs, 274-662 or equiv.
R1, R2, R3, R4	8-Ω 50-W resistors. Available from: MCM Electronics, 650 Congress Park Dr., Centerville, OH 45459-4072.
Cabinet	3″-X-5⅛″-X-5⅞″ metal cabinet, 270-253 or equiv.
Misc.	#22 hookup wire, solder, terminal strips, 6-32 bolts, nuts, etc.

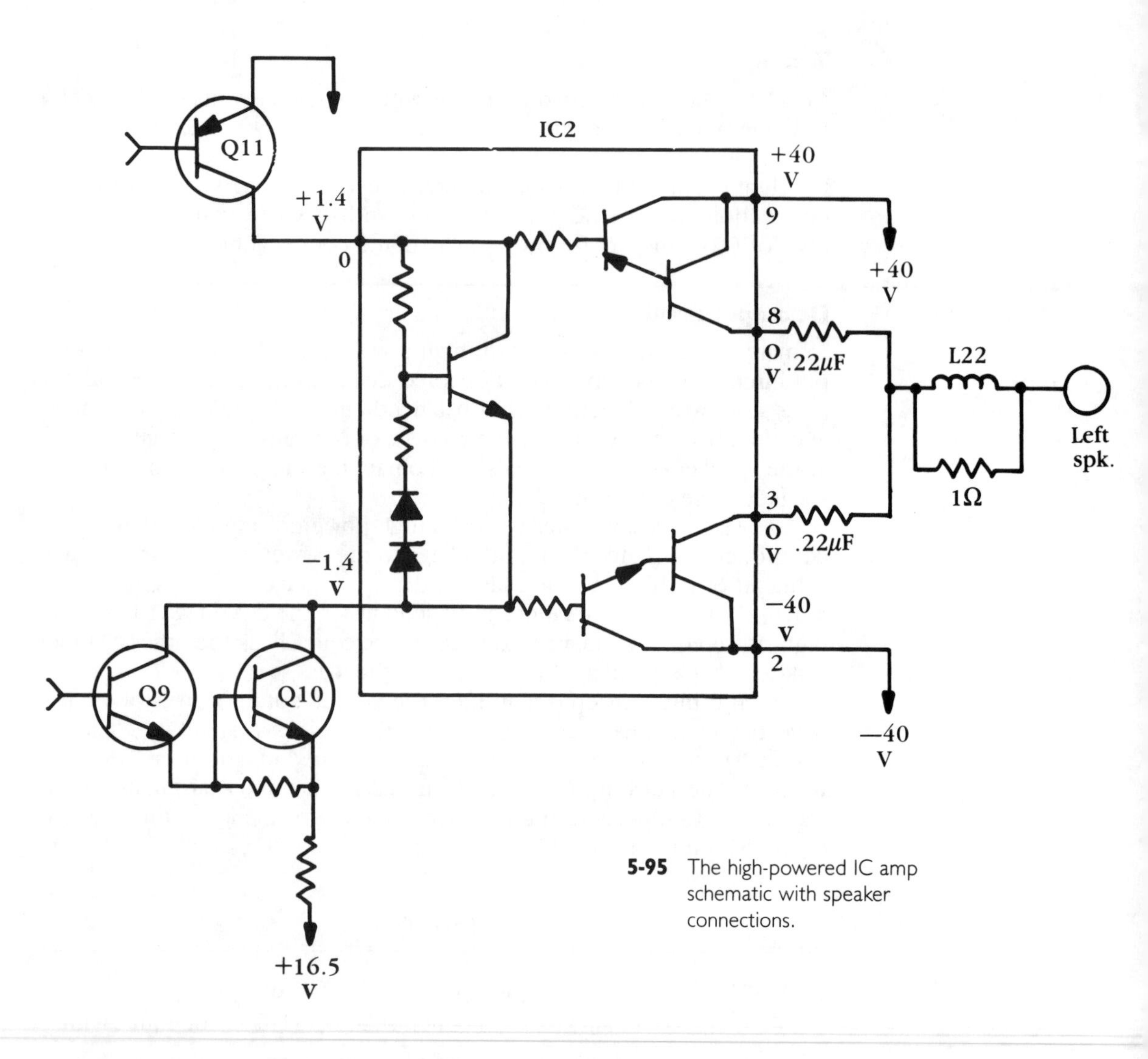

5-95 The high-powered IC amp schematic with speaker connections.

voltage across the speaker terminals. If you need a dummy load with an output higher than 100 watts at 4 ohms, a series/parallel arrangement of 8-ohm resistors can be connected.

TUNER SUBBER

Today, tuner subbers are difficult to find and very few are manufactured. If you are an electronic student or technician coming into the TV-service business, you might have to do without one. So, why not build your own? The tuner subber is injected into the circuit to determine if the TV tuner or the circuit is defective. The subber is still needed with the many different wafer, varactor, and modular tuning systems (FIG. 5-96).

5-96 The finished tuner subber beside a portable TV.

The tuner subber can be clipped into the i-f input stage or plugged into the i-f input. The subber is a substitute for the tuner in the TV set. The defective tuning system or tuner can be eliminated with the tuner subber. As long as the TV chassis has an i-f input or cable, the tuner subber can be used to check the defective tuner or system. This subber can be used on TVs that use a system control processor that controls a varactor tuner.

A few years ago, you could pick up new or rebuilt tuners for less than $15.00. Today, very few are even listed in surplus catalogs. However, there are many discarded transistorized tuners from color and B&W TVs. Often, a B&W TV portable is discarded if the picture tube or flyback transformer is defective, because the cost of the replacement would be beyond what the customer would want to pay. Just take one of these transistorized tuners out of a discarded TV and make your own tuner subber.

Dismounting old tuners

Be careful when removing the tuner so that you don't break any top components. The red lead is the B+ supply. The AGC lead is either white or gray. Notice which plug goes into the i-f cable of the TV set. Disconnect the UHF tuner and cable. You might find a B+ wire running from the VHF tuner to the UHF tuner. Cut the wire at the VHF tuner. Unplug the shielded cable from the VHF-to-UHF tuner.

You will need the VHF output cable or plug, the B+ red supply wire, the gray AGC voltage wire, and the VHF antenna lead-in wires with the old tuner. The smaller VHF tuner is ideal to place in a smaller cabinet. Of course, the standard solid-state tuner is fine to use as long as it works.

Save the VHF and fine-tuning knobs. If a VHF dial assembly is removable, place it on the new subber. If not, the dial can be numbered with rub-on letters and numbers. Often, the VHF tuner is held into position with a metal front brace or bracket assembly. Keep this assembly intact to mount inside the new case. In most cases, the VHF tuner bracket will have to be mounted with long metal or plastic spacers.

This particular transistorized VHF tuner came out of a discarded KMC B&W portable (FIG. 5-97). The front bracket was kept intact for mounting. The 2-to-13-channel VHF tuner was a standard model. Before trying to mount the tuner, remove the bottom cover and clean it with tuner spray. Each wafer section was sprayed individually while rotating the tuner shaft; no sense starting with a dirty tuner!

5-97 This VHF tuner was taken from a discarded B&W KMC portable to place in the tuner subber.

Preparing the front panel

The standard tuner fits snugly in a MB-3C instrument enclosure. Fit the tuner to the right side of front metal cover. Metal covers can be ordered from the MB-855 enclosure series. Center the tuner or mount it flat on the bottom cover. Mark the 4 front mounting holes with a large 7/8-inch hole for the fine-tuning knob. The larger hole can be cut with a circle cutter or by drilling several 1/8-inch holes in a round circle. Now break out the center piece and smooth the edges with a half-round file (FIG. 5-98).

Drill a 5/16-inch hole for R1. Mount SW1 on the rear of R1. At the center bottom, drill a 5/16-inch hole for the gold-plated phono chassis mount

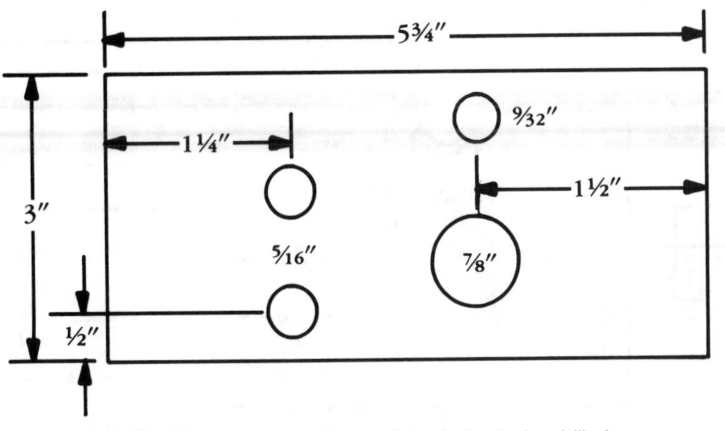

5-98 The front panel layout of the holes to be drilled.

jack. Use metal jacks and plugs to connect and disconnect the various plugs and cables. The standard i-f cable will plug directly in this type plug with the other plug connectors. The numbered dial assembly can be cemented to the front panel. In this case, three 1/8-inch holes were drilled to fasten the dial bezel to the front panel.

Cut four 3/4-inch metal spacers to hold the tuner back inside so the front knobs will line up. These metal or plastic spacers should be large enough for 1/4-inch metal screws. Make sure that the spacers are the exact length and that they can be ground level or at right angles. Since the tuner projects through the TV cabinet, it must be spaced in back of the metal front panel.

The circuit

Most of these transistorized tuners operate with 12, 15, or 18 volts dc. If in doubt, check the old TV tuner schematic for the supply voltage. In this case, 11.7 volts was supplied to the tuner from the TV. Most tuner subbers operate from batteries. Often, two 9-volt batteries are wired in series. In this model, the 12-volt source came from a regulated power supply. Since most TV chassis, while being serviced, are plugged into an isolation transformer, this subber operates from a stepdown transformer power source to prevent damage from a hot TV chassis.

SW1 operates when R1 is turned on. N1 is a 120-volt ac neon indicator that will light, showing that the subber is operating (FIG. 5-99). T1 is a 12.6-volt stepdown transformer tied to a full-wave bridge rectifier, D1. C1 filters the dc voltage that is applied to the input terminal of IC1. IC1 regulates the dc voltage applied to the solid-state tuner and R1 provides positive bias to the AGC circuit. With tubes, the AGC circuits are negative-biased, and positive bias runs to the solid-state tuner.

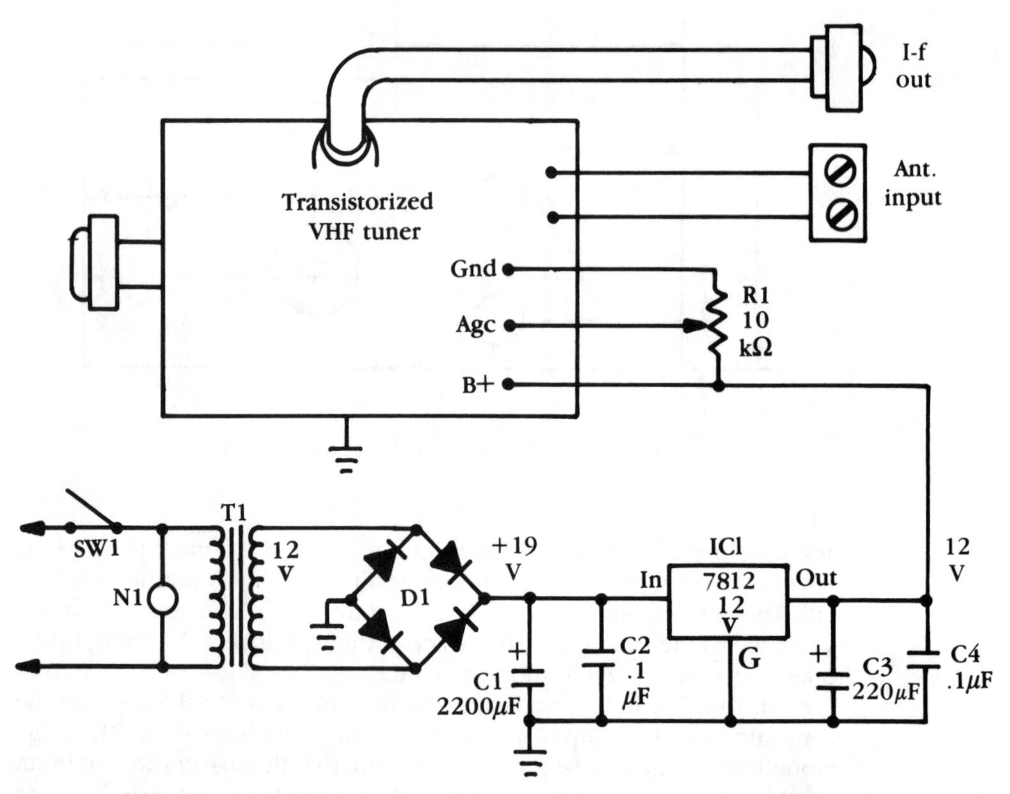

5-99 The circuit and connections of the low-voltage power supply and VHF tuner.

Board construction

Mount the small parts for the power supply on a general-purpose component pc board. Run a bare wire down one side for the common ground and another down the opposite side for the B+ components. Start with the bridge rectifier and leave the two center terminals stick outward for the transformer ac leads, 12.6 volts (FIG. 5-100). Next, mount C1 and C2, then IC1, C3, and C4. Solder each component into the circuit as you mount it. Solder a six-inch piece of flexible hookup wire for the B+ and ground wires to the tuner.

Cabinet construction

Since the VHF tuner takes up much of the cabinet room, mount the pc board toward the front on the bottom cover. Mount the small power transformer behind the pc board at the back of the case. Drill two 9/64-inch holes to mount the transformer. Drill four 9/64-inch holes to mount the pc board with 1/2-inch spacers. Mount the tuner to the front panel with four 1/4-inch metal screws.

Prepare the back plastic panel for mounting an antenna strip to con-

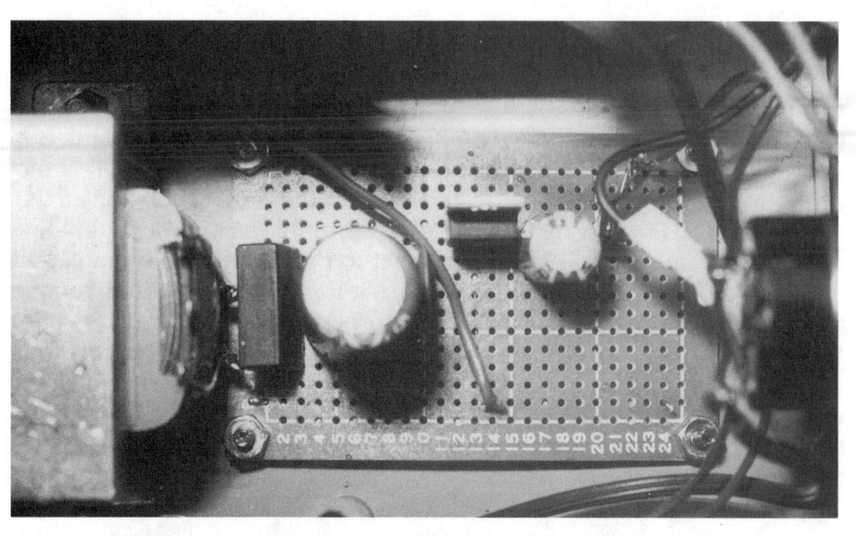

5-100 A close-up of the pc indexed board for the power supply.

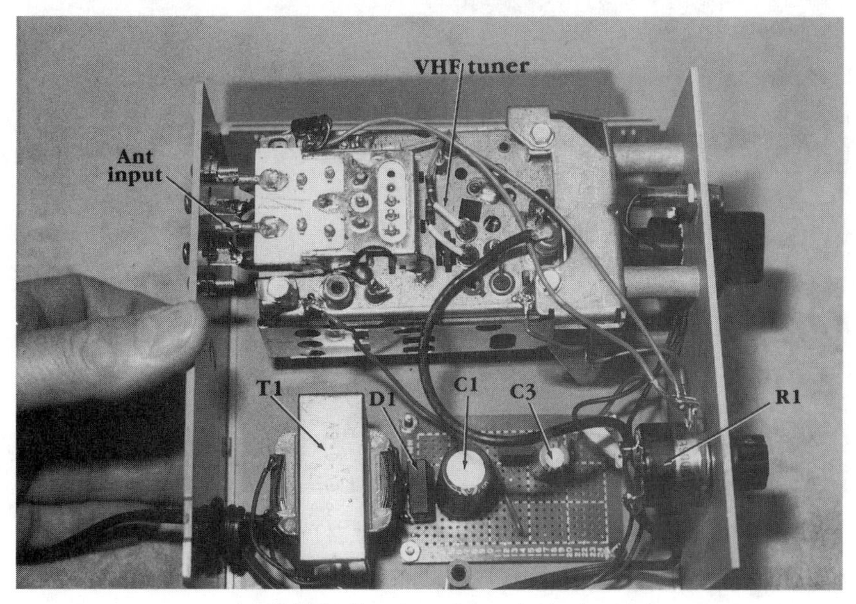

5-101 Top view of the wired unit.

nect the antenna. Drill two 1/8-inch holes for the strip and one 5/16-inch hole for the rubber grommet around the ac cord. Remember to tie a knot in the ac cord so it will not pull out (FIG. 5-101). To prevent scratches and damage to the front and rear covers, draw the required hole dimensions on a piece of paper and tape it to the sides.

Testing

Doublecheck all wiring on the pc board of the small power supply. Check the different connections between the pc chassis, transformer, and tuner. Connect the red or purple wire of the tuner to one side of R1 and the B+ power. Both the chassis and tuner are grounded to one side of R1. Run the center top of R1 to the white or gray wire of the VHF tuner. Solder the two antenna VHF tuner input wires to the terminal strip. Check the output jack and shielded wire from the tuner to the front panel input jack.

Turn on R1 and N1 should light. Check for 15 volts B+ on the tuner (FIG. 5-102). Most tuner AGC voltages are between +1.5 and 4.5 volts. Rotate R1 and see if the voltage changes. Set R1 at the low end of rotation. Connect a TV antenna to the back of the tuner subber. Insert a shielded input cable with two small alligator clips. Procedures for making tuner subber test probes and cables are given in chapter 6.

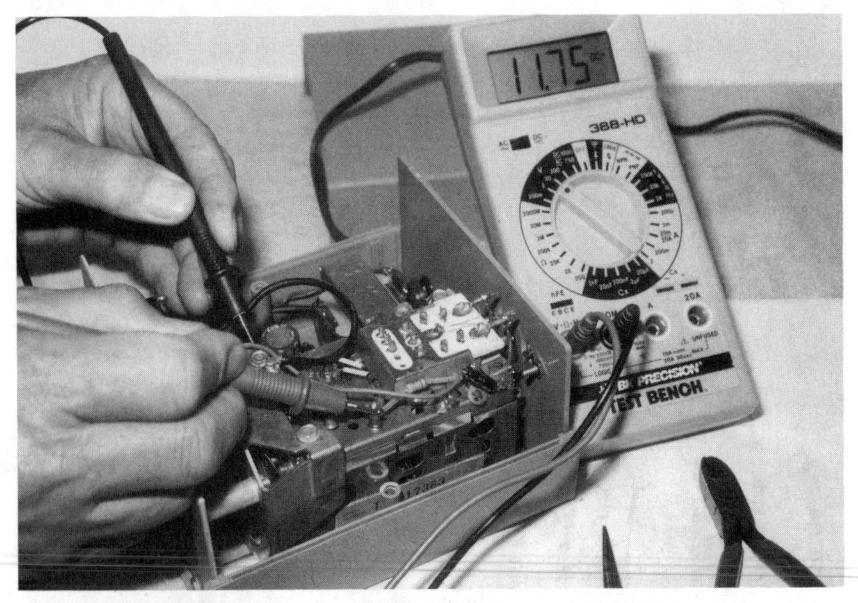

5-102 Test the unit with voltage measurements on VHF tuner.

Locating the defective tuner

If the defective chassis has no sound, no picture, and good raster symptom, you can assume that the tuner or the TV chassis are defective. One quick method to determine if the tuner, tuner module, or system-tuning control circuits are not working, is to substitute another tuner. In this case, the tuner subber is helpful. In today's ac/dc TVs (without the power transformer), plug it into an isolation transformer.

Leave the old tuner or tuning system connected as before, but unplug the i-f cable. If the i-f cable is soldered in, use the shielded cable with alli-

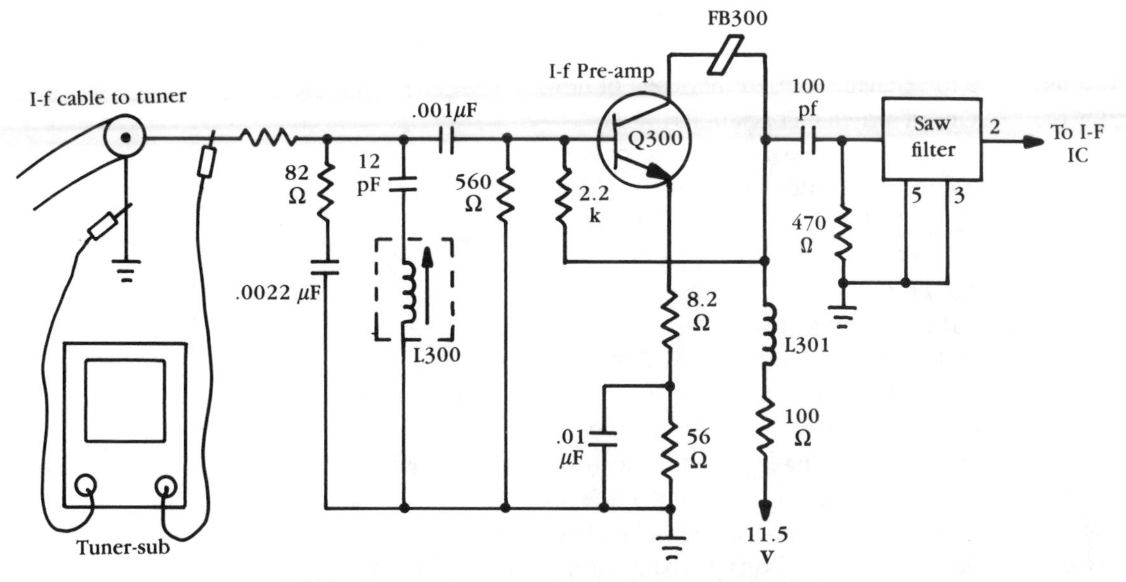

5-103 Connect the tuner subber with the soldered i-f cable.

5-104 A 75-to-300-ohm matching transformer must be used with 75-ohm antenna cable.

Parts list

Tuner	2-to-13-channel VHF tuner used or new.
SW1	On/off switch on back of R1.
T1	12.6-V 1-A secondary, 273-1352 or equiv.
D1	2-A full-wave bridge rectifier.
C1	2200-μF 35-V electrolytic capacitor.
C2, C4	0.1-μF 100-V ceramic capacitor.
C3	220-μF 35-V electrolytic capacitor.
IC1	7812 1-A 12-V regulator.
R1	10-kΩ linear control with SPST switch.
Case	ABS instrument plastic enclosure with satin finish aluminum front panel, MB-3C from All Electronic Corp.
Pc board	Multipurpose ringed-eyelet pc board, 276-150 or equiv.
TER-1	2 screw-strip terminals, 274-663 or equiv.
J1	Shield ¼″ metal phono jack, 274-346 or equiv.
Misc.	Power cord, nuts, bolts, hookup wire, solder, 1¼″ metal screws, four ¾″ metal spacers, etc.

gator clips. You can plug the i-f cable directly into the tuner subber, if one is found. In some TVs, a female jack is the input to the i-f stages. Simply unplug the tuner cable to the i-f jack and insert a shielded cable with male prongs on both ends to the TV chassis and the tuner subber.

Connect the alligator clips across the i-f cable where it enters the chassis if soldered into the circuit (FIG. 5-103). Make sure that the ungrounded alligator clip goes to the above ground or center terminal of the i-f cable. Clip the ground (shield) clip to the common ground at the i-f input stages. Now, test it.

Turn on the tuner subber and rotate the VHF tuner to one of the local channels. Make sure that the outside cable or antenna lead is connected to the tuner subber antenna terminals. If a 75-ohm cable runs from the antenna, place a matching transformer between the antenna terminals and the cable. Remember, the tuner subber has an 300-ohm antenna input (FIG. 5-104). Now, rotate the fine tuning of the tuner subber and tune in a station. Suspect a defective tuner or system if all stations tune in normally. Adjust R1 for the best picture. This tuner subber can be hooked to any color or B&W TV. If the chassis is still without sound or picture, check the i-f stages.

Chapter 6

Test probes and test leads

You can save a bundle of money by making your own or repairing test leads. Even a good set of test leads will not last forever. This chapter features information on making different test leads and cables for use around the service bench.

COMMERCIAL TEST LEADS

Most every new test instrument comes with a set of test leads. Many test instruments have banana-type plugs with test probes or hook-type clips (FIG. 6-1). A good pair of meter test leads costs from $4 to $15 if purchased separately. Most of the banana-type plugs fit inside a hole to protect the operator from voltages and accidental shock.

The commercial test leads for VOM and DMMs are available in 36- and 48-inch lengths. Replacement test leads to fit B & K, Bechman, Fluke, Simpson, and other companies that use long-sheathed banana plugs or jacks cost from $6 to $10. Often, these leads feature a probe tip at one end and some type of banana plugs at the other. Rf cables with plugs are available in 2, 5, 6, and 10-foot lengths. Scope probes with BNC connectors usually cost between $13 and $20.

ROLL YOUR OWN

You can make test leads in a few minutes and it's lots of fun. Just purchase flexible test lead wire (red and black) in 25- or 50-foot rolls. Add your favorite plug to one end and probe to the other and you're in business (FIG. 6-2). Try to avoid making junctions or soldered connections outside

6-1 Commercial test leads for VOM, DMM, audio, and TV applications.

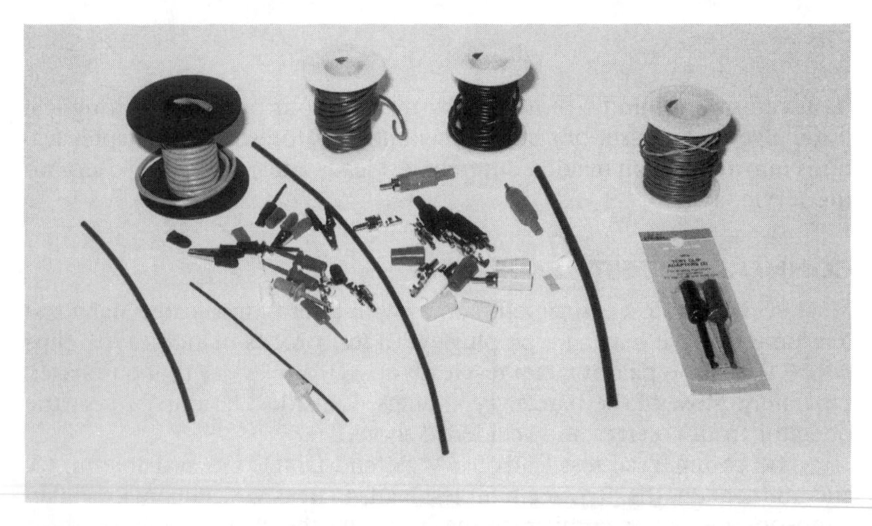

6-2 Roll your own test leads from rolls of flexible wire and solder the end fittings.

the plug areas. If two leads are to be joined at one end, use duct tape or shrink tubing to give it a professional finish. Banana plugs, nylon plastic plugs, test clips, phono plugs, and BNC fittings can be purchased at Radio Shack or practically any electronic parts store. Solder both the male and female plugs on the wires at each end.

METER AND TEST INSTRUMENT LEADS

To make new or repair old test leads, pick up the desired fittings for the test instrument and probe or clip leads at the other end. For most of the test equipment found in this book, leads with banana plugs at one end and test probes or alligator clips at the other are appropriate. You should have at least three different sets of test leads to fit most meter and test instruments: banana plugs at one end and probes, alligator, and insulated hook-type clips at the other (FIG. 6-3). Why not make a few sets to fit the ready-made test leads on the DMM.

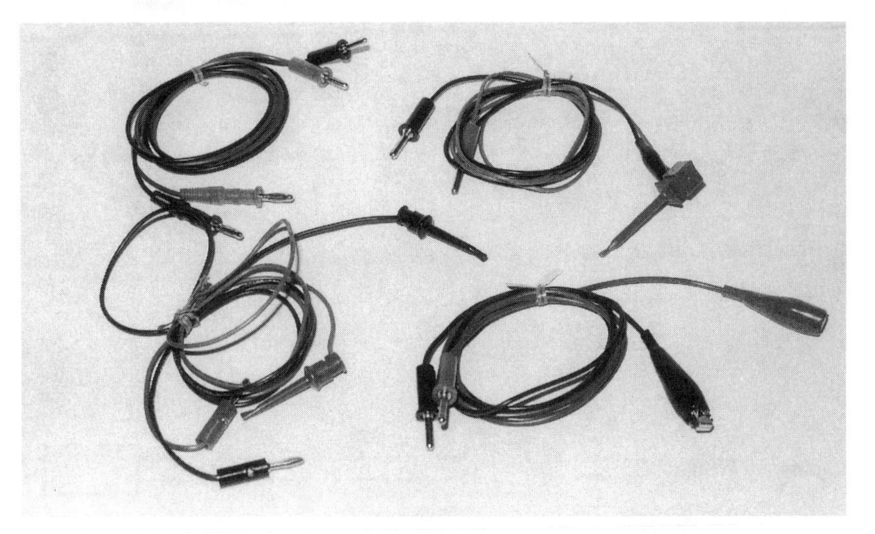

6-3 Three different test instrument leads with probe, alligator clip, and minihook-type test clips.

When you replace or repair test leads for the volt/ohmmeter, or other test instruments with a tip-jack-type plug at one end and probe at the other, select the soldered-in type. Many of commercial-type plugs and probes are molded right on the test lead. Replace these with a screw/solder-type fitting so if the cord breaks at the male plug or probe, the fitting can be removed and resoldered (FIG. 6-4).

BANANA TEST LEADS

Banana-type plugs and jacks are used throughout homemade electronic projects and commercial test equipment. You can buy banana-type plugs and jacks at any electronic outlet and they provide good, clean contacts. The banana plug seems to be tough and it does not easily break off within the test instrument jack. Test leads with unbreakable strain-relief banana plugs seem to withstand everyday wear and tear. Select gold-tip jacks and plugs for audio and TV applications.

Besides banana plugs on one end and alligator, probe, and hook-type clips on the other, make a couple of test leads with audio spade lugs. Also, make a set of test leads with banana plugs on each end (FIG. 6-5). Stacking standard banana plugs on each end accommodates the most complex cir-

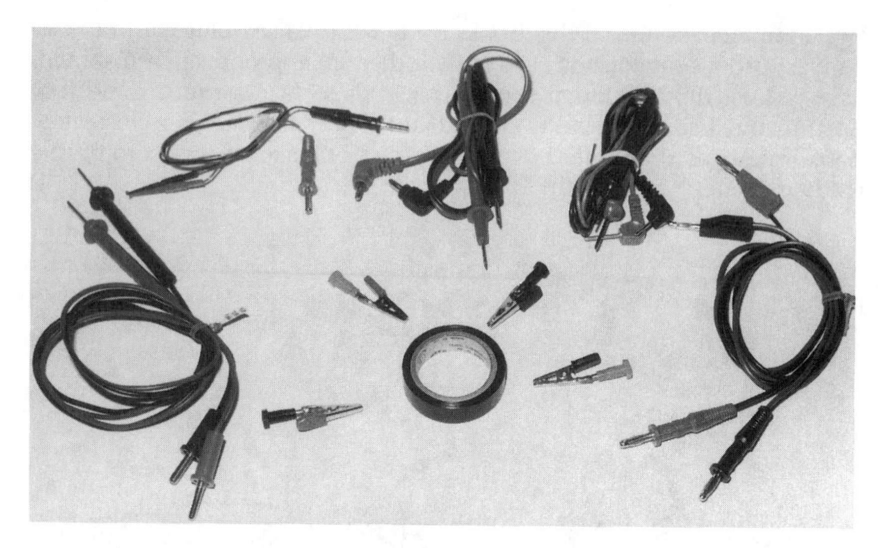

6-4 Screw-type probe and male tip ends used on the VOM and older-type instruments.

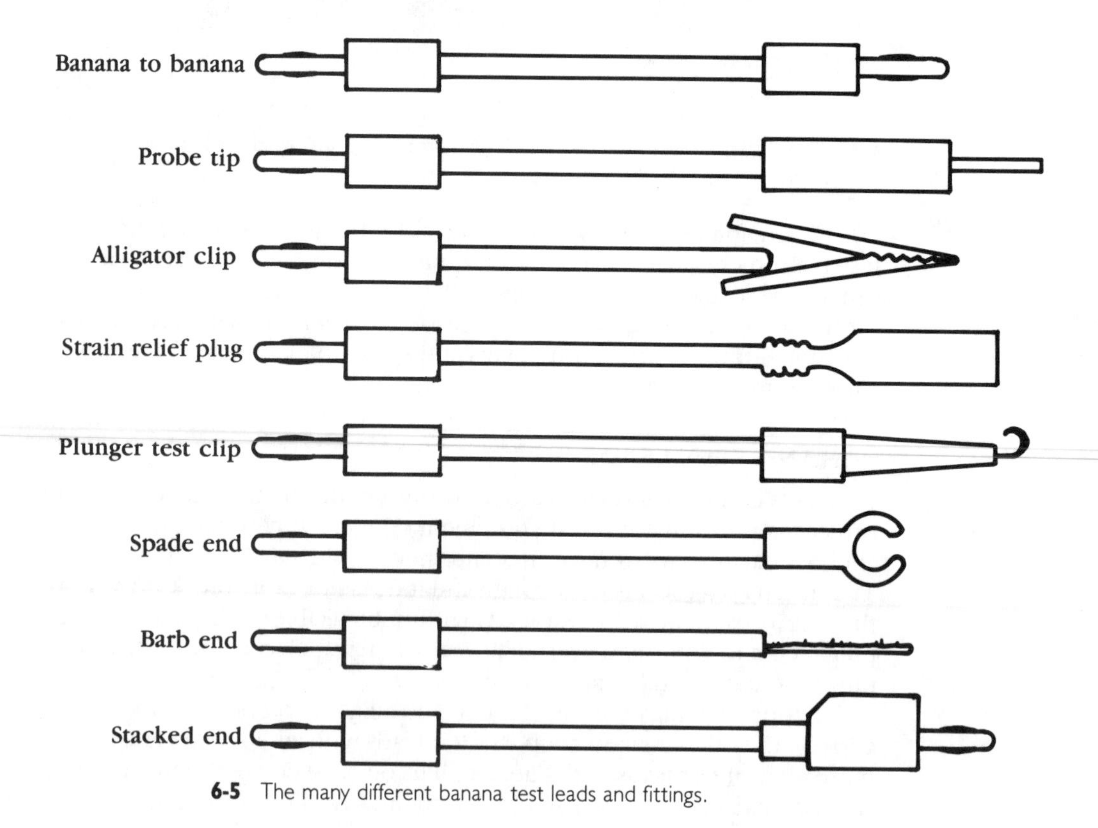

6-5 The many different banana test leads and fittings.

cuitry. Attach standard straight banana plugs to one end and angle-stackable plugs to the other.

AUDIO TEST LEADS

If you connect different test instruments to the amplifier input and output jacks, you will need several different combinations of test leads. Make two sets of input test leads that have male phono plugs on each end. Dual shielded RCA-type plugs can be used for audio input, video, VCR, CD, and television connections (FIG. 6-6). Make one set with banana plugs on one end and phono plugs on the other. Although gold-tip plugs are the best in video and audio work, they are also the most expensive.

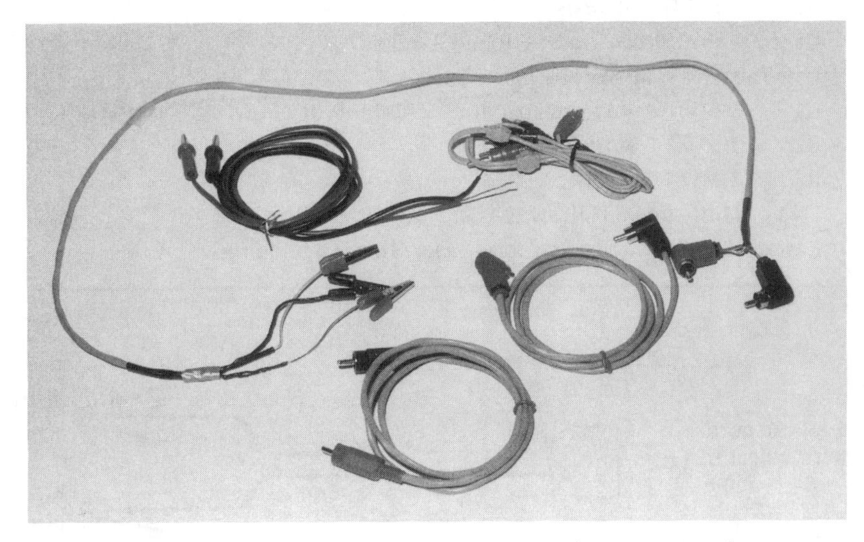

6-6 Several sets of audio test leads for input and output connections.

For speaker output instruments, make a set with banana plugs on one end and alligator clips on the other. On another set, place banana plugs on one end and bare soldered leads on the other end. Likewise, make another with U-spade lugs on one end with banana plugs on the other. Of course, a cable with the banana plugs on one end and clip/hook-type probes on the other, is handy when testing a component in the circuit. Make special audio input test leads with one set of banana plugs on one end and two sets of phono input plugs on the other. With this configuration, you can inject the signal into both channels at the same time.

AC CORDS

To check or apply 120 volts ac to a power transformer or to ac-input circuits, solder two insulated alligator claw-type clips to a regular ac cord.

VTVM probe parts list

C1	0.01-μF 2-kV ceramic capacitor
D1	1N34 crystal diode
R1	470-kΩ 1-W resistor
1	CTP-1 probe case, Available from: Global Specialties, P.O. Box 1405, New Haven, CT 06550
Demodulator scope probe parts list	
C1	270-pF 500-V ceramic or mylar capacitor.
D1	1N34 crystal diode.
R1	150-kΩ 1-W resistor.
R2	220-kΩ 1-W resistor.
Case	CTP-1 Probe Case; Global Specialties.
Low-capacity scope probe parts list	
C1	6-50 pF trimmer capacitors, 272-1340 or equiv.
R1	10-MΩ 1-W resistor.
R2	2-MΩ 1-W resistor.
SW1	Mini-slide STDP switch, 275-409 or equiv.
Case	CTP-2 gray plastic probe case; Global Specialties.

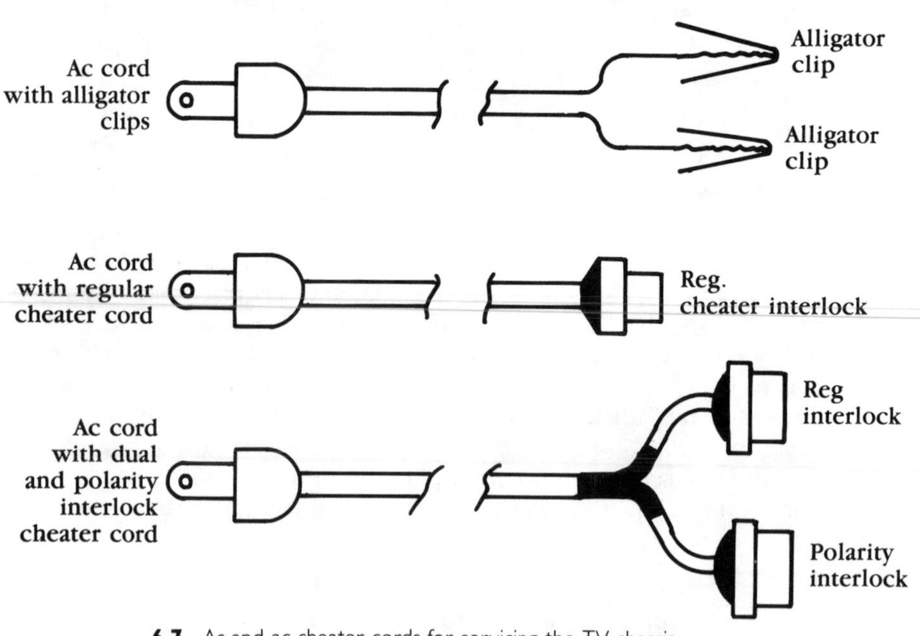

6-7 Ac and ac cheater cords for servicing the TV chassis.

For the TV interlock cord, connect a polarity-type plug and a regular cheater cord to plug into the TV (FIG. 6-7). Tape or use plastic heat-shrink tubing over the soldered junctions.

VTVM RF PROBE

Place the VTVM rf components in most any type of probe container. In this case, Global Specialties probe case CTP-1 was used to house the VTVM, the low-capacity scope and the demodulator scope probe. The probe case is an ideal housing for signal injectors, logic probes, small meters, voltage probes, resistance probes, continuity checkers, and signal tracers, with the probe tip, LED holder, and perfboard included (FIG. 6-8). The body of the probe case is only $5^1/_4$ inches by $^3/_4$ inches for easy handling.

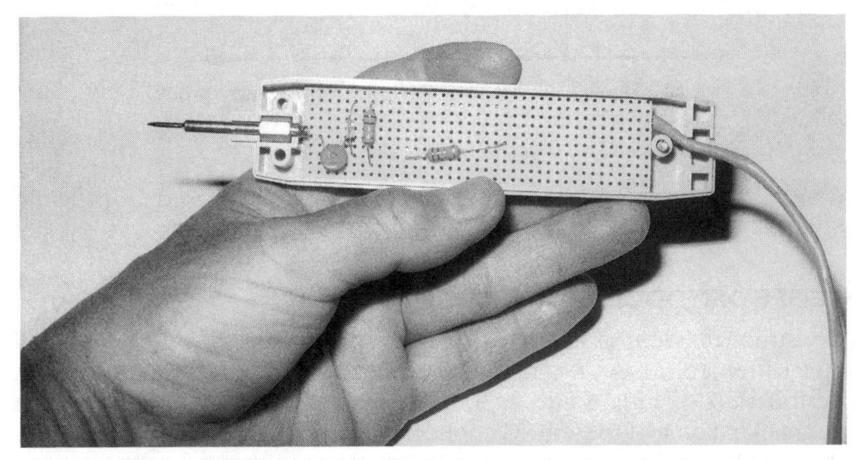

6-8 The VTVM probe fits in the palm of your hand.

Mount the three internal components on the enclosed perfboard. This board fits between the screw holds and is insulated from the plastic side for soldered connections. Solder the 0.01-μF ceramic capacitor, C1, to the metal probe tip. Use either a heatsink tool or the tips of long-nose pliers as a heatsink when soldering in the IN34 crystal diode. Check the continuity of the diode to see if about 350 to 550 ohms is found in only one direction. If a lower ohm measurement is found in both directions, the diode needs to be replaced (FIG. 6-9).

Attach both the shield and the shielded wire to the perfboard connections. When the plastic sides are screwed together, the plastic ends tighten against the shielded cable to secure it from pulling out. Place the correct male plug at the opposite end, which corresponds to the VTVM input jack. Often, a shielded-type microphone plug is found on older VTVMs and BNC connectors are used on the newer-type meter connec-

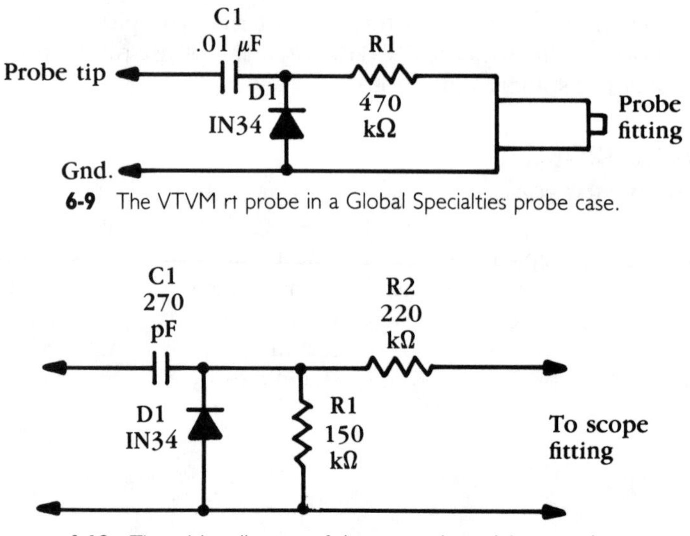

6-9 The VTVM rt probe in a Global Specialties probe case.

6-10 The wiring diagram of the scope demodulator probe.

tions. BNC and microphone connectors can be purchased from Radio Shack.

SCOPE DEMODULATOR PROBE

Construct the scope demodulator probe inside another CTP-1 Global Specialties probe case. Again, mount all small components on the 3/4-inch perfboard (FIG. 6-10). Solder a 30-inch shielded cable to the BNC scope connector. Some of these BNC connectors have solderless terminals. Trim the shielded cable according to the instructions on the rear of the mounting cord. Keep the braid flat and twist the outer braid clockwise. A stray or loose braid can cause shorts. Insert the center conductor into the back end of the connector, into the guide hole. Push in the cable and screw the connector clockwise until it stops (FIG. 6-11).

SCOPE LOW-CAPACITY PROBE

House the low-capacity probe inside a Global Specialties CTP-2 probe case or any other suitable probe case. This gray plastic case has provisions for mounting a mini-slide switch and a slot, through which C1 can be adjusted. The scope low-capacity probe can switch the input of the scope directly or through the probe components (FIG. 6-12).

Although this special probe case has no perfboard enclosed, you can easily cut a 3/4-×-4 inch piece. Mount all parts directly on the perfboard. Place the STDP slide switch on the board, directly under the slotted area.

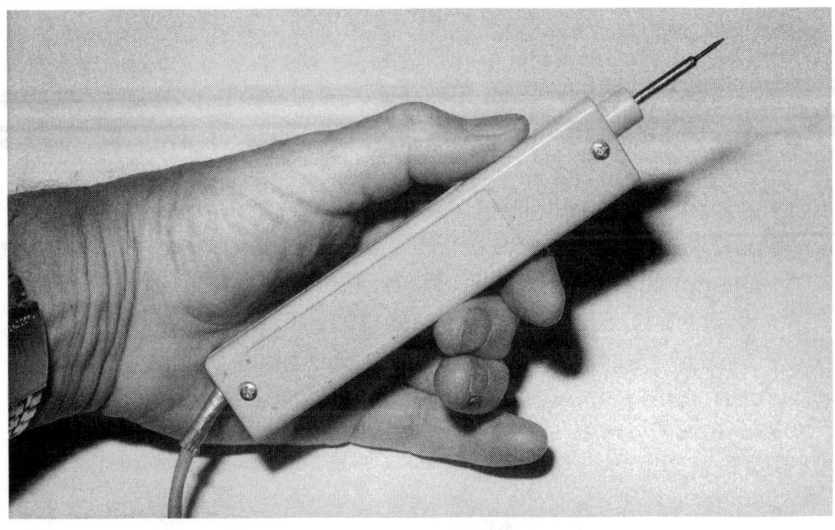

6-11 The finished VTVM rf probe and demodulator probe.

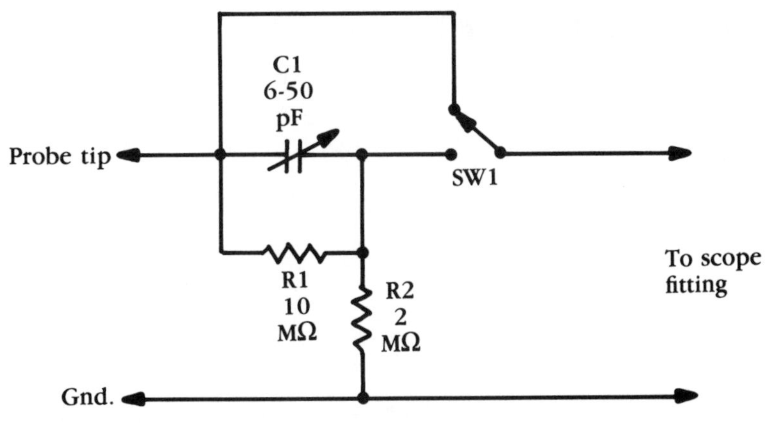

6-12 The wiring diagram of the low-capacity probe.

Solder and mount all parts on the board. Now, bolt the slide switch and perfboard together on the top panel. Mount C1 directly under the other slotted hole for easy adjustment. Solder the correct-fitting cable end to connect the probe to the scope input jack.

CONCLUSION

Building your own test probes can be quite fun and profitable. The few parts needed can be found at most electronic parts stores. If not, they can be ordered.

Test leads parts list

- Red and black banana plugs, 274-721 or equiv.
- Red and black claw-type alligator clips, 274-359 or equiv.
- Red and black mini-alligator clips, 270-1545 or equiv.
- Red and black flexible banana plugs, 274-730 or equiv.
- Red and black mini hook-type clips, 270-372 or equiv.
- Set of jumper test leads (−10), 278-1156 or equiv.
- Phono plugs with color-coded covers, 274-321 or equiv.
- Shielded-type phono plugs, 274-339 or equiv.
- Gold phono plugs, 274-250 or equiv.
- Red and black test probes.
- Ac power cords.
- Assorted spade-type lugs.
- Phone tip jacks.
- BNC male plugs.
- Roll of red flexible hookup wire.
- Roll of black flexible hookup wire.

Chapter 7

Basic construction

Building and wiring your own electronic projects can be a lot of fun, but make it safe with a few safety precautions. Anytime that you work with projects or tools that operate from the power line, it's possible to make a few mistakes. Play it safe and have fun (FIG. 7-1).

SAFETY FEATURES

Keep line cords on power tools in tip-top shape. Replace any broken ac plugs. Tape the cord if it is cut or old, or better yet, replace it. Keep away from metal posts, water pipes, and ground wires when using electric power tools. Inspect the 3-prong plug. Make sure the power tool is grounded (FIG. 7-2). Never cut off the metal ground or third terminal of the power tool if it will not fit into the extension cord or receptacle. Instead, buy a 3-prong fitting with ground wire (FIG. 7-3).

Inspect for poor ac cord or plug on your soldering iron. Keep the iron in a regular soldering-iron holder or on a metal tool. Prevent burning the table or workbench. Handle the soldering iron with care, you can accidentally burn your hands or arms if you move foolishly.

Know how to operate portable drills and sanders. Reread the instruction manual. If you improperly operate a portable drill, it can pierce a hole in your hand or arm. The sander might remove more than just the object that you intended to sand or grind. Your hands and fingers are irreplaceable, so keep them intact. Think before you use (FIG. 7-4).

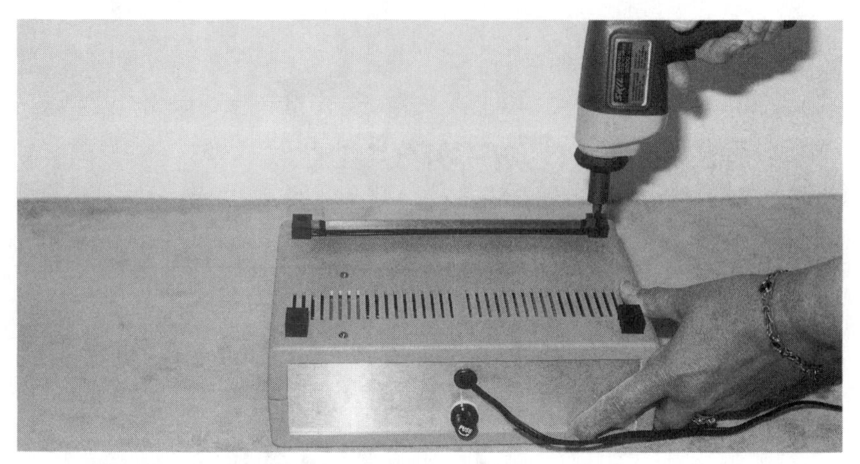

7-1 Be very careful when using power tools around electronic cabinets.

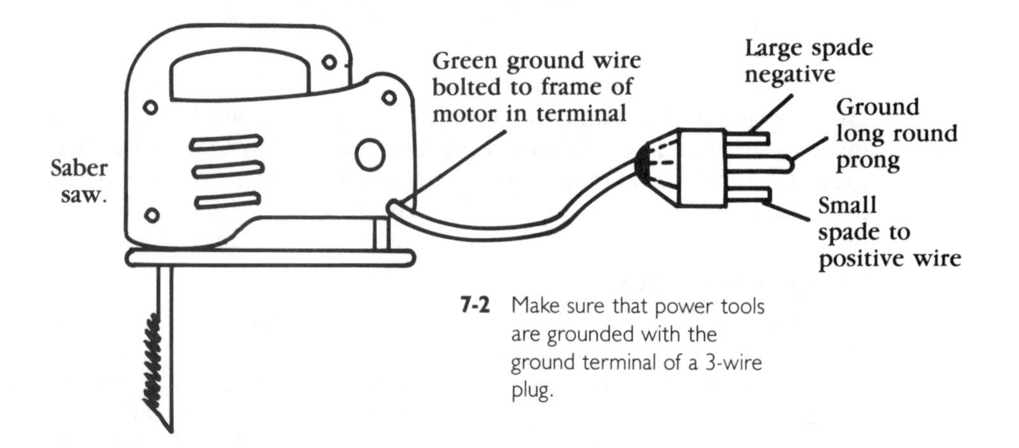

7-2 Make sure that power tools are grounded with the ground terminal of a 3-wire plug.

LIST OF HAND TOOLS

You might already have the necessary tools to build these electronic projects. Of course, power tools can speed up the construction process and the end result is a professional appearance. Generally, only three or four hand tools are required (FIG. 7-5).

- Side cutters
- Long-nose pliers
- Regular pliers
- Phillips screwdriver
- Small-blade screwdriver
- 35-watt soldering iron
- Hacksaw

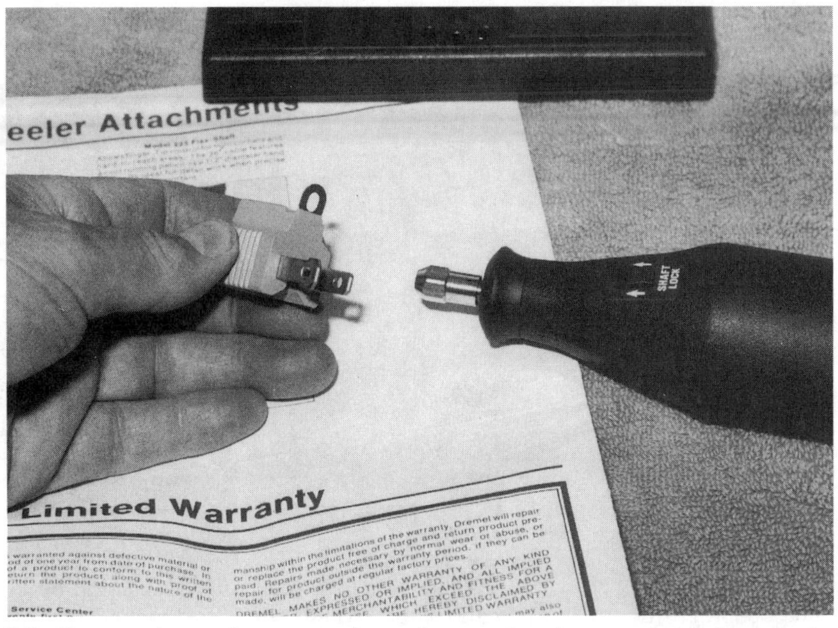

7-3 If your outlets are 2-prong, purchase a 3-prong plug adaptor and ground the wire lug to the center screw of the receptacle.

7-4 Always read the instruction manual of a new power tool for operation and safety procedures.

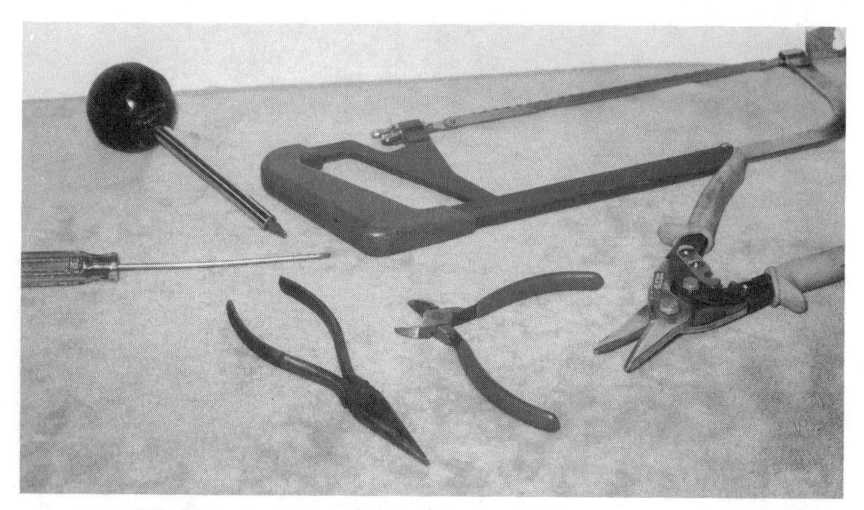

7-5 You only need a few hand tools to build electronic projects.

The small power or bench drill is ideal to make all those small holes. The bench grinder or sander can remove rough edges and make a smooth wood finish. Knibbling tools, punches, reamers, and circle cutters can easily make neat holes in metal chassis and cabinets. The handheld saber saw can quickly cut pc boards, perfboards, cabinet materials, and large rectangular metal holes (FIG. 7-6). The small battery-operated iron was designed to solder transistors and ICs.

7-6 Use the saber or hand saw to cut thin perf and pc boards.

DO'S AND DON'TS

- Do proceed with caution and follow all power in the operating manuals.
- Don't operate a power tool in the rain or in damp conditions.
- Do wear protective glasses when drilling, grinding, and sanding surfaces (FIG. 7-7).

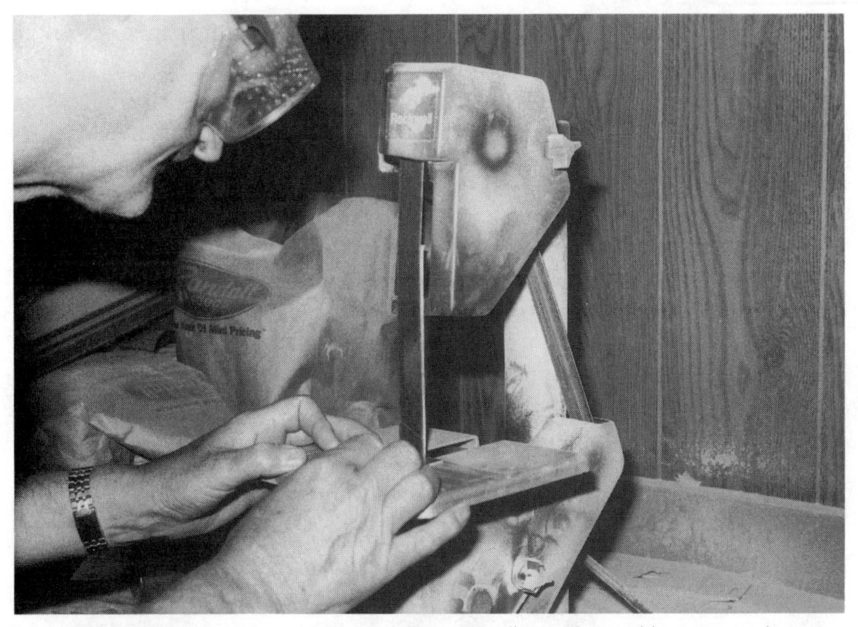

7-7 Wear safety glasses when grinding or sanding surfaces with power tools.

- Do use clamps to hold all pieces to be drilled with a power drill or drill press.
- Do be careful when using a hacksaw to cut thin plastic, metal, or round tubing.
- Do be careful when operating the soldering iron to prevent burns on arms, fingers, tables or projects.
- Don't let battery-operated power tools remain idle for lengths of time before charging the batteries.
- Don't use acid-core solder on electronic projects.
- Don't hold the soldering iron on transistor or IC terminal connections too long.
- Don't tug or pry on transistor or ICs.
- Do doublecheck all wiring at least twice before firing up the project.

- Do have a lot of patience when working with electronic projects, especially when you make a mistake.
- Do collect all parts and components before starting the electronic project.
- Don't forget to order those special parts that can't be found locally from mail order firms.
- Do dab enameled paint over each nut so it won't loosen.
- Don't forget to place rubber grommets or strain insulators on ac cords to prevent shorted and pulled out cords.
- Do keep bare wires off the metal chassis or from shorting against each other by connecting them to insulated terminal strips.
- Do have fun! Build your own test equipment for your workbench.

SELECTING THE CORRECT DRILL BIT

You can start out drilling with two or three drill bits. Most hardware stores have small drill kits, which contain 6 or 8 bits that can do the job (FIG. 7-8). The 1/16-, 1/8-, 1/4-, and 3/8-inch bits are most commonly used in electronic construction. Use the 1/16-inch bit to drill terminal mounting holes in the pc boards. The 1/8- and 9/64-inch bits are just right for bolt holes. For making holes for volume controls and tuning capacitors, choose the 1/4- and 3/8-inch bits. The plastic or metal drill bit template is nice to have on the bench, because it has correct hole sizes stamped on it. Keep drill bits sharp (FIG. 7-9).

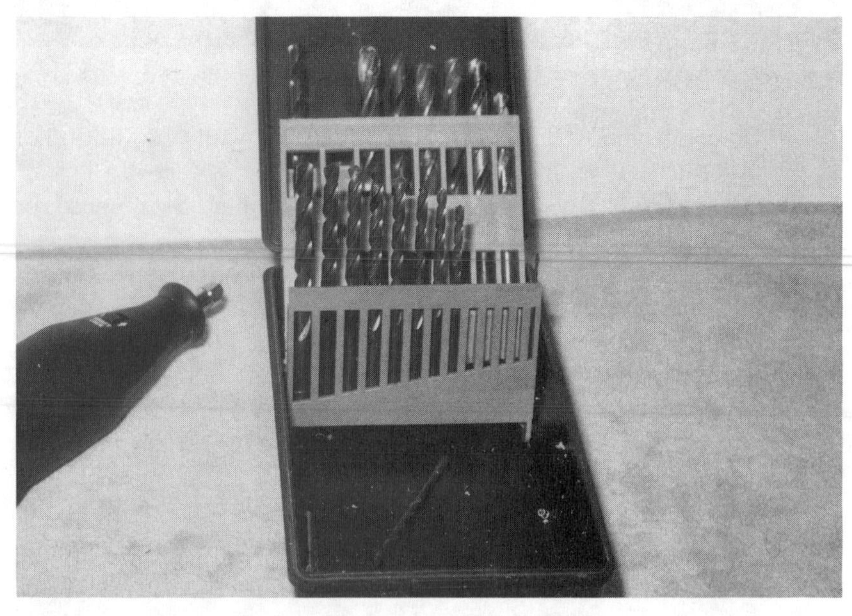

7-8 A set of power bits inside a plastic case.

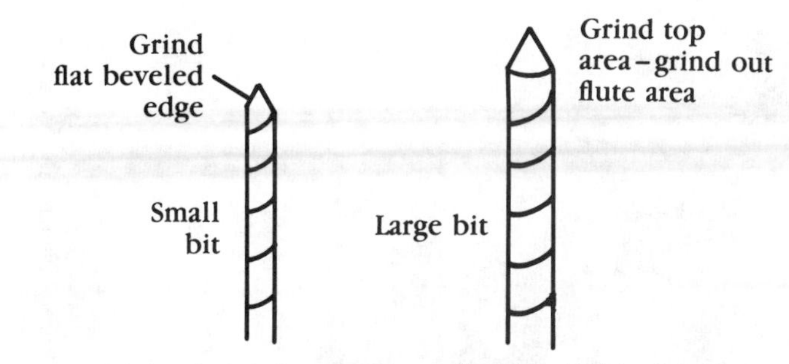

7-9 Sharpen small bits by grinding flat sides against the drill bit. Grind larger bits at the bevel end.

HOW TO DRILL HOLES

It takes a little longer to make holes in metal with the hand drill. The power hand drill can make them quickly, but it can also scratch or place additional holes before you realize what has happened. Mark each hole to be drilled with a center punch (FIG. 7-10). Be careful not to hit it too hard with the hammer or larger dent marks will appear in the light metal. Centerpunching keeps the bit from skidding and marring the finish (FIG. 7-11).

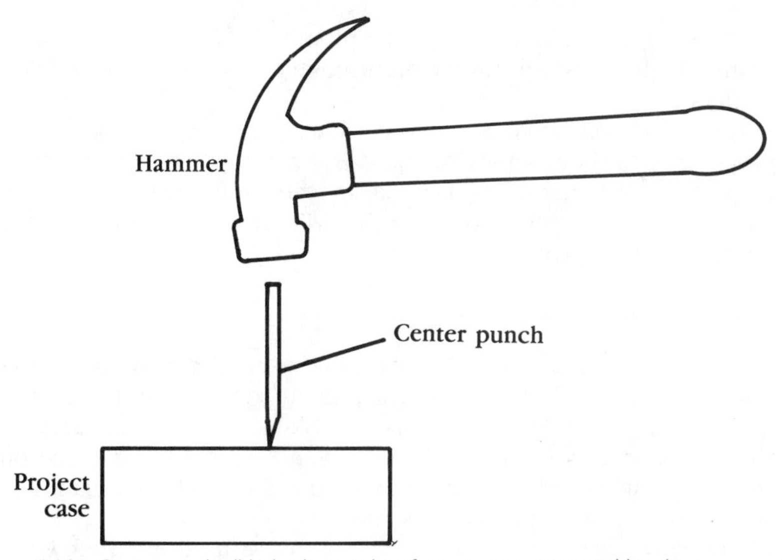

7-10 Centerpunch all holes in metal surfaces to prevent scratching the cases.

Be very careful when drilling plastic cases and project covers. Too much pressure can crack the plastic before the drilling is complete. Take it slow and easy; firmly hold down the plastic piece against a block of wood. When drilling the bottom area of the box, place it upside down or place a piece of wood inside the case to prevent it from cracking. When

7-11 Use the bench power drill to drill small holes in plastic or in pc boards.

plastic burrs form on top, stop. Let the plastic cool and push off the burrs with fingers. If necessary, run the bit through the hole once again to clean the edges.

Keep a light oil can handy when drilling large or deep holes in metal. Squirt oil into the hole while drilling. If you don't, you might quickly ruin the tip of the bit. Stop drilling if smoke appears or if the bit becomes overheated. Proper care and handling of the drill bits keep them alive to drill holes in your next project.

PROTECTING CABINETS AND CASES

Leave the plastic covers on expensive cabinets until you are ready to use them. Just one scratch can ruin the whole project (FIG. 7-12). When drilling holes in the top plastic case, place masking tape over the area to be drilled. This procedure prevents the drill and the plastic or metal burrs from scratching the cabinet. Always shut the drill off before removing it from the drilled piece, to prevent scratching the finish.

Lay down a protective cloth, towel, or mat. After drilling a few holes, lift the project and shake out any small pieces caught in the cloth. You might want to layout the holes on a separate piece of paper (FIG. 7-13). Then fasten the layout to the panel with masking tape. When drilling plastic, start with a small bit and then use a larger bit to finish the large hole, to prevent splitting around the hole area. Brush or pick off all small pieces before removing the paper.

7-12 Leave all cabinets wrapped until you are ready to use them.

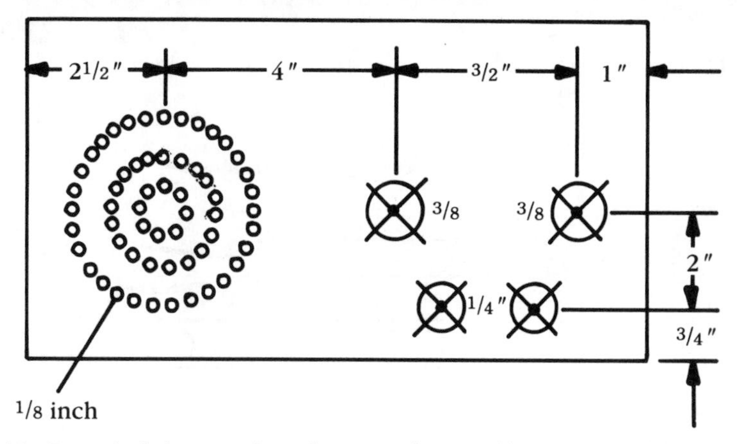

7-13 Draw the holes on a piece of paper and use masking tape to hold it in place.

SOLDERING

There are many ways to use a soldering iron. Always lay down a cloth or newspaper to prevent burning the bench or table (FIG. 7-14). Place the iron in a soldering iron holder or on a metal tool to prevent burn marks. Keep plastic-covered tools and the plastic case away from the soldering iron. When soldering inside the plastic case, keep the hot iron away from the plastic sides or risk melting them.

When soldering a connection, heat the area with the iron tip and then apply the solder to the material (FIG. 7-15). Make sure all connections are clean. Hold the iron on the connection long enough to let the solder

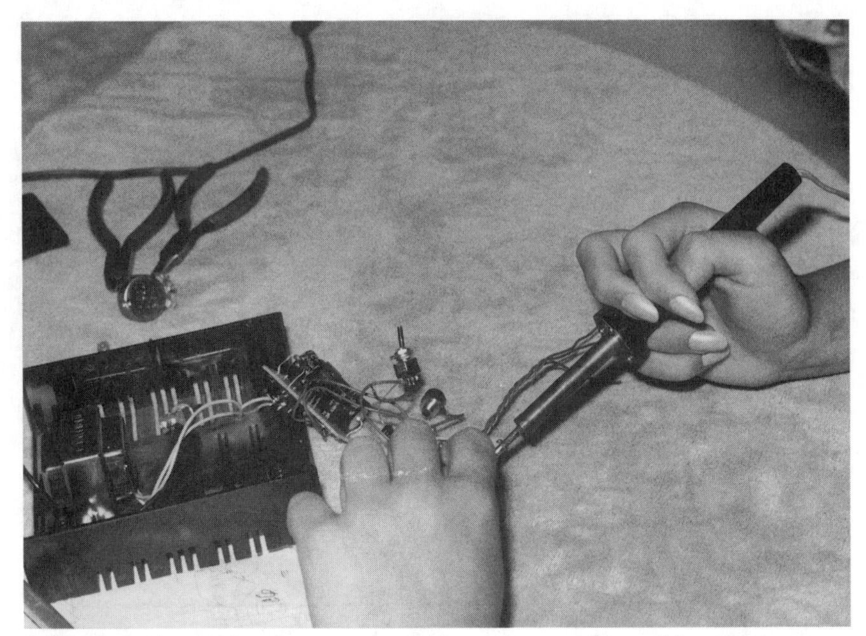

7-14 Use a drop cloth or newspaper to solder projects on. Be careful not to damage or burn the work area.

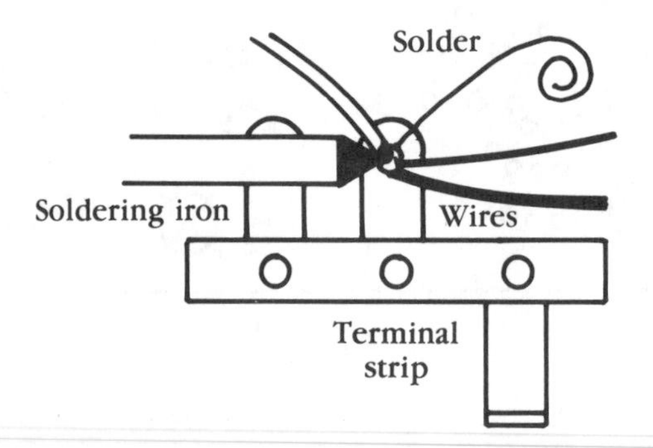

7-15 Rest the soldering iron point against the connection and melt solder to the material for a good solder joint.

flow. Don't melt the solder on the iron tip and then place it on the material to be soldered. This technique always produces a poor soldering joint. Pull on each wire to make sure it's solid. Wrap hookup wire around the terminals or connections before soldering. If need be, check the soldered connection with a handheld magnifying glass.

Use small light-duty or pc-work solder (0.05) on transistor or IC terminal connections. Solder from a larger gauge might bunch up and connect more than one terminal together. Use standard-size solder (0.062) for

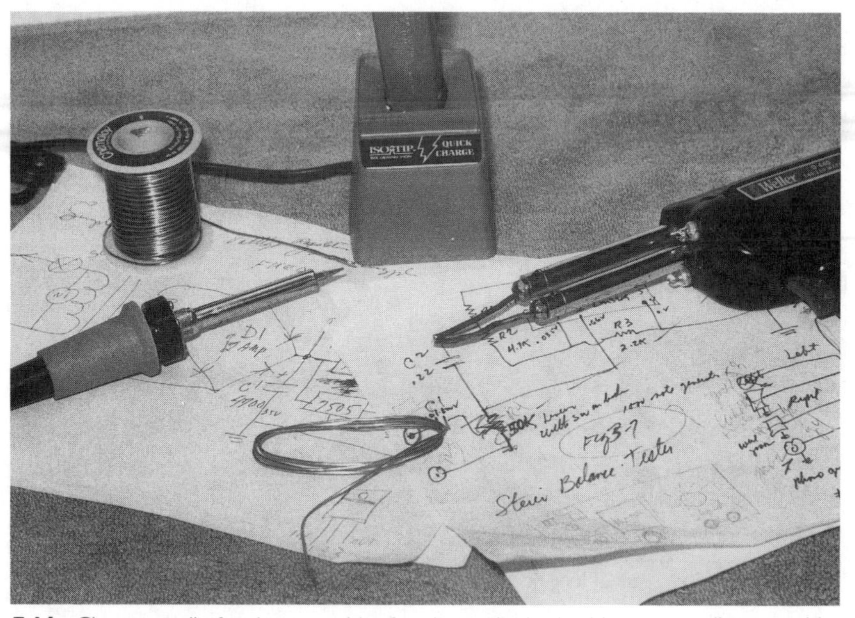

7-16 Choose a roll of rosin-core solder for electronic circuits. Use very small gauge solder for IC board connections.

other electronic connections (FIG. 7-16). Always use rosin-core solder, not acid-core, for electronic circuits.

SOLDERING HEATSINKS

When soldering small transistor and IC terminals, heatsinks should be used to help drain heat from components (FIG. 7-17). Sometimes the transistor can be damaged if no heatsink is used. Excessive heat from the connection goes into the terminal lead inside to the junction. Clip the heatsink tool at the base of the transistor lead to be soldered.

Forcep clamps or straight mini-forceps are ideal to clamp transistor leads. A forcep clamp looks like a pair of scissors with blunt nose ends (FIG. 7-18). These clamps lock onto the terminal and do not necessarily have to be held during the soldering process. Don't forget that your pair of long-nose pliers can be used as a solid-state heatsink.

Transistor and IC heatsinks

To keep the transistor or IC from overheating and destroying itself, heatsinks can be applied or solid-state parts can be mounted on heatsinks. Heatsinks come in many different sizes and shapes (FIG. 7-19). Usually, heatsinks are soldered or bolted to the metal ground.

Heatsinks are available for TO-3, TO-5, TO-18, TO-66, TO-82, and TO-220 case-type transistors. Some of these heatsinks wrap around the

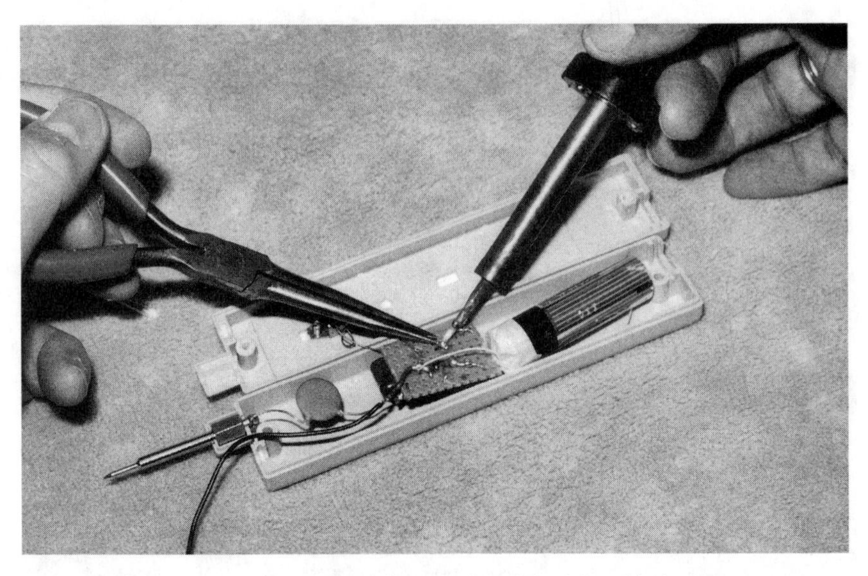

7-17 Use a pair of long-nose pliers as a heatsink while soldering ICs.

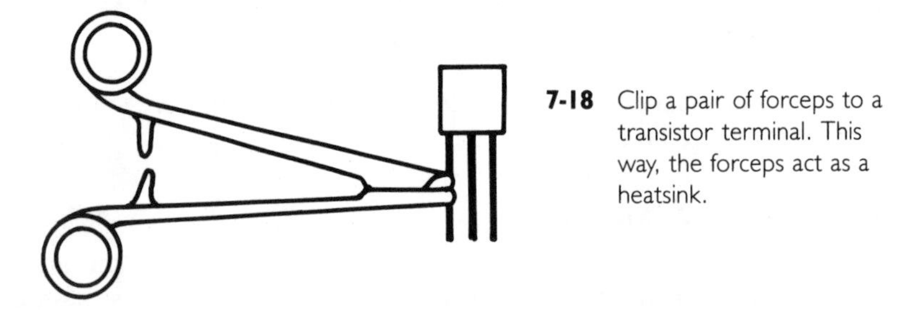

7-18 Clip a pair of forceps to a transistor terminal. This way, the forceps act as a heatsink.

transistor and others have the transistor bolted to it. High-wattage power output transistors require longer and wider heatsinks to dissipate the operating heat. IC heatsinks mount on top or come with the heatsink already attached.

NEAT WIRING

Try to keep the hookup wiring as neat as possible. Of course, all leads in electronic circuits should be directly wired. Sloppy wiring, with loops and extra wire, should be avoided inside the instrument case (FIG. 7-20). Twist the wires connected to the volume control, input wiring, and microphone hookups. Shielded wiring should be used on high-gain audio input amplifier circuits. Do not leave the bare ends exposed, except on common ground leads.

7-19 Several different heatsinks for use with the smallest transistor up to high-wattage output transistors.

SOLDERING IRONS

Always use a 35-watt iron or less when soldering transistors and IC components. The 15-to-30-watt pencil irons are great. Replacement tips are needed after years of soldering. When one tip gets burned badly, replace it with a new tip. Some pencil soldering irons change from 15- to 30-watts with a flick of a switch.

7-20 Keep the wiring as direct and neat as possible. Twist input and high-gain audio wires together to prevent hum pickup and improve neatness.

7-21 Use a solder gun for large terminals and surfaces.

Special irons, such as the 100-, 200-, and 300-watt guns, are used on larger melting surfaces (FIG. 7-21). These irons work best for soldering #10 and #8 wire, removing parts soldered to a metal base, and soldering ground connections. The pistol-grip soldering guns might have different wattages applied to the soldering tips. Replace the tip when it becomes burned into; these wire tips are often copper or plated copper.

Battery-powered soldering iron

The battery-powered soldering iron is small, lightweight, and it can be taken anywhere. It is ideal for connecting transistors and IC terminals (FIG. 7-22). Of course, the battery iron will not remove components from large grounded areas or connect large power line wires. Its battery charges when it is returned to the cradle. If the battery iron becomes cold after making several connections, it needs to be recharged. After 15 or 20 minutes of recharging, the iron can be used again. Today, a small portable gas-powered soldering iron can operate for up to 60 minutes per tank.

Temperature-controlled irons

The temperature-controlled iron operates on a set temperature between 350 and 850 degrees. The small dual-wattage iron can be switched between 15 and 25 watts. Larger soldering stations handle all precise and delicate soldering operations in the most demanding applications. The selected tip temperature is automatically controlled (FIG. 7-23). These variable-temperature soldering stations might contain an off/on switch, power-line insulation, a grounded iron tip for CMOS usage, micropoint,

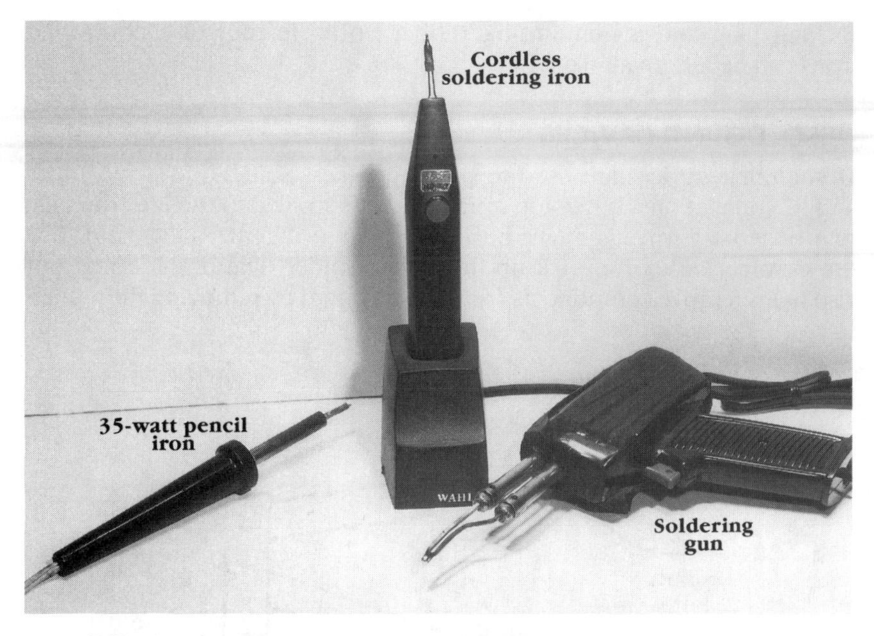

7-22 Use small cordless battery irons to solder transistors and IC terminals.

7-23 The temperature-control soldering iron can operate for many hours applying a constant temperature. Also, these irons are grounded for IC, MOS, and microprocessor replacement.

locking plug and a nonburning rubber cord. Temperature controlled irons can be left on all day.

DESOLDERING DEVICES

Desoldering devices are used to pick up excess solder from transistors and IC connections. When too much solder is applied to the IC terminals, two or more terminals might become soldered together (FIG. 7-24). Use the desoldering tool to pick up the excess solder. The desoldering tool also helps remove components from the pc board by removing the solder.

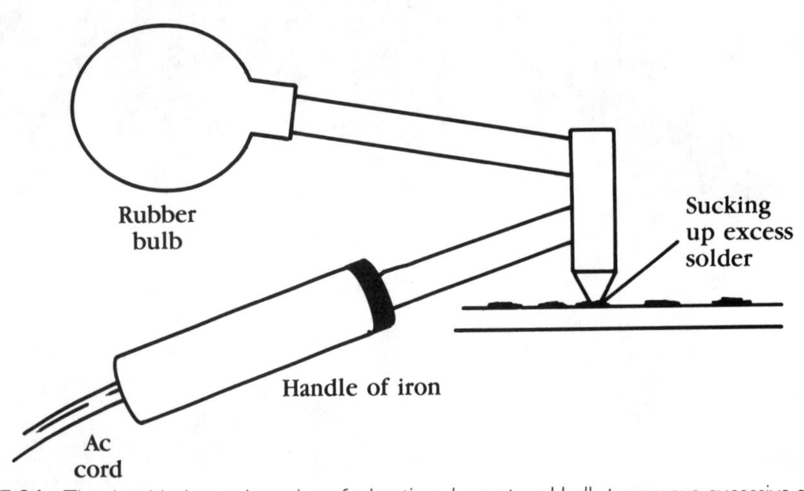

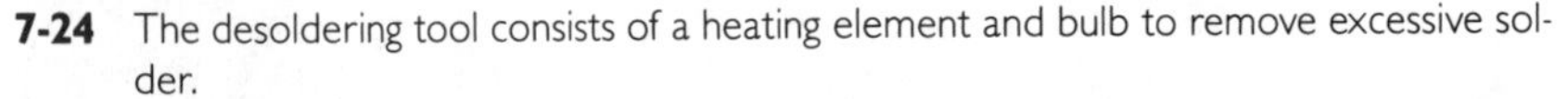

7-24 The desoldering tool consists of a heating element and bulb to remove excessive solder.

These kits come as a desoldering adapter, bulb, and desoldering iron. The suction bulb is depressed and the heated excess solder is sucked inside. The solder wick, which is made up of copper mesh with soldering paste, is used the most by electronic technicians. It is cheap, comes in various widths, and does a clean job (FIG. 7-25). Cut off the end of the mesh when it becomes saturated with solder.

CUTTING PERFBOARDS AND PC BOARDS

Cut the small perf or pc board from a larger piece of stock with a common hacksaw or saber saw (FIG. 7-26). Place the board in a bench vise or alongside the bench for cutting. Mark the small perf or solder-ringed-hole universal pc boards with a pocket knife, place it in a vise at the point of breakage, and snap it. You can even cut holes or slots in the pc boards with a metal blade placed in the bench scroll saw (FIG. 7-27). Grind the rough edges on a sander or bench grinder.

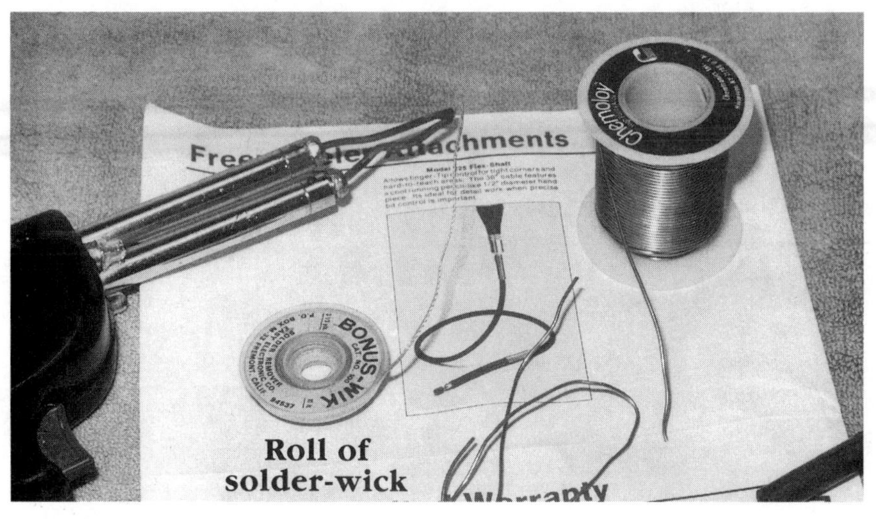

7-25 Solder-wick is an inexpensive mesh-type copper roll, available in different widths.

7-26 Perf or pc boards can be cut with a common hacksaw or saber saw.

Cutting larger holes

Besides the bench scroll saw, the hand knibbling tool neatly cuts up to #18 steel, 1/16-inch copper, aluminum, and plastic. These tools can be purchased at local hardware or electronic mail order firms (FIG. 7-28). After drilling a starting hole, the knibbler can cut round and square holes in minutes. The tapered reamer can enlarge small holes up to 1 inch in diameter. Circular and rectangular punches can be used with the hammer

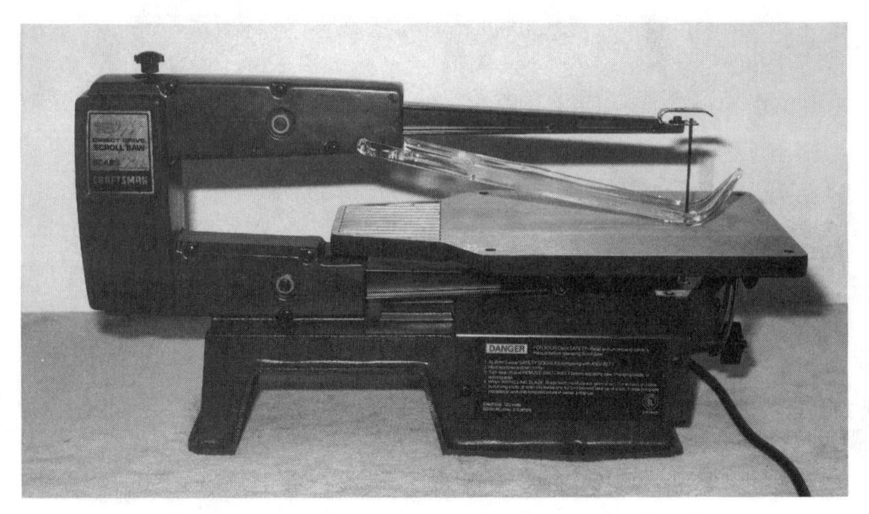

7-27 Circuit boards can be cut on the scroll saw, if handy.

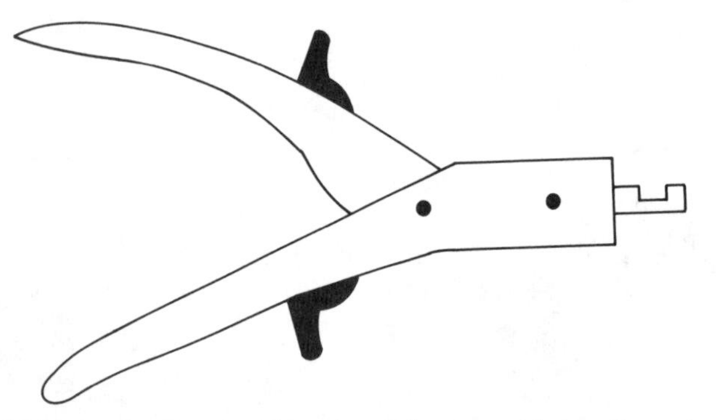

7-28 The nibbling hand tool can quickly cut round or square holes after a starting hole is made.

to make holes in the metal chassis. There are many different types of punch and die-socket sets to make holes in metal with a metal-working press, however, this method is rather expensive for project builders.

Power tools

The electric drill and saber saw found in most homes are rather inexpensive (FIG. 7-29). Ac-powered screwdrivers and drills have dropped in price. Even portable battery-operated drills and screwdrivers are quite inexpensive nowadays. Battery-operated tools make quick work in drilling holes and securing bolts, nuts, and screws in building projects (FIG. 7-30). These battery-operated tools operate from a 3-to-10.8-volt source.

7-29 Most people already have an electric drill or saber saw on the workbench.

7-30 The cordless power drill and screwdriver are ideal when building electronic projects.

Keep cordless tools charged

The cordless drill or screwdriver can operate from a 3-to-7-volt source. Usually, small nickel-cadmium batteries are wired in series to produce the required operating voltage (FIG. 7-31). When the screwdriver begins to

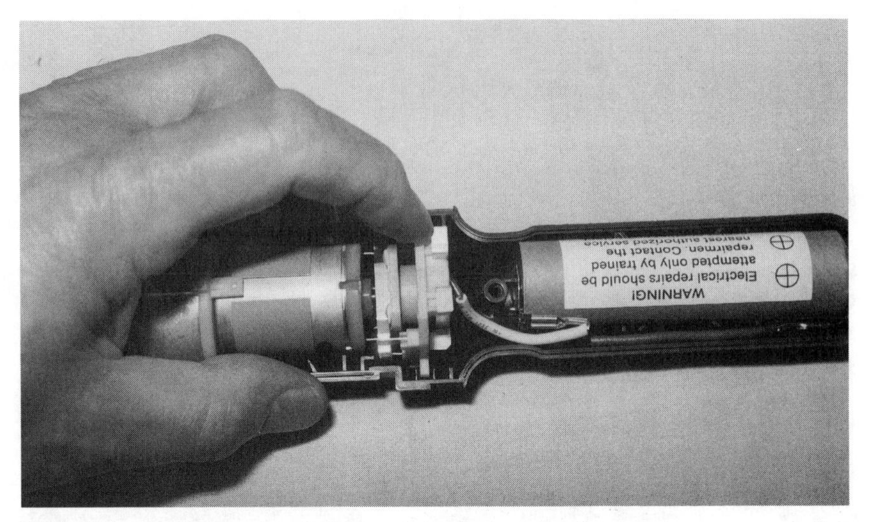

7-31 This large battery is actually two batteries in series to operate a power cordless screwdriver.

slow down, the battery needs to be recharged. All cordless tools come with a small battery charger that plugs into the ac power line. Some chargers shut off when the battery is fully charged and others are charged in 2 to 4 hours (FIG. 7-32).

Suspect one or more dead batteries if the battery pack will not charge. If it takes twice as long to charge the batteries, you know one or more of the batteries is getting weak. Check the total voltage of the batteries with a DMM or VOM if the batteries will not charge (FIG. 7-33). Remember, nickel-cadmium battery voltage is less than that of a flashlight cell—usually, 1.12 or 1.35 volts. Check the batteries if the cordless tool

7-32 Charge the cordless, battery-operated tool by plugging it into the ac power line.

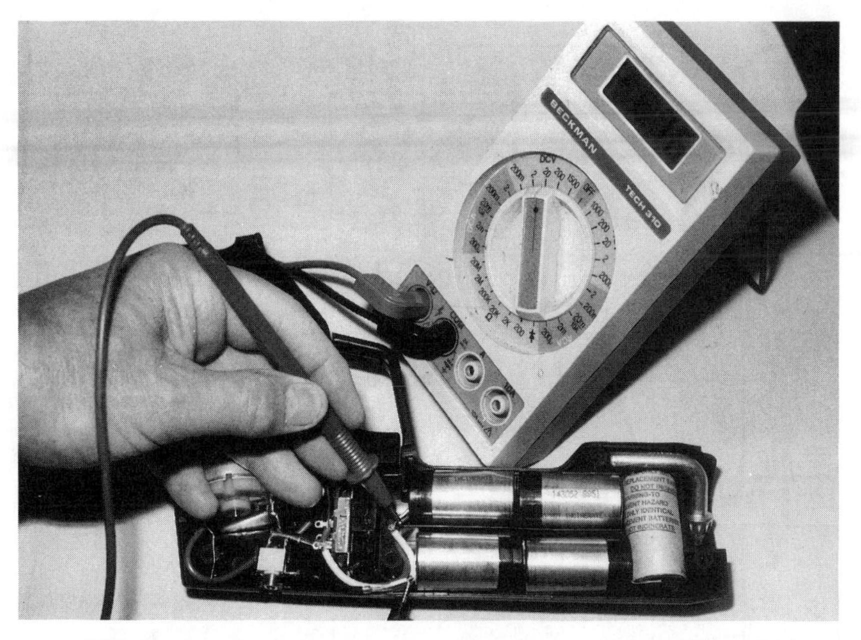

7-33 Check small batteries when they will not charge with the DMM or VOM.

will not operate after recharging. Suspect broken wiring or a defective switch if the cordless tool does not work when batteries are charged (FIG. 7-34).

Cross off wiring on the schematic

To know where you are at in a schematic, simply cross off the circuit with a pencil as you wire each component (FIG. 7-35). This way, you know what is left to connect. Copy or redraw the original schematic from a book or magazine in ink. Of course, with a direct copy, you can't make any mistakes sketching the circuit.

MAKE YOUR OWN DIALS

You can make your own dials and scales with rub-on transfers and lettering. Today, very few dials are made to place on dark cabinets. You might locate a few dark rub-ons, but these do not show up well on dark cases. So, why not make your own dials and scales (FIG. 7-36)?

Purchase a low-priced protractor, drawing compass, ruler, india ink and pens, or direct-etching dry transfers. Dry transfer panel labels and etching transfers can be picked up from Radio Shack and many other electronic stores. Dry transfer lettering can also be purchased at art and hobby stores. If you go the direct-dry transfer route, you do not need india ink or a pen.

By making a white background, you can place black lettering on it.

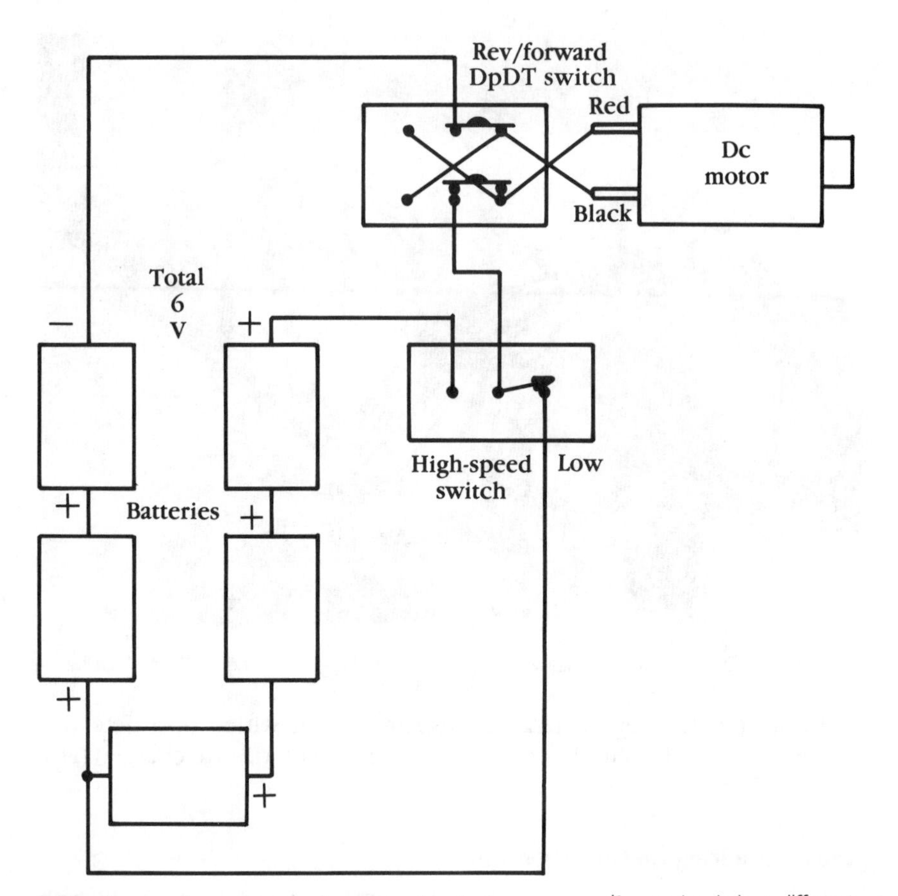

7-34 The cordless power tool consists of batteries, a reverse/forward switch, a different-speed switch, and a low-voltage motor.

Glue or cement the finished product to the new surface or place a small sheet of stick-on clear plastic over each dial for protection. Clear stick-on plastic can be obtained at camera stores or U-SEAL-IT™ plastic can be obtained from vending machines.

First, lay out the meter dial or control dial with a pencil on at least #25 bond white paper. For the dial, draw three different 1/2 circles. The first half circle should be close to the dial knob. Now, place the second circle 3/8 or 1/2 inch from the first circle. Place the last 1/2 circle 1/4 inch from the second 1/2 circle.

Now draw the grid or division lines in with a protractor and pencil. These lines should be equal and angled around the 1/2 circle. Place these lines between first and second 1/2 circles. Draw or press the letters or

7-35 Cross off the schematic wiring with a pencil as you make each connection.

7-36 Use india ink, a pen, dry transfer lettering, and dry etching symbols to make your own dials and meter scales.

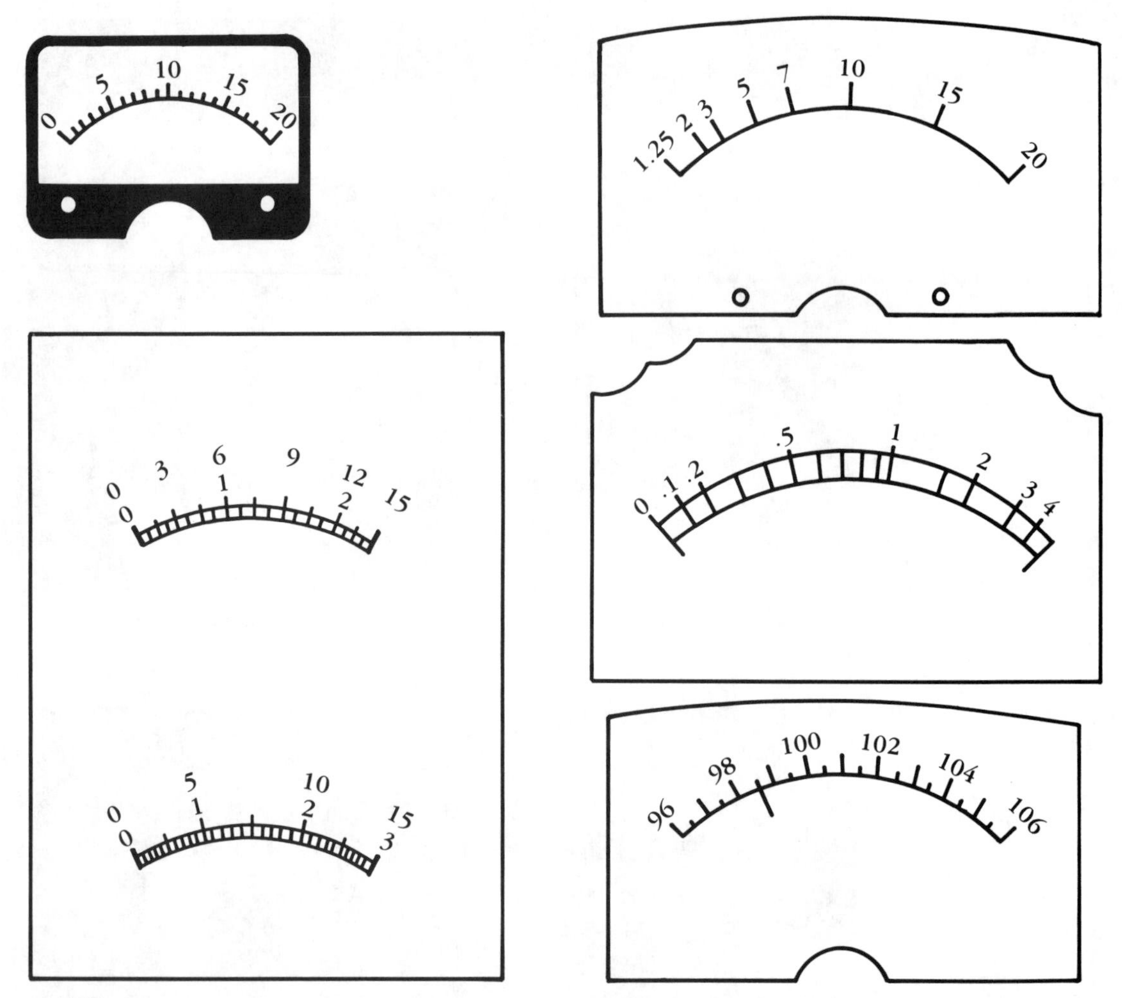

7-37 Draw grid or division lines with india ink and a pen or by using dry-etching transfer lines and patterns.

numbers between the second and third circle. Space each grid or division line evenly around the 1/2 circle (FIG. 7-37).

Select dry etching transfer symbols for the grid lines. Actually, the dry transfer symbols are normally used for etching pc boards. Use the thin lines for the division indicator. Place a 3/8- or 1/2-inch direct-transfer line over each division or grid on the dial scale. Rub on a black line between the first and second 1/2 circle. Try to be neat and don't run over the scale. If a dry transfer line sticks out over the second line, cut or scrape it with a razor blade. Remember, the dry direct transfers can't be moved after they have been rubbed down. After the dial is completed, if you have any black lines where not needed, dab on white typewriter fluid. Place the letters or numbers between second and third 1/2 circle.

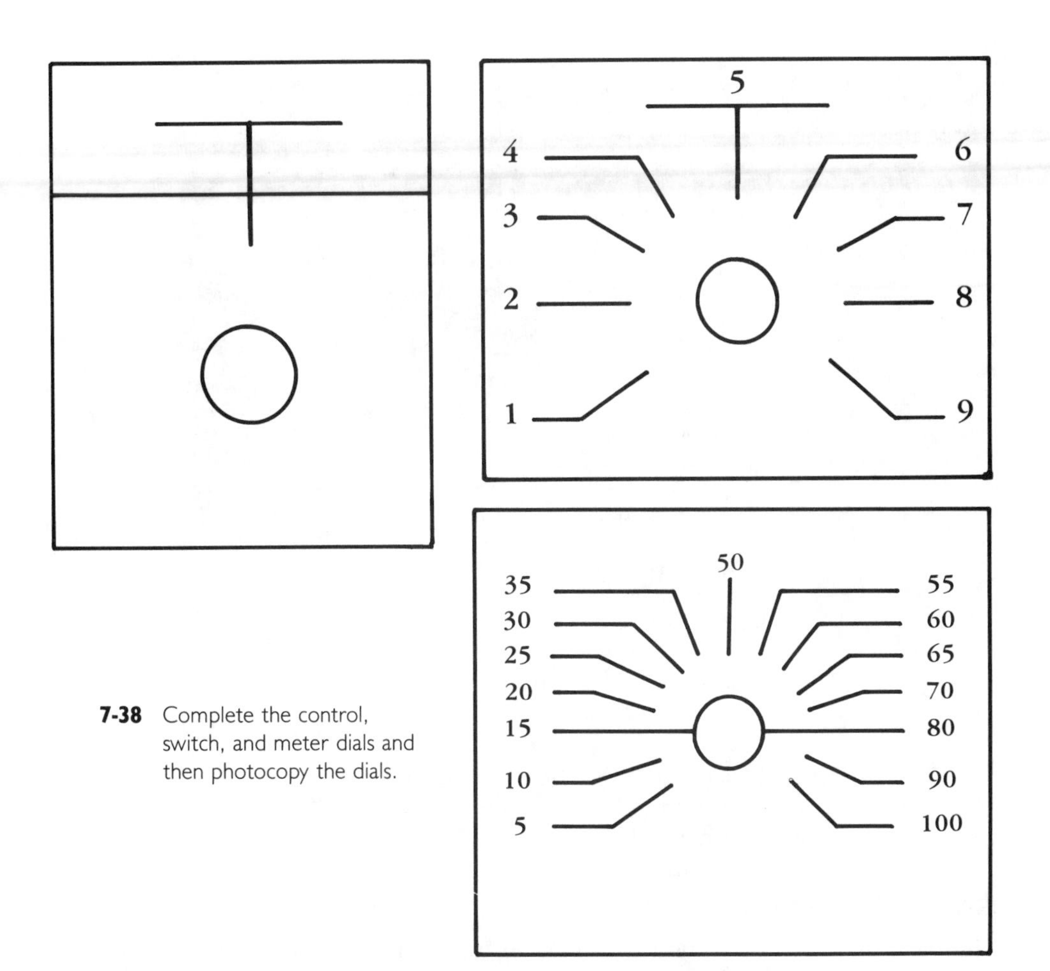

7-38 Complete the control, switch, and meter dials and then photocopy the dials.

You can make each dial in the same manner or photocopy each type. Many mistakes do not show after being copied. Now you can make as many dials as you need without spending hours working (FIG. 7-38).

When making switch dials, be sure that the division or grid line is where the switch stops each time. Place a piece of paper over the switch and draw a line for each switched position. You can do the same for meter voltage, current, or ohm scales. For instance, if you are making a voltmeter dial, draw the grid lines evenly and place the volt numbers where each voltage will be measured with another DMM or VOM.

Mount the dial after cutting if either round, rectangular, or square. Place airplane glue or regular glue in four different places to hold the dial in place. Place the dial horizontally with the matching component. Cut a piece of clear adhesive plastic and leave about 1/8 inch overlap for each side of the dial. Be careful to place the plastic evenly, because it will not come off after it is stuck into position (FIG. 7-39). Letter and mark the parts on the front panel before mounting any components.

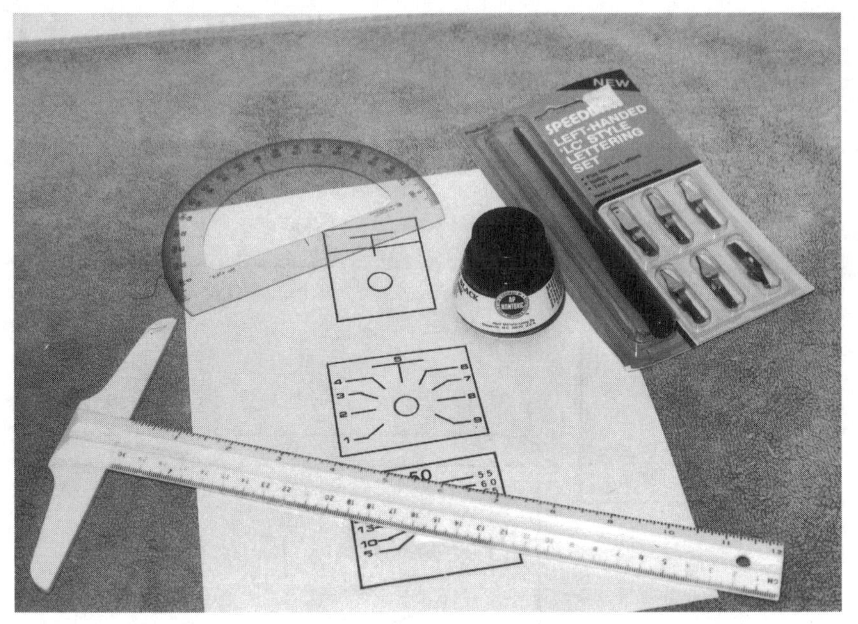

7-39 The tools needed to make dials for your project.

Cut holes in the dial if it goes 280° around the component knob assembly. The dial will stay bright and last for a long time. The case will be protected from finger marks for years.

ISOLATION TRANSFORMER

The electronic technician uses an isolation transformer when servicing the ac/dc chassis of a radio, TV, or amplifier. Anytime a transformer is not used in a test instrument or in a unit to be serviced, connect it to an isolation transformer (FIG. 7-40). The isolation transformer might be able to select different ac voltages. This transformer isolates the chassis from the

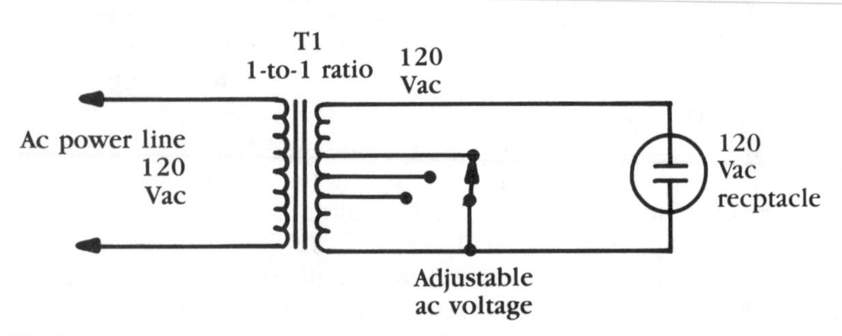

7-40 For safety protection, always connect the ac chassis to an isolation transformer before attaching the test instrument.

tester. If not the direct ac voltage might be hot and destroy line fuses or solid-state components in the chassis.

TROUBLESHOOTING PROJECTS

Nothing is more heartbreaking or discouraging than finishing an electronic project that won't operate. Besides doublechecking the wiring, project connections can be checked with transistor, IC, voltage, current, and resistance measurements. You will need a good DMM or VOM to take voltage, current, and ohmmeter measurements (FIG. 7-41).

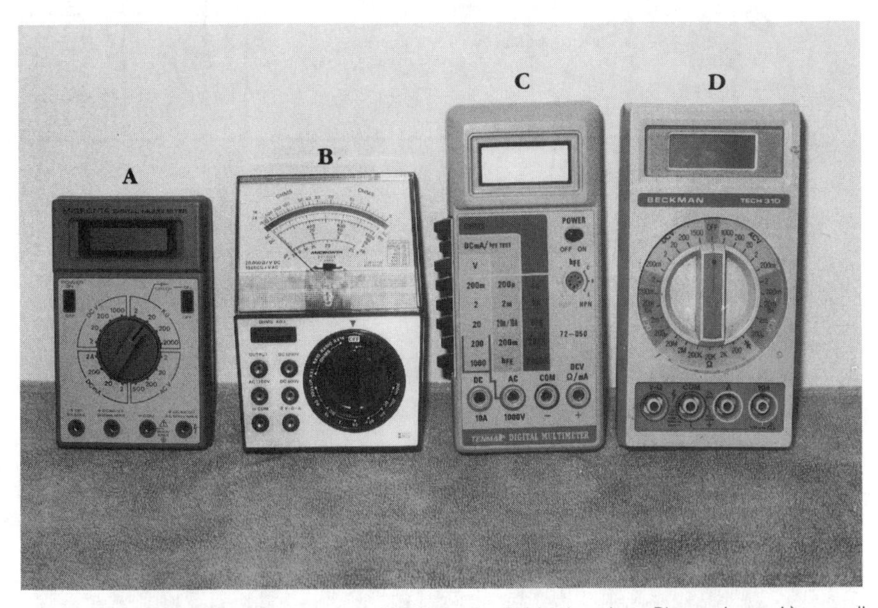

7-41 Use a VOM and DMM to troubleshoot the finished project. Pictured are: A) a small DMM, B) a VOM, C) a push-button DMM, and D) a DMM/diode tester.

The DMM is ideal for it will accurately measure low voltage, current, and resistance. The base bias voltage on a transistor is 0.03 or 0.06 volts. Low-bias resistors, less than 1 ohm, can be measured with the DMM (FIG. 7-42). Current meter readings in low milliampere measurements can be taken with the DMM. Also, you can test diodes, transistors, and capacitors with some DMMs. Some DMMs have audible sounds for different test measurements.

After checking the wiring, take critical voltage measurements (FIG. 7-43). Low voltage might occur as a result of a leaky or shorted component. Then, make a current test. The current meter can be inserted across the switch, between the battery and the circuit, or in the leg of each circuit (FIG. 7-44). A quick overall current check in a battery-operated project can be made by connecting the dc voltmeter across the switch terminals.

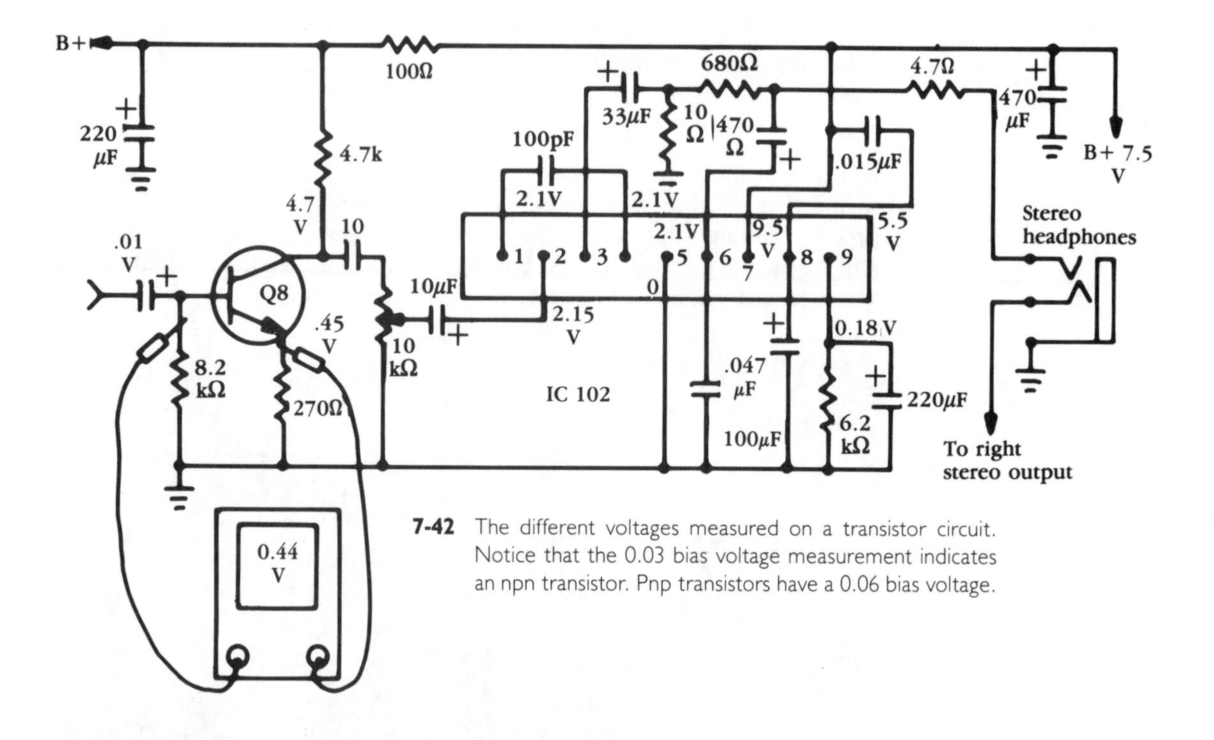

7-42 The different voltages measured on a transistor circuit. Notice that the 0.03 bias voltage measurement indicates an npn transistor. Pnp transistors have a 0.06 bias voltage.

7-43 Several projects with lettered dials.

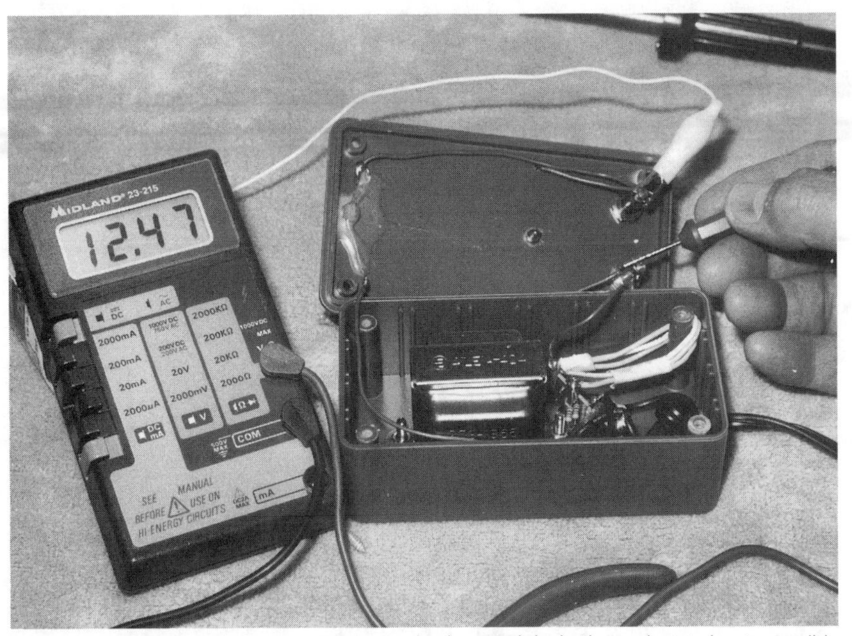

7-44 Take accurate voltage measurements, both ac and dc, in the project to locate possible defects.

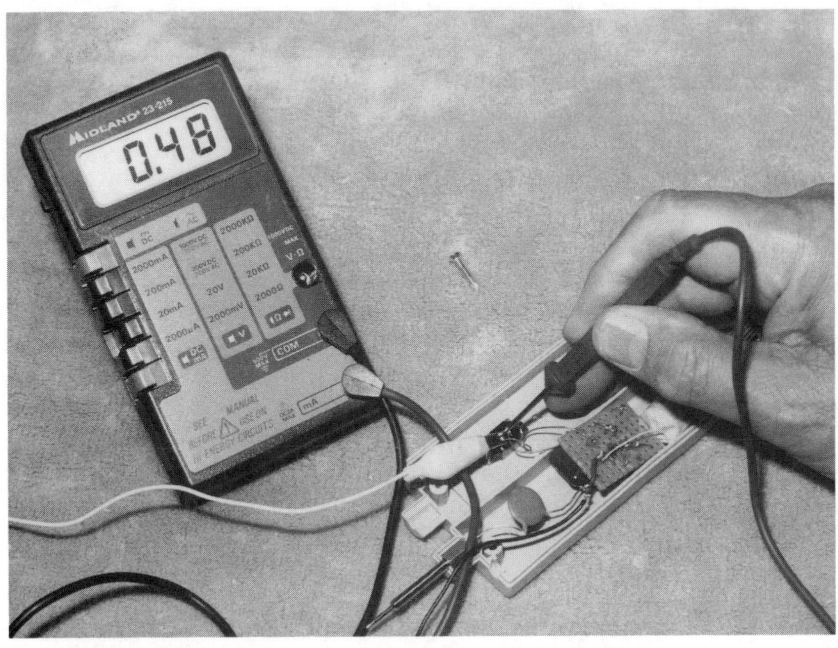

7-45 Check the current across the switch terminals of a dc-powered project.

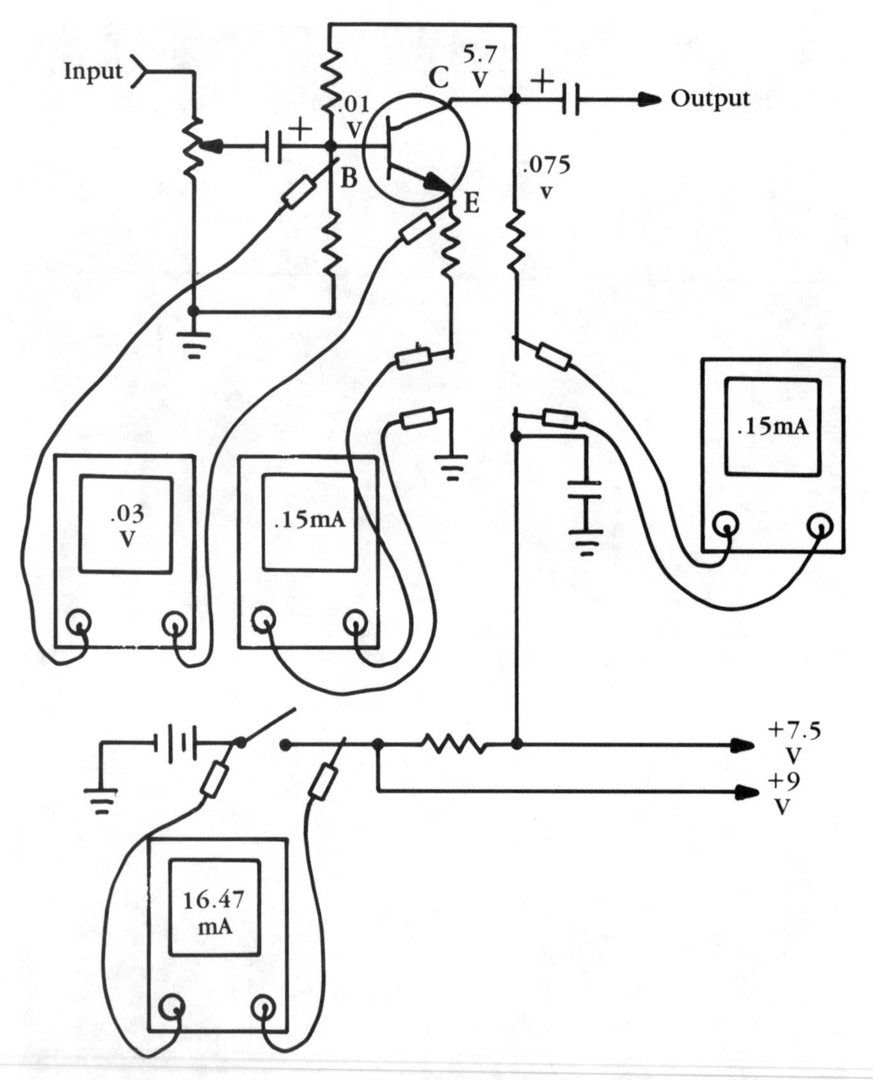

7-46 Check the current in the different areas of a solid-state circuit.

Do not try this in the ac switch circuits. A high-current pull might indicate a shorted part or leaky transistor/IC current. Don't overlook misplaced or incorrectly soldered wires, which could cause a higher-than-normal current measurement (FIG. 7-45).

After locating a suspected transistor or IC with low-voltage measurements, check the transistor with a tester. Measure the resistance of each terminal pin of the suspected IC to common ground. A resistance measurement under 1 kΩ on the IC voltage supply terminal (VCC) might occur as a result of a leaky IC. Of course, the measured voltage will also be

7-47 Label the finished project with a tape writer machine.

lower on this pin. Take a resistance measurement from each IC pin to ground to locate the leaky component. Many ICs have been replaced, even though they were normal, because a leaky part tied to one of the pins was defective. Practically all of the test equipment in this book have troubleshooting procedures at the end of the text.

THE FINISHED PROJECT

The completed electronic project can be improved by placing it in a fancy cabinet. Dress up the cabinet by placing tape writing labels by each terminal or control (FIG. 7-46). These tapewriter kits come with different colored labels. The letters can be made horizontally or vertically. After the label is printed, cut it off on the machine, and stick it on the project case.

Project identification can be made with dry transfer lettering and numbers. The surface must be clean and flat. Anyone can stick transfer lettering on dials and controls with a little patience and care. Place the correct letter in position and rub it on with the blunt point of a pencil. Spray it with clear lacquer or poster spray to prevent damaging the labels. Correctly place the transfer lettering, spray the top of the case, and finally, mount the controls or components on the case or cabinet (FIG. 7-47). Before mounting parts on the front panel, apply letters and numbers. Spray on a clear matte finish so that the letters will not become scratched.

Chapter 8

Obtaining components

Try to locate all of the components for a project before you start to build circuit boards. Electronic parts can be purchased locally, through mail-order firms, and through manufacture servicing depots (FIG. 8-1). You might find surplus electronic parts will fit the bill or by looking a little closer, you might find the parts in your junk box.

LOCAL ELECTRONIC DISTRIBUTORS

Most every town has a local electronic parts distributor or a Radio Shack store. Sometimes a larger Radio Shack store, located in cities, will have everything needed to construct most of these test instruments. You might find that some electronic distributor will not sell to ''every Tom, Dick or Harry,'' but give the locals a try, anyway.

Don't forget to check your local radio/TV establishment. Sometimes when a special part is needed, they can obtain the component through their electronic sources. Don't be surprised if they even have one on hand. Just take a part list, description, and component number to obtain the correct part. Many transistors, ICs, resistors, capacitors, and controls can be purchased from your local radio/tv technician.

MAIL-ORDER PARTS

Although you might want to start at once on one of these electronic test instruments, you probably can't purchase all the parts locally. Why not try ordering from one of the mail-order firms. You might find the parts at a much lower price. Sometimes you can find unused cabinets, boxes, and

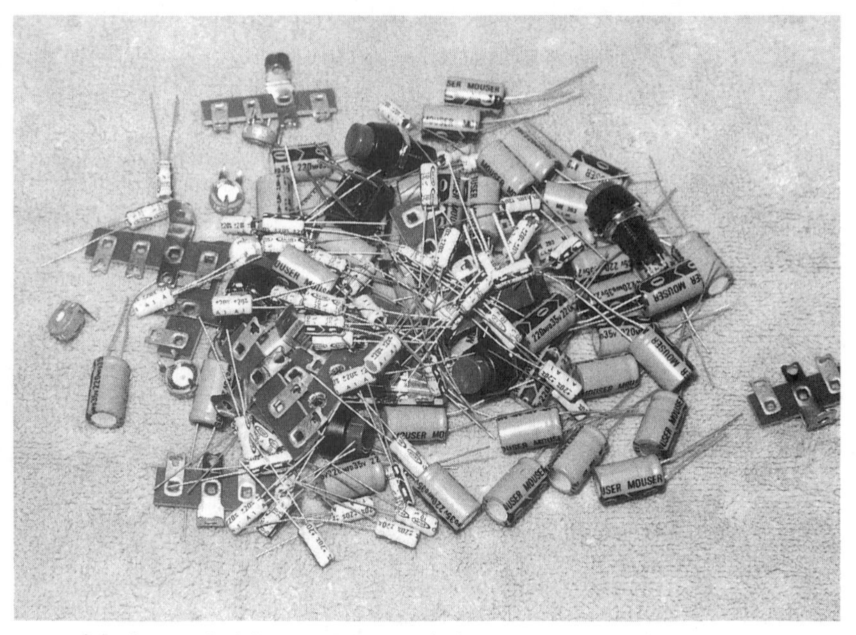

8-1 Locate all of the necessary parts before you start to build any project.

8-2 Several excellent commercial-looking cabinets from Radio Shack, Global Specialities, and Hosefelt Electronics mail order firms.

transformers that were made for another commercial manufacturer (FIG. 8-2). These new components work nicely in building your test equipment projects.

The best bet is to write to these companies and order catalogs if you don't already have any. Some mail-order firms have no minimum order, and others might have a $10 or $25 minimum order. Some electronic out-

8-3 Try mail order firms to locate those electronic parts. You might find that these parts are available at a lower price and can be shipped within 24 hours.

lets will accept check, money order, Mastercard, VISA, COD, and cash (FIG. 8-3). Most of these firms ship within 24 hours. You might also be able to call in your order by phone for fast delivery. Some of the various electronic outlets who supply electronic components are:

ALL ELECTRIC CORP.
P.O. Box 567
Van Nuys, CA 91408

AMERICAN DESIGN CORP.
815 Fairview Ave.
P.O. Box 220
Fairview, NJ 07022

BCD ELECTRO
P.O. Box 830119
Richardson, TX 75083

BUDGET ELECTRONICS
P.O. Box 1477
Mareon Valley, CA 92337

CIRCUIT SPECIALISTS
P.O. Box 3047
Scottsdale, AZ 85271-3047

CONSOLIDATED ELECTRONICS
705 Waterviet Ave.
Dayton, OH 45420-2599

D&C ELECTRONICS
P.O. Box 3203
Scottsdale, AZ 85271-3203

DIGI-KEY CORP.
701 Brooks Ave. So.
Box 677
Thief River Falls, MN 56701

ELENCO ELECTRONICS, INC.
150 W. Carpenter Ave.
Wheeling, IL 60090

FORDHAM RADIO
260 Motor Parkway
Hauppauge, NY 11788

GLOBAL SPECIALITIES
70 Fulton Terrace
New Haven, CT 06512

HOSEFELT ELECTRONICS, INC.
2700 Sunset Blvd.
Steubenville, OH 43952

H&R CORP.
401 E. Erie Ave.
Philadelphia, PA 19134

INTERNATIONAL MICROELECTRONICS
P.O. Box 170415
Arlington, TX 76003

JAMECO ELECTRONICS
1355 Shoreway Rd.
Belmont, CA 94002

MCM ELECTRONICS
858 E. Congress Park Dr.
Centerville, OH 45459-4072

MICRO MART
508 Central Ave.
Westfield, NJ 07090

MOUSER ELECTRONICS
National Circulation Center
P.O. Box 699
Mansfield, TX 76063

PARTS EXPRESS
390 E. First Street
Dayton, OH 45402

RADIO SHACK
1400 One Tandy Center
Fort Worth, TX 76102

SOLID STATE SALES
P.O. Box 74D
Somerville, MA 02143

PART SUBSTITUTION

Many different electronic components can be substituted for other parts, such as transistors, ICs, diodes, transformers, resistors, and capacitors (FIG. 8-4). Cabinets and cases can also be substituted.

Transistors

Simply look into semiconductor replacement guides from RCA, GE, Sylvania, and TAB Books for the needed transistor. If you cannot locate this special transistor locally or through mail-order companies check one of these guides for a substitute.

For example, you may need a substitute for an MPS 222A transistor. The semiconductor manual lists the GE-20, Sylvania ECG 123A, and RCA SK3122 as proper replacements. For a 2N3053 transistor, the different solid-state replacement manuals list the RCA SK 3024, Sylvania ECG 128, and GE-241. Check these replacement part manuals for correct transistor substitutions. You might find that your local electronic technician has these transistors in stock. Check transistors with the diode test position on the DMM, on the transistor tester, or on the tester that you can build from this book (FIG. 8-5).

IC components

Often, IC substitutions are a little more difficult, but thousands of ICs have replacements (FIG. 8-6).,

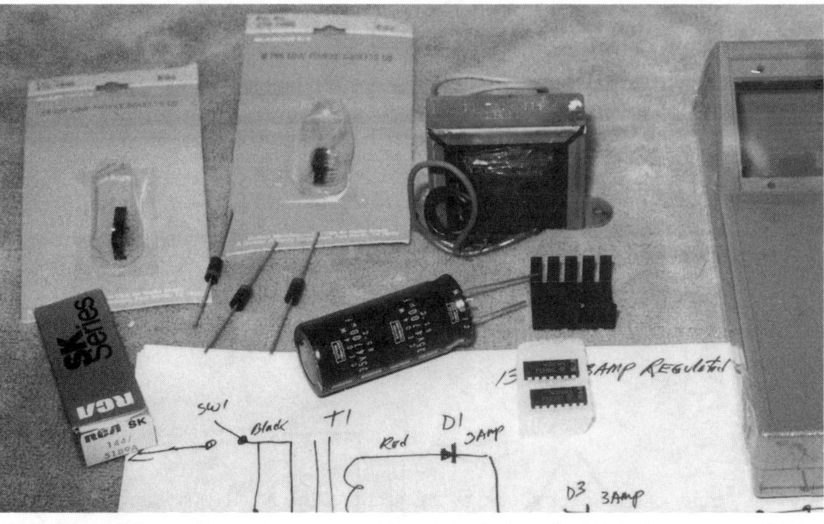

8-4 Many different electronic parts may be subbed when the originals are not available.

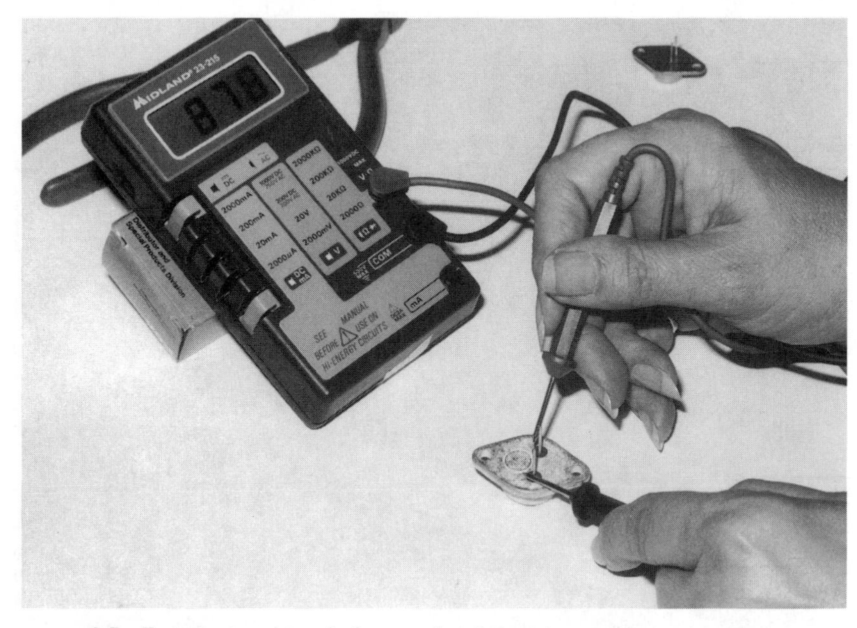

8-5 Test new transistors before you install them in your electronic project.

For example, maybe you cannot locate an LM 386, audio amp IC. When you check the solid-state replacement manual, you will find that the RCA SK 9210 is an exact replacement. So maybe you need another audio amp, an LM 380, and you cannot find one locally. The RCA solid-state replacement manual lists the SK 3228 as the exact replacement. Most

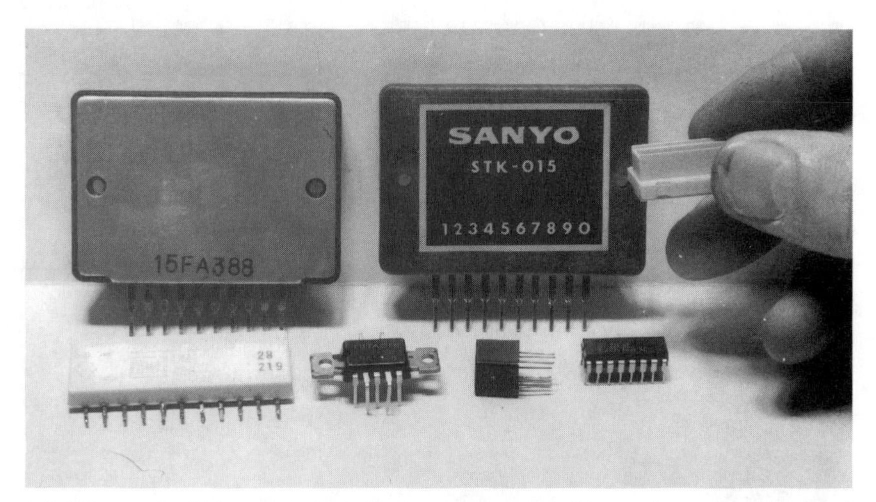

8-6 Thousands of ICs are substituted by the electronic technician each year. Try this route if you cannot locate a special IC.

8-7 This electronic technician is installing a universal IC in a police scanner.

of these semiconductor replacement manuals list different makes of transistor replacements, such as the RCA SK line that can be substituted with GE, Phillips ECG (Sylvania), and Zenith. The direct substitutions are listed in the front of the semiconductor replacement manuals (FIG. 8-7).

Transformers

Many different power transformers can be substituted, if space and weight is not a problem. Watch out for correct voltage and current requirements (FIG. 8-8). If you need a 12.6-volt 300-mA power transformer and cannot locate one, try a 12.6-CT 450-mA or a 12.6-CT 1.2-amp power transformer. Simply cut off the center tap lead and tape it up. The voltage is the same and these replacements furnish more current than you need. If you have room for larger transformers, they will work.

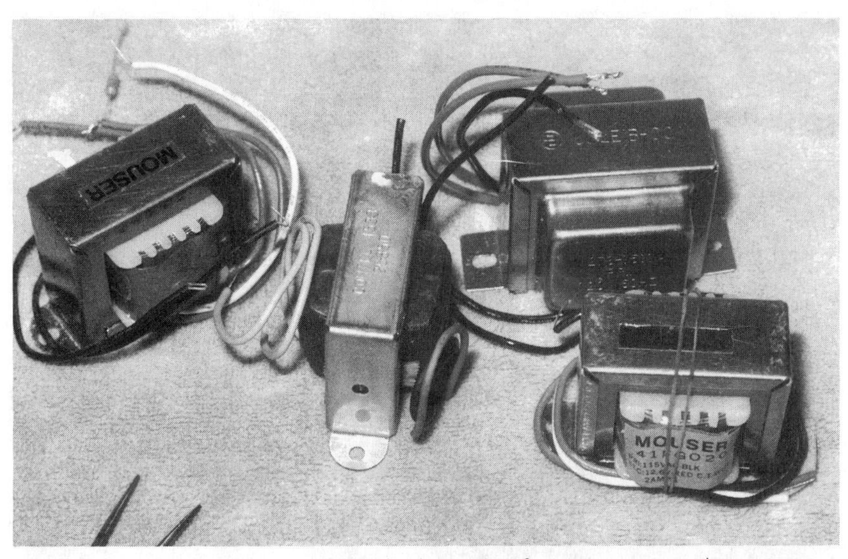

8-8 Substitute a power transformer with a larger transformer or try a surplus component.

Do not replace a 12.6-volt 1.2-amp transformer with a 12.6-CT 450-mA job. It will run warm and could burn up because the current-carrying capacity is too low. You can always substitute a larger current transformer for a smaller one, but not vice versa.

Or maybe you need a 6.3-volt ac power transformer, but you cannot find one locally. So, maybe you can locate a 12.6-volt CT transformer or have one in the junk box. Make the 12.6-volt transformer work by using only $^1/_2$ of the winding (FIG. 8-9). The same holds true when you need higher secondary voltage. Let's say you need 18 volts at 1 amp. Maybe you can locate a 6.3-volt 1.5-amp transformer and a 12.6-volt 2-amp power transformer. Simply wire the primaries in parallel and the secondary windings in series to obtain 18 volts. Of course, more room is needed and the project is a little heavier (FIG. 8-10).

Capacitors

Somewhat like resistors, capacitors can be wired in parallel and in series to reach the correct voltage and capacitance. Let's say you need a 470-μF

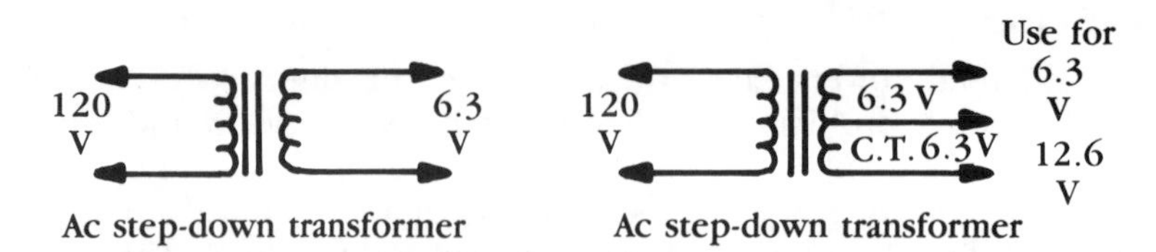

8-9 In this case, a 12.6-volt CT power transformer is substituted for a 6.3-volt transformer by using only half of the secondary winding.

8-10 This electronic technician has installed a 12.6-volt 1.5-amp transformer for a burned-out transformer in a stereo 8-track player.

35-volt electrolytic capacitor for one of your experiments and you already have two 220-μF at 35-volts. Simply parallel both capacitors (FIG. 8-11). If you need a 10-μF at 35 volts and have a 22-μF at 25 volts, use it, it will work okay.

If the project calls for a 3300-μF 35-volt capacitor and none is available, you can parallel a 1000-μF 50-volt with a 2200-μF 35-volt electrolytic

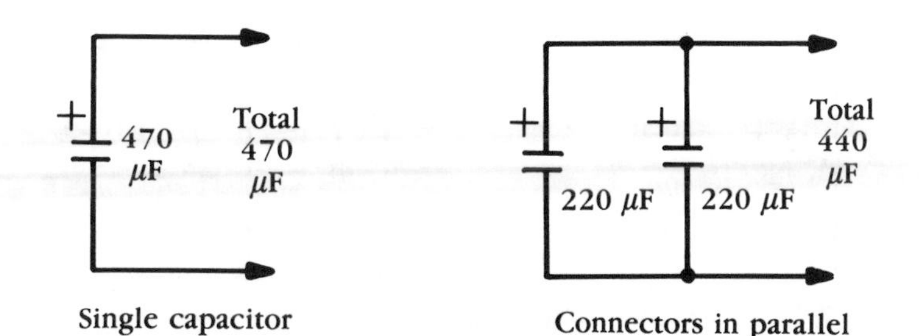

8-11 Simply parallel electrolytic capacitors to increase their capacity. Make sure the working voltage is the same or higher.

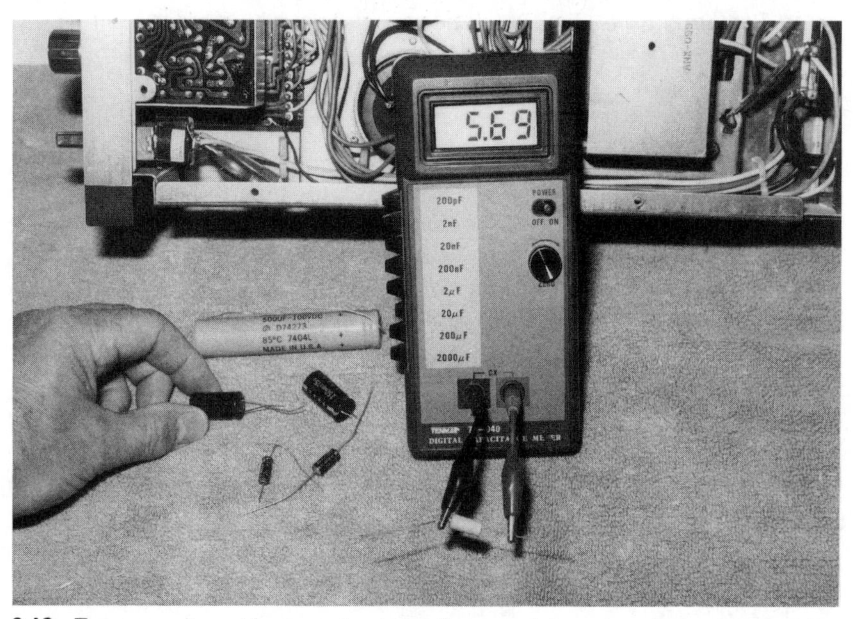

8-12 Test a capacitor with a capacitor tester if you don't know its value or if you need to check it for the correct capacity.

capacitor and never notice the difference. Remember, the capacity of electrolytic capacitors increases when in parallel and the exact or higher voltage ratings can be used for the original. If you do not know the value of a capacitor, check it on a capacitor tester (FIG. 8-12).

The same steps apply to small bypass capacitors. Paralleled bypass capacitors increase capacity. Make sure that the operating voltage breakdown is the same or higher. If the project calls for a 0.01-μF 50-volt capacitor and you have a 0.022-μF at 100 volts, use it, you will not hear the difference (FIG. 8-13). If you need a 2.2-μF 35-volt bypass capacitor and have several 1-μF 35-volt capacitors on hand, simply parallel two 1-μF capacitors. In circuits that call for a high-Q ceramic disc or a silver mica

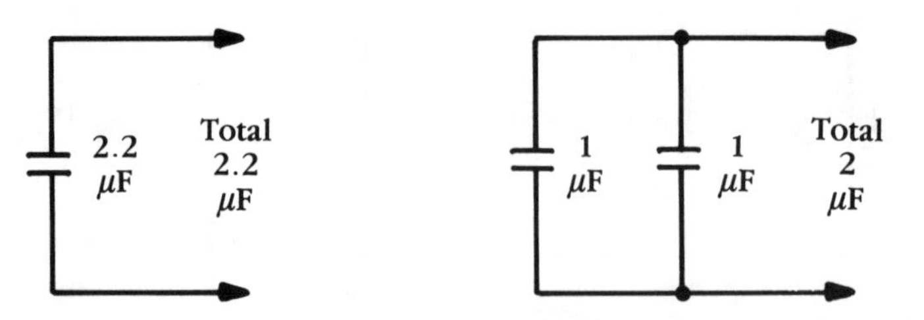

8-13 Like the electrolytic, the bypass capacitor can be paralleled to increase its capacity.

capacitor try to obtain the originals. You cannot substitute ordinary bypass capacitors for high-Q components, designed for low loss in rf circuits. Remember, a 5-μF electrolytic capacitor can be replaced with a 4.7-μF capacitor.

Resistors

Small resistors can be placed in parallel for the least resistance and in series for higher resistance (FIG. 8-14). Watch for total resistance and wattage. Most 1/8-, 1/4-, and 1/2-watt resistors will work in most test equipment projects. If the part list calls for a 1-kΩ 1/8-watt resistor and you have a 1-kΩ 1/2-watt resistor, use it if plenty of room is available. Never substitute a 100-ohm 1-watt resistor for a 100-ohm 10-watt resistor. The resistance is

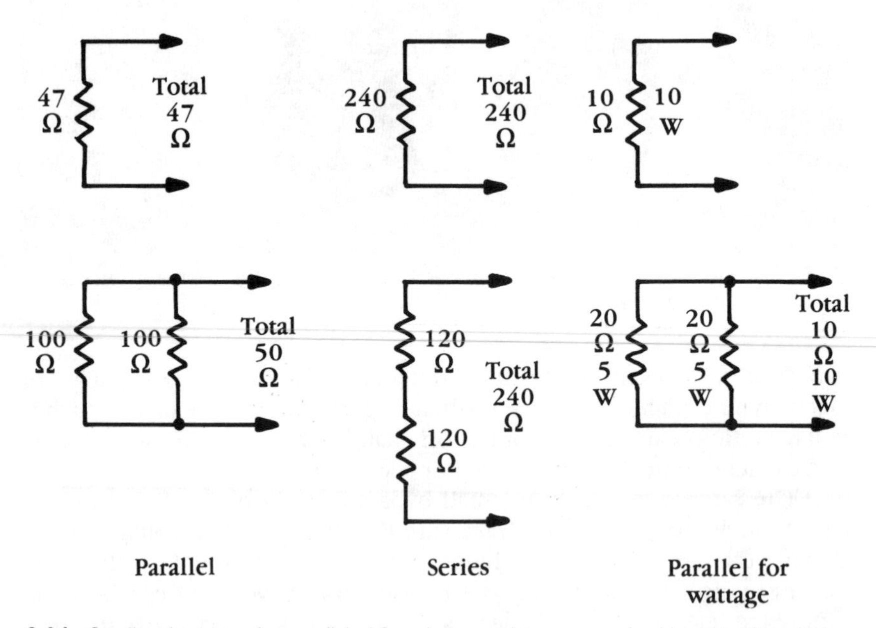

8-14 Small resistors can be paralleled for a lower resistance, or wired in series for a higher resistance. Never replace a 10-watt resistor with a 1-watt type. Check the correct resistance with the DMM.

8-15 With variable controls, use audio-tapered controls in audio circuits and linear variable controls in other circuits.

correct, but the wattage is not. It might smoke and burn up as a result of improper wattage.

In audio circuits, you can use a 10-kΩ audio taper variable control for a 5 kΩ without any problems (FIG. 8-15). The same holds true for miniature pc pots. Always use a linear taper control in circuits other than audio circuits. Use audio taper controls, for controlling volume, in audio circuits. Remember, many of the various parts can be substituted without any problems if the voltage, current, and resistance values are equal.

CONCLUSION

Most any electronic component can be easily located if you know where to look. If you cannot find the part locally, try mail-order firms. Do not overlook surplus stores for critical components. Substitute parts can be found locally or in the junk box.

Dress your favorite test equipment with new cabinets or cases. Spend as much on the cabinet as you do the components. A sleek, professional appearance is hard to beat. Besides, you can build test equipment for much less than companies sell these instruments.

Index

I

L

M

N

O

P

R

S

T

U

V

W